REPORT OF

A MAGNETIC SURVEY OF SOUTH AFRICA

REPORT OF

A MAGNETIC SURVEY OF SOUTH AFRICA

BY

J. C. BEATTIE, D.Sc. (Ed.)
PROFESSOR OF PHYSICS, SOUTH AFRICAN COLLEGE, CAPE TOWN

PUBLISHED FOR
THE ROYAL SOCIETY
AND SOLD BY THE CAMBRIDGE UNIVERSITY PRESS
FETTER LANE, LONDON, E.C.
1909

Price Twenty Shillings net

CAMBRIDGE
UNIVERSITY PRESS

University Printing House, Cambridge CB2 8BS, United Kingdom

Cambridge University Press is part of the University of Cambridge.

It furthers the University's mission by disseminating knowledge in the pursuit of education, learning and research at the highest international levels of excellence.

www.cambridge.org
Information on this title: www.cambridge.org/9781107427464

First published 1909
First paperback edition 2014

A catalogue record for this publication is available from the British Library

ISBN 978-1-107-42746-4 Paperback

PREFACE

THE magnetic survey of South Africa—the results of which are given in the following pages—was carried out with the aid of grants from the Royal Society, the British Association for the Advancement of Science, and the governments of Cape Colony, the Transvaal, the Orange River Colony, Natal, and Rhodesia.

The observations were made by Professor J. T. Morrison and the writer, with assistance at one time and another from Mr S. S. Hough, Professor A. Brown, Professor L. Crawford and Mr V. A. Löwinger. The period over which the observational work extends is from 1898 to 1906. The greatest number of stations—about 300—was occupied in 1903 when the writer had leave of absence for a year. The other stations were occupied during the college vacations of the different years.

The reduction of the astronomical observations was carried out by Mr Löwinger and the writer, with the assistance of several computers of the Royal Observatory at the Cape. The reductions of the magnetic observations were made mainly by the writer, with the assistance of Professor Morrison, Miss Lucy Stapleton, Mr F. D. Hugo and Mr C. Craggs.

The calculations involved in the discussion of the results were carried out by the writer, with the assistance of his wife.

The region surveyed extends from L'Agulhas on the south to the Victoria Falls on the north, and from Saldanha Bay on the west to Beira on the east. More than 400 stations have been occupied, and at a number of these—about twenty—the observations have been repeated at intervals in order to obtain data for the secular variations of the different elements.

It gives the writer great pleasure to be able to thank those—and they are many—who have directly and indirectly helped in the carrying out of the survey. In particular, thanks are due to Viscount Milner, Sir Walter Hely Hutchinson, Sir David

Gill, Hon. J. W. Sauer, Hon. T. W. Smartt, Mr L. Mansergh and Mr A. Bell of the Public Works Department of Cape Colony, the railway managers and the postmasters-general in the different South African Colonies, the surveyors-general in the Cape Colony, in the Transvaal, and in Natal, and His Majesty's Astronomer at the Cape.

Thanks are also due to Professor Brown for reading over the manuscript and for valuable suggestions with regard to the arrangement of the matter of the Report.

The writer also wishes specially to thank Dr Chree for the sympathetic communications he has had from him as representing the Committee of the Royal Society to whom the publication of the work was referred, for many valuable suggestions, and for his kindness in undertaking the final revision in England of the proof-sheets.

The field books used in the survey and the reduction sheets for each station are in the Physical Laboratory of the South African College, and can be consulted there.

J. C. BEATTIE.

PHYSICAL LABORATORY,
SOUTH AFRICAN COLLEGE,
CAPE TOWN.

May, 1908.

[The Secretaries of the Royal Society desire to record their thanks to Dr C. Chree, F.R.S., whose careful and laborious revision, freely given, has contributed very substantially to the accurate printing of the text and particularly of the maps, in England, in the absence of the Author.

They desire also to express their acknowledgments to the officials of the Cambridge University Press, for the pains which they expended on the efficient execution of the work.]

J. L.

April, 1909.

CONTENTS

CHAPTER		PAGE
I.	Introductory and historical	1
II.	General account of the survey including distribution of stations	11
	Instruments used	11
	Determination of latitude and longitude	12
	,, magnetic declination and the probable error	15
	,, dip and the probable error	16
	,, horizontal intensity and the probable error	17
III.	Daily variation of declination	20
	,, ,, horizontal intensity	26
	Secular variation of declination	27
	,, ,, dip	30
	,, ,, horizontal intensity	32
IV.	Comparison of instruments	34
V.	Summary of declinations, dips and horizontal intensities reduced to the epoch 1st July, 1903	44
	Values of magnetic elements at district centres	52
	Variation of magnetic elements with latitude and with longitude	53
	Values of declination at intersections of degrees of latitude and of longitude	55
	Values of the dip at intersections of degrees of latitude and of longitude	56
	Values of the horizontal intensity at intersections of degrees of latitude and of longitude	57
	Values of the total intensity at intersections of degrees of latitude and of longitude	58
	Values of the vertical intensity at intersections of degrees of latitude and of longitude	59
	Values of the northerly intensity at intersections of degrees of latitude and of longitude	60
	Values of the westerly intensity at intersections of degrees of latitude and of longitude	61
	Summary of observed and of calculated values of declination, northerly, and westerly intensities and their anomalies	63
	Summary of observed and of calculated values of dip, horizontal, vertical, and total intensities and their anomalies	72
	Natal and Cape Colony disturbances	85
	Transvaal and Orange River Colony disturbances	106
	Bechuanaland and Rhodesian results	122
APPENDIX A.	Specimen of astronomical observations and reductions	*1*
,, B.	Specimen of declination observations and reductions	*6*
,, C.	Specimen of dip observations	*7*
,, D.	Specimen of horizontal intensity observations and reductions	*9*
,, E.	Summary of magnetic results obtained	*12*
,, F.	Summary of results obtained at Durban by Mr E. N. Nevill, F.R.S.	*227*
INDEX		*229*

LIST OF MAPS IN TEXT

No.	Page	
1	29	Secular variation of Declination.
2	31	Secular variation of Dip.
3	33	Secular variation of the Horizontal Intensity.
4	86	Natal and Transkei. Horizontal Intensity and Declination Anomalies.
5	87	,, ,, Vertical Disturbances, Ridges, Valleys, etc.
6	89	,, ,, Lines of equal Horizontal Intensity.
7	93	Central, Southern and Eastern Cape Colony. Vertical Disturbances, Ridges Valleys, etc.
8	96	Central, Southern and Eastern Cape Colony. Lines of equal Horizontal Intensity.
9	99	South-Western Cape Colony. Vertical Disturbances, Ridges, Valleys, etc.
10	101	Horizontal Intensity and Longitude.
11	102	Declination and Longitude.
12	104	North-Western Cape Colony. Vertical Disturbances, Ridges, Valleys, etc.
13	105	,, ,, Horizontal Intensity and Declination Anomalies.
14	107	Transvaal. Vertical Force Anomalies, etc.
15	108	North Transvaal. Ridges, Valleys, etc.
16	110	,, ,, Horizontal Intensity and Declination Anomalies.
17	111	Transvaal. Lines of equal Horizontal Intensity.
18	113	,, Horizontal Intensity and Declination Anomalies.
19	115	Orange River Colony. Vertical Intensity Disturbances, Ridges, Valleys, etc.
20	117	,, ,, ,, Horizontal Intensity and Declination Anomalies.
21	119	Declination values at different longitudes along parallel 25° 45′ S.
22	120	,, ,, ,, ,, 27° 20′ S.
23	121	,, ,, ,, ,, 28° 18′ S.
Unnumbered	9	Isogonic Lines for 1825, 1850 and 1875.

LIST OF CHARTS

CHART I. Distribution of stations.

" II. Isogonics.

" III. Isoclinals.

" IV. Lines of equal horizontal intensity.

" V. Lines of equal total intensity.

" VI. Lines of equal vertical intensity.

" VII. Lines of equal northerly intensity.

" VIII. Lines of equal westerly intensity.

" IX. Geological map prepared by Mr A. W. Rogers, Director of the Geological Survey of the Cape of Good Hope.

Available for download from www.cambridge.org/9781107427464

CORRIGENDA

(*Pages in Italics refer to the Appendices.*)

PAGE

11, line 2. Plate I should be Chart I.

25, line 13. Klipplast should be Klipplaat.

50. Latitude of Cotswold Hotel should be 30° 42′·7, not 32° 42′·7.

55. Name of Station 81 should be Deelfontein Farm, not Deelfontein.

„ „ „ 82 „ Deelfontein, not Deelfontein Farm.

67. Latitude of Gamtoos River Bridge should be 33° 55′·2, not 33° 15′·2.

69. Declination at Glenallen should be 47° 59′·5, not 47° 56′·5.

72. Latitude of Grange should be 29° 37′·9, not 29° 7′·9.

75. Longitude of Groenplaats should be 28° 33′·8, not 28° 3′·8.

„ „ Grootfontein should be 21° 15′·0, not 19° 15′·0.

96. Declination at Kenhardt should be 26° 22′·4, not 26° 15′·4.

101. The description under Map 10 applies to 10 (*a*) only; in 10 (*b*) Dip is plotted against longitude.

133. Movene should be numbered 224A, not 223A.

134. M'Phateles should be M'Phatele's.

147. Pivaans should be Pivaan's.

189. Number of Station Twelfelhoek should be 333, not 331.

„ „ „ Tweepoort „ 331 „ 332.

190. „ „ Twee Rivieren „ 332 „ 333.

212. Latitude of Wolvefontein should be 33° 19′·0, not 23° 19′·0.

219. Longitude of Station 390 should be 29° 44′·8, not 29° 4′·8.

The positions given for Kenilworth and Beaconsfield, near Kimberley, do not agree with those indicated in Chart I.

CHAPTER I

INTRODUCTORY AND HISTORICAL

THE fact that the earth is a magnet has been known now for over three hundred years. A study of the earth from the magnetic point of view has to be carried out as it would be in the case of other magnets. It may be said that every magnet produces in the space around it an effect depending on the amount and the distribution of its magnetism; the portion of space so influenced is called the magnetic field. To specify this at any point it is necessary to give the strength and the direction. In the case of the earth it is customary to specify its field by three independent quantities which are determined by observing the behaviour of other magnets when exposed in suitable ways to its influence.

The three elements usually determined by travellers and observers not at a permanent magnetic observatory are the declination,—sometimes called the variation,—the horizontal intensity, and the dip or inclination.

The declination at a point is the angle between the geographical north and south, and the magnetic north and south lines passing through the point. The latter is determined by suspending a suitable magnet horizontally in the earth's field by a torsionless suspension, in such a way that the position of its magnetic axis can be accurately determined. The declination is usually given in degrees and it may be east of true north, as it is at the present day in Melbourne, or west of it as in Cape Town, or it may be zero.

In the accompanying diagram O is a point where the declination is 20° W. NOS is the geographical north and south line supposed drawn on the ground, WOE the east and west line. MOM_1 is the position taken at O by the magnetic axis of the properly suspended magnet, M being its positive or north seeking end. The angle NOM is the declination.

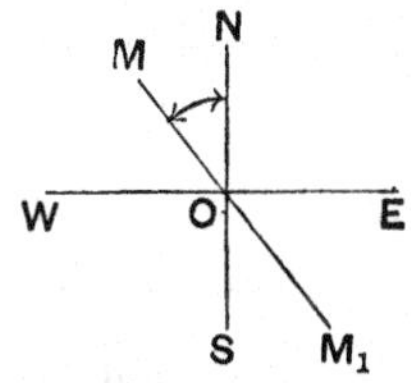

Fig. 1.

A rough determination of the declination may be made in the following way. A stick is placed upright with one end in the ground so that it casts a shadow. About three hours before midday draw a circle with the point of fixture as centre and the length of the shadow as radius; as the day advances the shadow first shortens then lengthens, mark on the circle the point where the shadow cuts it in the afternoon. Bisect the angle between the two points of contact. The bisecting line is the line *NOS* of Fig. 1. Take out the stick and hold a magnet suspended horizontally so that its point of support is over the point which formed the centre of the circle. The position the magnet takes up will be the line *MOM*, of the diagram. The vertical plane containing the axis of the magnet and the point *O* is the magnetic meridian.

The dip or inclination is the angle which a suitable magnet free to move about a horizontal axis through its centre of gravity and perpendicular to the magnetic meridian makes with the horizontal. The value of the dip is usually given in degrees and ranges from 0° at the magnetic equator to ±90° at the magnetic poles. In Fig. 2 HOH_1 is a horizontal line, *O* is the projection of the axis of rotation of the magnet, MOM_1 is the position of equilibrium of the magnet and the magnetic meridian is in the plane of the paper. The angle *MOH* is the angle of dip.

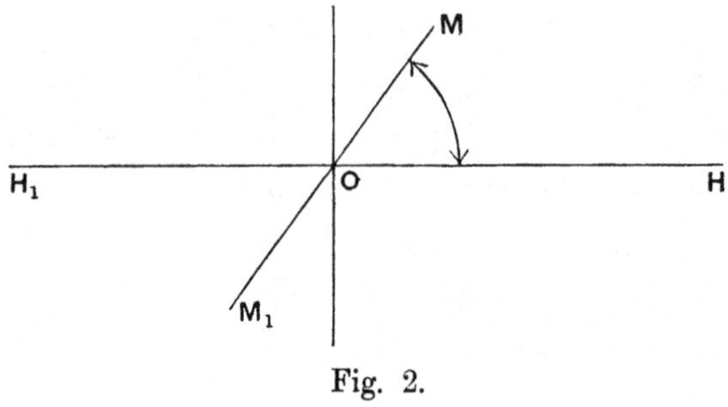

Fig. 2.

The angles of declination and of dip at any point determine the direction of the earth's field at that point. There remains the intensity. The horizontal intensity at a point is the number of lines of magnetic force per unit area about that point, the unit area being taken perpendicular to the magnetic meridian. The horizontal intensities at two places can be compared by allowing the same magnet suspended horizontally to vibrate under the action of the earth's field, first at one place then at the other.

The earth's field can be specified in other ways; or in other words we may give the magnetic data at a given point at a given time in terms of other quantities than the three mentioned above. In Fig. 3 let *OM* represent the position of equilibrium of a magnet in the earth's field, *Y* be the zenith, *N* and *W* the geographical north and west respectively. Then *YOML* is the magnetic meridian, *YON* the geographical meridian, *NOL* the angle of declination (*D*), *MOL* the angle of dip (θ). If *OL* represent in magnitude the horizontal intensity (*H*) of the earth's field, we evidently have the northerly component (*X*) represented by *ON*, the westerly (*Y*) by *OW*, the vertical (*Z*) by *LM*

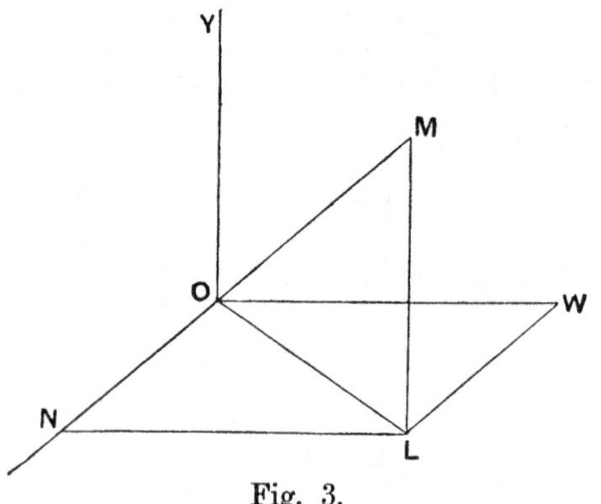

Fig. 3.

and the total (T) by OM, and these different quantities are connected by the following equations.

$$H = T\cos\theta,\ \ X = H\cos D,$$
$$Z = T\sin\theta,\ \ Y = H\sin D.$$

Observation has shown that the magnetic data for a given point have different values at different times. The changes are usually classified under four heads (a) secular, (b) annual, (c) diurnal variations, and (d) irregular variations called magnetic storms.

The secular variation is the change in a magnetic element such as the declination which goes on from year to year. Its value is obtained by subtracting the value of the element at a place for a given year from that of the succeeding year at the same place. The value must in both years have been corrected for all other variations and disturbances. The secular variation may be such as to decrease, or to increase the absolute value of the element, it may in the course of time change from an increase to a decrease as has been the case with the secular variation of declination in Cape Town.

The annual variation depends upon the season of the year. It is obtained by making a series of observations at regular intervals at one place for a series of years. The mean of all the values in any one year gives the mean value of the element for that year. The mean is also taken for each month and the monthly means are corrected for secular variation. The difference between the monthly and the yearly means is the annual variation.

The daily variation of a magnetic element depends on the time in the twenty-four hours at which the observation is taken. Theoretically it is sufficient to take throughout the whole of one day a series of observations at regular intervals in order to determine this variation. The mean of all the observations is the mean value of the element for the day, the difference between this and the actual value at any hour corrected for disturbances is the daily variation at that hour. In Cape Town in the summer the value of the declination is on the average 4′ greater at 9 a.m. than the mean of the day, at 3 p.m. its value is below the mean by about 2′.

In addition to considering the value of the magnetic elements at one place at different times, it is necessary for a complete knowledge of the magnetic state of the earth to have the value of these elements at different places at the same time. It is for this purpose that magnetic surveys are undertaken; in such a survey the values of the magnetic elements are determined at different places and usually at different times. These values are then corrected for secular, annual and daily variations, and allowance is made for any magnetic disturbance which may have influenced the experimentally determined results. In this way the values of the elements over an extended region may be determined at a given date, known as the epoch of the survey. The results of such a survey are shown in maps. A map on which all the points having the same declination are joined is an isogonic map, one on which all the places having the same dip are joined is called an isoclinic map. Maps showing lines of equal intensity are called isodynamic maps.

In so far as South Africa is concerned, the magnetic results obtained to within a few years ago were confined to isolated observations at places on the coast. For a short period a magnetic observatory was maintained at the Cape of Good Hope, and what little is known of the daily and the annual variations in South Africa has been learned from the results obtained there and published by Sabine under the title *Magnetical and Meteorological Observations at the Cape of Good Hope*, Vol. I. Magnetism.

The earliest recorded observation of a magnetic element in this part of the world was made as long ago as 1595; in that year the declination was determined at Mossel Bay by C. Houtman*, while on a voyage to the East Indies. From that time to the present day observations of declination have been carried out at irregular intervals by seamen and travellers at different points on the coast, by surveyors in different parts of the interior, and in particular of recent years a number of results have been obtained by Mr J. J. Bosman, Geodetic officer to the Cape government and by Mr H. G. Fourcade late of the Cape Forestry Department.

The earliest dip observation was made at the Cape in 1751 by La Caille, no observations of this element seem to have been made away from the coast until recent years.

The first determination of the horizontal intensity is as recent as 1840.

The observations made at the magnetic observatory were carried out by a detachment of Royal Artillery stationed at the Royal Observatory of the Cape of Good Hope from 1841 to 1846. After that date the work was carried on fairly regularly under Sir Thomas Maclear till the magnetic house was burnt down in 1853. After that date declination observations were still carried on till 1869. Since then, despite many attempts to resuscitate it, the Cape magnetic observatory has been in abeyance. There is no doubt that such a station would be invaluable for the more complete study of earth magnetism in South Africa, and the matter is at the present day so far advanced that the government of the Cape Colony recognises the necessity of such an observatory and doubtless at a suitable time it will give the necessary financial aid for its establishment.

* *Royal Magnetical and Meteorological Observatory at Batavia*, Vol. XXI. Supplement, p. 78.

The following table contains a list of the recorded observations made in South Africa by seamen, travellers, and surveyors.

Year	Place	Lat.	Long.	Declination	Dip	Intensity in C.G.S. units	Observer	Authority
		° ′	° ′	° ′	° ′			
1595	Mossel Bay	34 10 S.	22 0 E.	0	—	—	Houtman	(1)
1605	Saldanha	33 0 ..	17 52 ..	0 30 E.	—	—	Davis	(2)
1609	L'Agulhas	[34 50 ..	20 0] ..	0 12 W.	—	—	Keeling	(2)
1614	Saldanha	33 0 ..	17 52 ..	1 30 ..	—	—	Pring	(2)
1614	Table Bay	[33 56 ..	18 26] ..	1 45 ..	—	—	Daunton	(2)
1616	Saldanha	—	—	0 40 E.	—	—	Adams	(1)
1617	„	—	—	0 40 W.	—	—	Doubling	(1)
1620	Table Bay	—	—	1 50 ..	—	—	De Beaulieu	(1)
1622	„	—	—	1 50 ..	—	—	—	(1)
1622	L'Agulhas	—	—	2 0 ..	—	—	—	(2)
1636	Table Bay	—	—	1 0 ..	—	—	Caen	(1)
1656	Cape Town	33 56 ..	18 26 ..	6 15 ..	—	—	Bogaerde	(1)
1667	Table Bay	—	—	7 15 ..	—	—	—	(2)
1671	Robben Island	33 48 ..	18 24 ..	7 30 ..	—	—	Padtbrugge	(1)
1675	Table Bay	—	—	8 28 ..	—	—	Leydeker	(2)
1675	L'Agulhas	—	—	8 0 ..	—	—	—	(2)
1677	Table Bay	—	—	7 40 ..	—	—	Jongekoe	(1)
1678	Robben Island	—	—	7 10 ..	—	—	Goudtsmit	(1)
1687	Table Bay	—	—	8 30 ..	—	—	Leydeker	(2)
1699	„	—	—	11 0 ..	—	—	„	(2)
1699	Robben Island	—	—	11 55 ..	(11th May)	—	Zeeman	(1)
	„	—	—	12 55 ..	(12th May)	—	„	(1)
1699	Hout Bay	34 4 ..	18 20 ..	10 14 ..	—	—	—	(1)
1702	Table Bay	—	—	12 50 ..	—	—	Leydeker	(2)
1706	„	—	—	13 40 ..	—	—	„	(2)
1708	„	—	—	14 0 ..	—	—	„	(2)
1712	Algoa Bay	34 10 ..	26 20 ..	17 10 ..	—	—	Pietersz	(1)
1724	Simon's Bay	—	—	16 12 ..	—	—	Mathews	(2)
1724	Table Bay	—	—	16 27 ..	—	—	„	(2)
	„	—	—	16 18 ..	—	—	„	(2)
1751	Cape Town	33 55 ..	18 24 ..	19 15 ..	43 0 S.	—	La Caille	(2)
1752	„	—	—	—	43 7·5 ..	—	„	(2)
	„	33 21 ..	18 55 ..	19 0 ..	—	—	„	(2)
1753	„	—	—	19 0 ..	—	—	„	(2)
1768	„	—	—	19 30 ..	—	—	Wallis	(2)
	„	—	—	19 30 ..	—	—	Carteret	(3)
1770	Simon's Bay	34 8 ..	—	19 10 ..	44 26 ..	—	Ekeberg	(2)
1772	Cape of Good Hope	—	—	20 26 ..	45 37 ..	—	Bayley	(2)
1774	Simon's Bay	—	—	—	44 28 ..	—	Ekeberg	(2)
1774	Cape of Good Hope	33 56 ..	18 23 ..	21 33 ..	—	—	Bayley	(2)
1775	„ „	33 56 ..	18 23 ..	21 14·5 ..	45 19 ..	—	„	(2)
1775	Simon's Bay	—	—	—	46 26 ..	—	Abercrombie	(2)
1780	„	34 20 ..	18 29 ..	22 16 ..	46 47 ..	—	Cook	(2)
1783	Table Bay	—	—	22 23 ..	—	—	Lodberg	(2)
1788	„	—	—	23 16 ..	—	—	Bligh	(2)
1789	„	—	—	26 0 ..	—	—	—	(2)
1791	Simon's Bay	—	—	25 40 ..	48 30 ..	—	Vancouver	(2)
1792	Table Bay	—	—	24 30 ..	47 25 ..	—	Dentrecasteaux	(2)
1798	Cape Town	—	—	26 12 ..	—	—	Albrechtsen	(2)
1804	Table Bay	—	—	25 4 ..	—	—	Bönsöe	(2)
1818	Cape Town	—	—	26 31 ..	50 47 ..	·3245 (T.)	Freycinet	(4)
1836	Cape of Good Hope	34 11 ..	18 26 ..	28 30 ..	52 35 ..	—	Fitzroy	(3)
1837	„ „	—	—	—	52 54 ..	—	Wrokham	(5)
	„ „	—	—	—	52 26 ..	—	Bethune	(5)

(1) *Royal Magnetical and Meteorological Observatory at Batavia*, Vol. XXI. Supplement.
(2) *Magnetismus der Erde Anhang*, by Hansteen.
(3) Sabine, *loc. cit.* (4) Hydrographic Department, Admiralty.
(5) *Phil. Trans.* 1840.

Year	Place	Lat.	Long.	Declination	Dip	Intensity in C.G.S. units	Observer	Authority
		° ′	° ′	° ′	° ′			
1839	Cape of Good Hope	—	—	29 9 W.	53 6 S.	—	Du Petit Thouars	(6)
1839	Simon's Bay	—	—	29 13 ..	—	—	" "	(7)
1840	"	34 11 S.	18 26 E.	—	53 8 ..	·3482 (T.)	Ross	(6)
1841	R.O., Cape Town	33 56 ..	18 29 ..	29 0·2..	53 9 ..	—	R.A. Detachment	(6)
1842	" "	—	—	29 6 ..	53 12 ..	·2096 (H.)*	"	(6)
1842	Simonstown	34 11 ..	18 26 ..	29 8 ..	53 4 ..	·2097 (H.)	Belcher	*
1843	R.O., Cape Town	—	—	29 5 ..	53 19 ..	·2089 (H.)	R.A. Detachment	(6)
	" "	—	—	—	—	·3498 (T.)		
1844	" "	—	—	29 6·2..	53 36 ..	·2069 (H.)	"	(6)
	" "	—	—	—	—	·3470 (T.)		
1844	Simonstown	—	—	—	—	·3478 (T.)	Clark	(8)
1845	"	—	—	29 15 ..	53 44 ..	·3478 (T.)	"	(8)
1845	R.O., Cape Town	—	—	29 7·4..	53 31 ..	·2082 (H.)	R.A. Detachment	(6)
	" "	—	—	—	—	·3495 (T.)		
1846	" "	—	—	29 9·2..	53 33 ..	·2080 (H.)	—	(6)
1847	" "	—	—	29 12·4..	53 41 ..	·2077 (H.)	—	(4) & (6)
1848	" "	—	—	29 14 ..	53 47 ..	·2072 (H.)	—	(4) & (6)
1849	" "	—	—	29 16·4..	53 52 ..	—	—	(4) & (6)
1850	" "	—	—	29 18·8..	53 58 ..	·2066 (H.)	—	(4) & (6)
1851	" "	—	—	29 20·9..	54 2 ..	—	—	(6) & (9)
1852	" "	—	—	29 22·9..	54 4 ..	·2059 (H.)	—	(9)
	" "	—	—	—	—	·3506 (T.)		
1853	" "	—	—	—	54 9 ..	·2056 (H.)	—	(9)
	" "	—	—	—	—	·3511 (T.)		
1854	" "	—	—	—	54 19·6..	·2050 (H.)	—	(9)
	" "	—	—	—	—	·3516 (T.)		
1855	" "	—	—	—	54 24·5..	·2048 (H.)	—	(9)
	" "	—	—	—	—	·3517 (T.)		
1856	" "	—	—	—	54 23·9..	·2044 (H.)	—	(9)
	" "	—	—	—	—	·3511 (T.)		
1857 Oct.	" "	—	—	29 34·4..	54 36·4..	·2050 (H.)	Officers of the Novara	(4)
1857 Jan.	" "	—	—	—	54 23 ..	·2044 (H.)	—	(9)
	" "	—	—	—	—	·3511 (T.)		
1858 Mar.	" "	—	—	—	54 29·3..	—	—	(9)
1860	" "	—	—	29 41·8..	—	—	—	(9)
1861	" "	—	—	29 44·8..	—	—	—	(9)
1862	" "	—	—	29 50·3..	—	—	—	(9)
1863	" "	—	—	29 52·1..	—	—	—	(9)
1864	" "	—	—	29 53·9..	—	—	—	(9)
1865	" "	—	—	30 0·1..	—	—	—	(9)
1866	" "	—	—	30 2 ..	—	—	—	(9)
1867	" "	—	—	30 1·7..	—	—	—	(9)
1868	" "	—	—	30 1·9..	—	—	—	(9)
1869 Jan.	" "	—	—	30 1·5..	—	—	—	(9)
1871 Aug.	" "	—	—	—	55 45·4..	—	Stone	(9)
1871 Sept.	" "	—	—	—	55 34·9..	—	"	(9)
1872 Nov.	Tulbagh	33 17 ..	19 8·7..	30 25 ..	—	—	A. Moorrees	(14)
1873 Nov.	R.O., Cape Town	—	—	30 4 ..	55 56·3..	·1989 (H.)	—	(10)
	" "	—	—	—	—	·3551 (T.)		
Apr. 1873	Cold Bokkeveld, Tafelberg	32 57 ..	19 25 ..	30 24 ..	—	—	A. Moorrees	(14)
Apr. 1873	Cold Bokkeveld, Leeuwfontein	33 15 ..	19 28 ..	30 30 ..	—	—	"	(14)
1874	R.O., Cape Town	—	—	30 6·4..	56 0·8..	·1978 (H.)	—	(4)
1874 Apr.	Port Nolloth	29 15·5..	16 52 ..	28 55·6..	53 22·8..	·2014 (H.)	Stone	(11)
1874	Klipfontein	29 14·3..	17 41·3..	28 23·3..	53 21·9..	·2035 (H.)	"	(11)
1874	Oo'kiep	29 36·3..	17 53·3..	28 21·5..	53 22·3..	·2030 (H.)	"	(11)
1874	Orange River	28 53·1..	18 14 ..	28 27·4..	53 49·8..	·2009 (H.)	"	(11)

(4) Hydrographic Department, Admiralty. (5) *Phil. Trans.* 1840.
(6) *Magnetical and Meteorological Observations at the Cape of Good Hope*, Vol. I., Magnetism.
(7) *Phil. Trans.* 1849. * *Phil. Trans.* 1843. (8) *Phil. Trans.* 1846.
(9) Beattie and Morrison, *Transactions of the South African Philosophical Society*, Vol. XIV.
(10) *Challenger Report Narrative*, Vol. II. (11) *Proc. R. S. L.*
(14) Surveyor General's Department, Cape Town.

Year	Place	Lat.	Long.	Declination	Dip	Intensity in C.G.S. units	Observer	Authority
		° ′	° ′	° ′	° ′			
1875 June	Worcester	33 39 S.	19 27 E.	29 48 W.	—	—	J. J. Bosman	(14)
1875 Aug.	"	33 42 ..	19 40 ..	30 10 ..	—	—	"	(14)
1875 Oct.	"	33 33·4..	13 31·5..	29 50 ..	—	—	"	(14)
1875 Sept.	Hex River Poort	33 32·2..	19 32·5..	29 57 ..	—	—	"	(14)
1876 July	Kenhardt	29 20 ..	21 9 ..	28 16 ..	—	—	A. Moorrees	(14)
1876 June	Kenilworth, Cape	—	—	30 16 ..	—	—	J. J. Bosman	(14)
1877	R.O., Cape Town	—	—	30 7 ..	—	—	—	(4)
1877	" "	—	—	29 54 ..	—	—	—	(4)
1885 Dec.	Walfisch Bay (Pelican Pt.)	22 55·7..	14 25·9..	25 54 ..	—	—	Stapff	(4)
1886 Jan.	Mara	23 12·2..	14 35·3..	25 44 ..	—	—	"	(4)
1886 Jan.		23 17·2..	14 44·5..	25 36·5..	—	—	"	(4)
1886 Jan.	Salt River	23 32·4..	14 58·9..	25 51·8..	—	—	"	(4)
1886 Jan.	Nada	23 37 ..	14 58·5..	25 47·2..	—	—	"	(4)
1886 June	Wepener	29 41 ..	27 3 ..	28 40 ..	—	—	J. J. Bosman	(14)
1886 Nov.	Mogwoding	27 41 ..	24 40 ..	25 43 ..	—	—	Bosman & Moorrees	(14)
1888	R.O., Cape Town	—	—	29 54 ..	57 15 S.	—	—	(4)
1888 Feb.	Steekdoorns	27 21 ..	24 18 ..	26 15 ..	—	—	Bosman & Moorrees	(14)
1888 Mar.		27 29·5..	24 15·2..	26 16 ..	—	—	" "	(14)
1888 Mar.		27 20·9..	24 10·7..	26 6 ..	—	—	" "	(14)
1888 Apr.		27 28·3..	24 9·5..	26 17 ..	—	—	" "	(14)
1888 June		27 33·3..	24 3 ..	26 16 ..	—	—	" "	(14)
1888 Oct.	Mooifontein	27 17·2..	24 1·3..	26 14 ..	—	—	" "	(14)
1888 Oct.	Klippan	27 23·3..	24 34·6..	26 2 ..	—	—	" "	(14)
1888 Oct.	Biesjesdal	27 23·8..	24 30·3..	26 1 ..	—	—	" "	(14)
1890	R.O., Cape Town	—	—	29 36 ..	57 15 ..	·1916 (H.)	Preston	(12)
		—	—		—	·3542 (T.)		
1891 Mar.	Lapanie	27 58·1..	22 39 ..	26 46 ..	—	—	Bosman & Moorrees	(14)
1891 Mar.	Gahoe	27 58·3..	22 31·6..	26 21 ..	—	—	" "	(14)
1891 Mar.	Tnooi	28 11·8..	22 22·6..	26 23 ..	—	—	" "	(14)
1891 Apr.	Harreeboom	28 22·4..	21 53·2..	27 3 ..	—	—	" "	(14)
1891 Apr.	"	28 11·4..	21 53 ..	27 18 ..	—	—	" "	(14)
1891 May	Van Roois Vlei	28 32·3..	20 58·8..	27 33 ..	—	—	" "	(14)
1891 May	Gilbert	28 21·3..	20 48·3..	27 36 ..	—	—	" "	(14)
1891 May	Toeslaan	28 19·8..	20 38·1..	27 46 ..	—	—	" "	(14)
1891 June	Biesjespoort	28 25·8..	20 34·8..	27 44 ..	—	—	—	(14)
1891 June	"	28 11·3..	20 27·3..	27 50 ..	—	—	—	(14)
1891 June	Aries	28 12·8..	20 3·2..	27 55 ..	—	—	—	(14)
1891 Nov.	Walfisch Bay (Pelican Pt.)	20 56 ..	14 30 ..	24 48 ..	—	—	—	(4)
1892 July	"	—	—	25 0·8..	—	—	—	(4)
1892 Apr.	Beira	19 51 ..	34 51 ..	17 58 ..	—	—	—	(4)
1893	Simon's Bay	34 11·6..	18 26·6..	29 22 ..	—	—	Baird	(4)
1894 Nov.	Walfisch Bay	—	—	—	51 37 ..	—	—	(4)
1894 Dec.	Simon's Bay	—	—	29 16 ..	—	—	Baird	(4)
1894	R.O., Cape Town	—	—	29 24 ..	57 52 ..	·1901 (H.)	Combe	(4)
1894 May	Reuben Point	25 58·8..	32 36 ..	23 58·5..	—	—	Kiddle	(4)
1895 Jan.	Cape Point	34 21·3..	18 29·5..	29 14·3..	58 6·6..	·1889 (H.)	Combe	(4)
1895 Jan.	Simon's Bay	34 11·6..	18 26·8..	29 20·9..	58 3·4..	—	"	(4)
1895 Jan.	R.O., Cape Town	—	—	29 18 ..	57 52 ..	·1900 (H.)	Finlay	(9)
	" "	—	—	—	—	·3572 (T.)		
1897 Oct.	" "	—	—	29 8 ..	58 15 ..	·18755 (H.)	—	(4)
1897 Oct. & Nov.	"	—	—	29 2 ..	58 7 ..	·18835 (H.)	Finlay	(9)
1899 Jan.	Lourenço	25 58·9..	32 36 ..	23 11·4..	58 54·1..	·35660 (T.)	—	(13)
	Marques (Reuben Pt.)	—	—	—	—	·19782 (H.)		
1900 May	Kosi River	26 53·5..	32 54 ..	21 37 ..	—	—	—	(4)
1900 May	Oro Point	26 50 ..	32 55 ..	22 16 ..	—	—	—	(4)
1904 Apr.	Walfisch Bay	—	—	23 46 ..	—	—	—	

(4) Hydrographic Department, Admiralty.
(9) Beattie and Morrison, *Transactions of the South African Philosophical Society*, Vol. XIV.
(12) *U. S. C.* and *G. S. Bull.* 23. (13) *Ter. Mag.* Vol. VII. p. 197.
(14) Surveyor General's Department, Cape Town.

The above results are of very different orders of accuracy. The earlier ones in particular, up to the early part of the 19th century, are liable to error from the fact that the determinations of longitude were unsatisfactory, the compasses were unsuitable and were not comparable with one another, and it is improbable that the observations taken at different times in the same place were taken at exactly the same position, a condition which later experience has shown is very necessary in observations for secular variation.

The declination results supplied to me by Mr J. J. Bosman and obtained by Mr A. Moorrees and himself, were found by determining the true meridian astronomically with a five-inch theodolite to which a magnet needle was attached. A mark was then set up at as long a range as practicable in the direction of the true meridian, and the magnetic bearing of this point observed. Several readings were thus made and the mean taken. This was the method adopted at the villages of Kenhardt, Tulbagh, and Wepener, at Mogweding in Bechuanaland and at Kenilworth in the Cape Division. At each of the remaining places where Bosman and Moorrees observed, the declination was determined by observing the magnetic bearing of some line, the geodetic azimuth of which was, either previously or subsequently, determined by triangulation.

The declination at the Royal Observatory of the Cape of Good Hope from April 1841 to July 1846 was observed hourly, from September 1846 to March 1852 five times daily, and from October 1860 to January 1869 twice daily.

The dip was observed as a rule twice weekly from 1841 to 1846 and from 1852 to 1857.

The intensity was observed hourly from 1843 to 1850 and five times daily from 1852 to 1857. In the intensity observations however there are several gaps in different years.

The results for the different elements given in these years is the mean of all the observations for each year. Further details for these periods will be found in Sabine's *Magnetical and Meteorological Observations, Cape of Good Hope,* Vol. I., and in the *Transactions of the South African Philosophical Society,* Vol. XIV., "Magnetic Elements at the Cape of Good Hope," by Beattie and Morrison.

Finally there still remains of the earlier observations a number taken by surveyors in different parts of Cape Colony. These results have been obtained for me by Mr J. J. Bosman from diagrams in the Surveyor General's office at Cape Town. The observations of the magnetic bearings were as a rule made by the surveyors with the object of indicating the direction of the magnetic north on their plans and diagrams. It has not been thought necessary to give the whole of the results obtained in this way; the main facts are shown on the three following maps. The declinations shown on them were obtained by taking all the observations for twelve years before and after the particular epoch and reducing these to the epoch by help of the secular variation derived from the observations themselves and from the observations at the Cape of

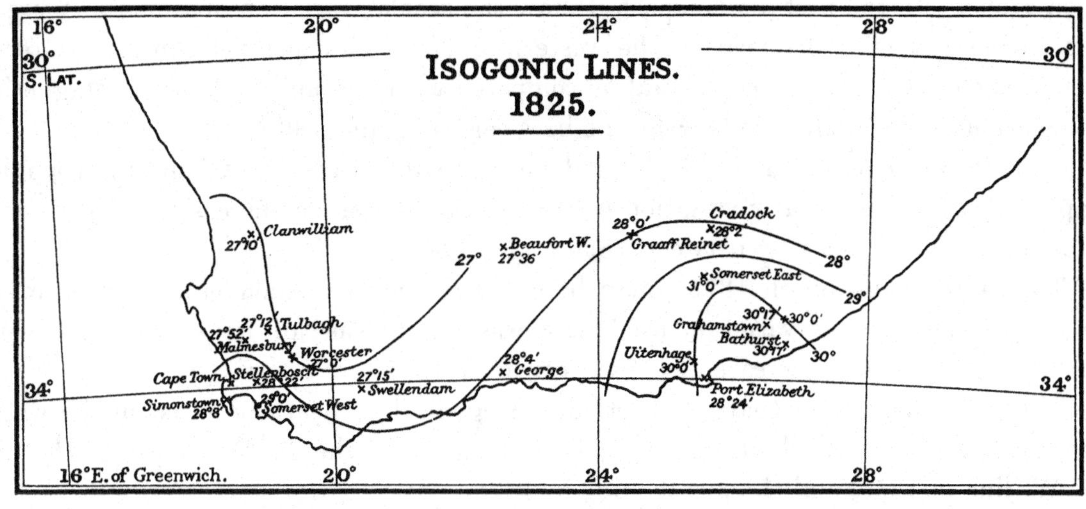
ISOGONIC LINES.
1825.
16°
20°
24°
28°
30°
S. LAT.
30°
34°
34°
16°E. of Greenwich.
20°
24°
28°
Clanwilliam
27°10′
Beaufort W.
27°36′
27°
28°0′
Graaff Reinet
Cradock
28°2′
28°
Somerset East
31°0′
29°
27°12′
Tulbagh
27°52′
Malmesbury
Worcester
27°0′
Cape Town
Stellenbosch
28°22′
27°15′
Swellendam
28°4′
George
Simonstown
28°8′
29°0′
Somerset West
30°17′
30°0′
Grahamstown
Bathurst
30°17′
30°
Uitenhage
30°0′
Port Elizabeth
28°24′

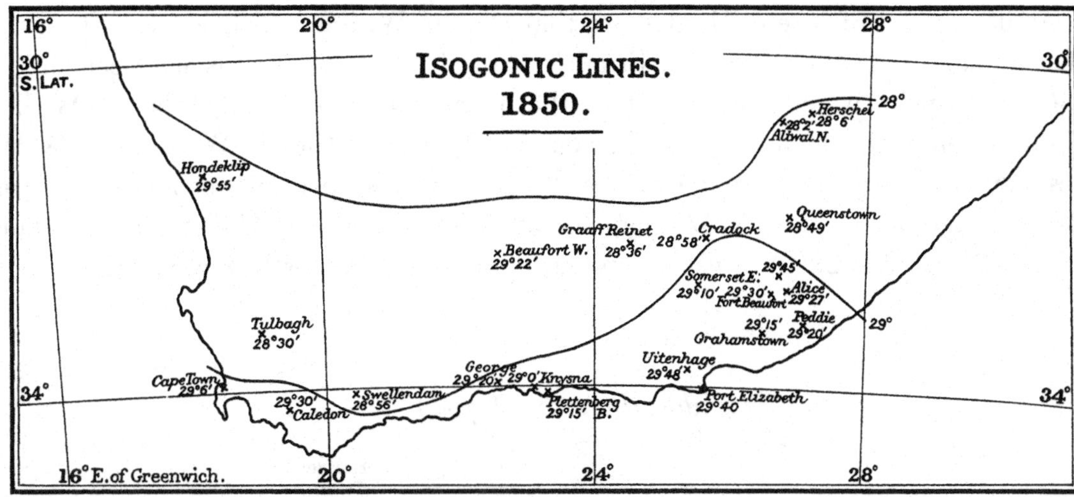
ISOGONIC LINES.
1850.
16°
20°
24°
28°
30°
S. LAT.
30°
34°
34°
16°E. of Greenwich.
20°
24°
28°
Herschel
28°
28°2′
28°6′
Aliwal N.
Hondeklip
29°55′
Queenstown
28°49′
Graaff Reinet
28°36′
Cradock
28°58′
Beaufort W.
29°22′
29°45′
Somerset E.
29°10′
29°30′
Alice
29°27′
Fort Beaufort
Tulbagh
28°30′
29°15′
Peddie
29°20′
Grahamstown
29°
Uitenhage
29°48′
George
29°20′
29°0′
Knysna
Cape Town
29°6′
29°30′
Caledon
Swellendam
28°56′
Plettenberg B.
29°15′
Port Elizabeth
29°40

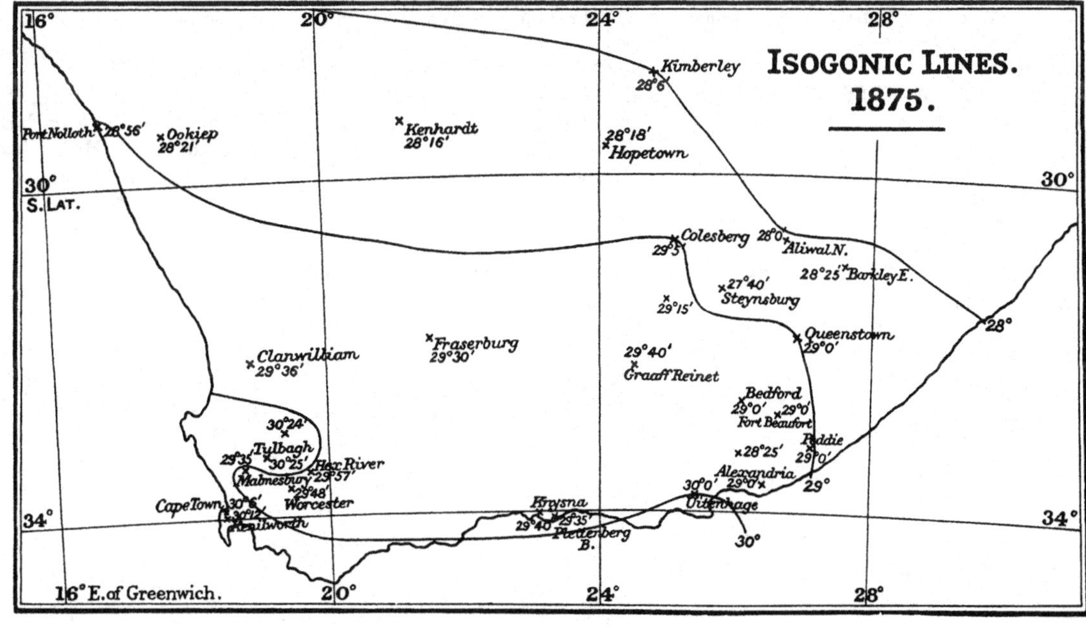
ISOGONIC LINES.
1875.
16°
20°
24°
28°
Kimberley
28°6′
Port Nolloth
28°56′
Ookiep
28°21′
Kenhardt
28°16′
28°18′
Hopetown
30°
S. LAT.
30°
Colesberg
29°5′
28°0′
Aliwal N.
28°25′
Barkley E.
27°40′
Steynsburg
29°15′
Queenstown
29°0′
28°
Clanwilliam
29°36′
Fraserburg
29°30′
29°40′
Graaff Reinet
Bedford
29°0′
29°0′
Fort Beaufort
30°24′
Tulbagh
29°35′
30°25′
Hex River
29°57′
Malmesbury
29°48′
Worcester
Peddie
28°25′
29°0′
Alexandria
30°0′
29°0′
29°
Uitenhage
Cape Town
30°6′
30°12′
Kenilworth
Knysna
29°40′
29°35′
Plettenberg B.
30°
34°
34°
16°E. of Greenwich.
20°
24°
28°

Good Hope. In the first map—epoch 1825—the secular variation obtained and used was an increase of 4′ per year in the western and one of 3′ per year in the eastern part of the colony. This map should be compared with Erman's map of the Magnetic Declination drawn from the observations between 1827 and 1830.

The second map—epoch 1850—should be compared with the Admiralty Isogonic map for 1858. The secular variation used in this case was an increase of 3′ per year up to 1850, and 2′ per year between 1850 and 1863.

The third map—epoch 1875—may be compared with Neumayer's Isogonic map for 1885. The secular variation used here was an increase of 1′ per year between 1863 and 1875, and a decrease of 1′ per year after 1875.

The three maps show that the particular isogonic line near the coast in the west and south is nearly parallel to the coast in each case, viz. the 28° W. line in 1825, the 29° W. line in 1850, and the 30° W. line in 1875.

Another point of interest is the fact that the 30° W. line which is evident on the S.E. mainland in the 1825 map,—Erman's map shows this also—is not seen in the 1850 map, a point which is borne out by the Admiralty map of 1858; the 30° W. line appears to have drawn off to sea again; on the other hand the 29° W. and the 28° W. lines have extended much more to the N. and the distance between them is considerably increased. In the 1875 map the 30° W. line again reaches to the mainland and the 29° W. and 28° W. lines extend still farther to the north in each case.

Addendum to Chapter I.

Date	Place	Lat.	Long.	Declination	Dip	Horizontal Intensity in C.G.S. units	Observer	Authority
		° ′	° ′	° ′	° ′			
1874 July Aug.	R.O. Cape Town	33 56 S.	28 28·8 E.	29 58·9 W.	56 2·4 S.	·19872	Perry	R.S. Proc. vol. 27
1874 Sep.	„		28 29·0 ..		56 6 ..	·2056	Officers 'Gazelle'	die Forschungsreise S.M.S. 'Gazelle'
1898 Jan.	Groote Laagte	21 14·1 ..	21 25·0 ..	22 0·0 ..			Passarge	Passarge's Die Kalahari
1906 Jul.	Inhambane Ponta de Belana	23 48·2 ..	35 22·1 ..	17 16·9 ..	58 14·5 ..	·19690	Chaves	Chaves
1906 Jul.	Boane	26 2·5 ..	32 19·8 ..	19 47·3 ..	59 28·0 ..	·18998	..	..
1906 Aug.	Olvrida	18 2·3 ..	36 56·1 ..	13 12·1 ..	53 11·8 ..	·22185	..	..
1906 Aug.	Beira	19 49·2 ..	34 50·0 ..	15 21·5 ..	54 41·4 ..	·21458	..	..
1906 Aug.	Quelimane	17 52·7 ..	36 52·9 ..	13 4·5 ..	52 56·7 ..	·22409	..	..

CHAPTER II

GENERAL ACCOUNT OF THE SURVEY INCLUDING DISTRIBUTION OF STATIONS, INSTRUMENTS, METHODS OF OBSERVING, AND ERRORS OF OBSERVATION

Distribution of stations and itinerary. The distribution of the stations at which observations were made is given in Plate I. It is only necessary to add that most of the observations were made between December 1902 and February 1904. In this period stations were occupied in the south of Cape Colony from Ladismith to Gamtoos River; in the north of Cape Colony from De Aar to Mafeking; in Bechuanaland, Matabeleland, Mashonaland, and Portuguese East Africa; in Natal; and in the Orange River Colony. Previous to this a number of observations had been carried out in the central parts of Cape Colony in July 1900; in the south west of Cape Colony in January, February and July 1902. Observations were taken between Bulawayo and the Victoria Falls and at repeat stations in the east of Cape Colony in July 1904; in December 1904 and January 1905 observations were taken in the west of Cape Colony and in January and February 1906 in the Transkei and in Bechuanaland.

Instruments. The instruments used were two dip circles by Dover No. 9 and No. 142, each with four needles; two unifilar magnetometers by Elliot, one No. 73 the property of the London Royal Society, and the other No. 31 belonging to the Royal Observatory of the Cape of Good Hope. The constants of the unifilars were determined at Kew and have not been redetermined in South Africa. These constants in so far as they are necessary for the reduction of the results are given in table 1; it will be noticed that the thermometer corrections are not given, these have been applied in the process of reduction and the temperatures given in the results are the corrected temperatures.

It has not been thought necessary to give the Kew certificates of the various needles used on the different dip circles. Experience shows that testing a needle in England on a part of the axle never used in South Africa has little value for South African purposes. The standard of comparison used was an arbitrary one, viz. the mean of the needles of the particular dip circle used.

In addition to the above magnetic instruments a five-inch theodolite, No. 1084, by Cooke & Son, York, and a chronometer—Reid & Son, No. 1078—were used.

TABLE 1.

Magnetometer Constants.

Magnet	Temperature correction, t deg. C.	Induction coefficient	Angular value of one scale division
Collimator 31	$0{\cdot}000601\,t + 0{\cdot}0_5\,241\,t^2$	6·532	2′·21
„ 73 A	$0{\cdot}000281\,t + 0{\cdot}0_5\,118\,t^2$	4·730	1′·75

Deflection apparatus 31 C, angular value of one scale division		1′·065
„ „ 73 C, „ „ „ „		1′·04

	Distance from centre (centimetres) at 0° C.			
Apparent	25	30	35	40
Brass Bar 31 true	24·995	29·995		39·990
„ 73 „ (corrected for bending)	24·993	29·993	34·994	39·994

Magnetometer	Log $\pi^2 K$ at 0° C.	Auxiliary cylinder of brass: Length	Diameter	Mass
31	3·46423	9·570 cm.	1·013 cm.	65·976 grm.
73	3·47406	9·536 „	0·995 „	62·803 „

Quantities observed and methods of determining them. The latitude and the longitude of each station were determined; and the magnetic elements observed were the declination, the dip, and the horizontal intensity.

Determination of latitude and of longitude. The observations from which the latitude was calculated were taken with a five-inch theodolite. The altitudes of the sun and the chronometer times of the observations were taken usually thrice in the ten minutes before noon, a second set of three was taken in the ten minutes after noon. The vertical circle of the theodolite was reversed between the ante-meridian and the post-meridian observations; in the first set the apparent upper and in the second the apparent lower limb was observed in transit over a horizontal wire in the theodolite.

The observations for the calculation of longitude were those of the altitude of the sun and of the corresponding chronometer times. They were taken about three hours east or west of the meridian. Four pointings were made with the vertical circle of the theodolite in one position, the first on the preceding, the second on the following, the third on the following, and the fourth on the preceding limb of the sun. The vertical circle was then reversed and a similar set of four pointings taken. Finally the axle of the theodolite telescope was reversed and the whole repeated.

In all observations of altitude, either for latitude or for longitude, the level was read for each pointing and the reading corrected accordingly*.

* The level correction in seconds of arc was found by taking

$$\tfrac{1}{2}\,(\text{left-hand}-\text{right-hand readings})\;12''\tan\alpha$$

where α was the altitude at the time of observation.

In order to obtain the reading for the sun's centre in the longitude observations it was necessary to combine an observation of a preceding with one of a following limb and to eliminate the index error of the circle a pair of circle direct (right) with a pair of circle reversed (left) observations.

After the readings had been combined in the above manner and the level correction applied, the reduction proceeded as follows. An approximate value of the latitude was obtained by combining a circle right and a circle left circum-meridian altitude and using the formula

$$\phi = z \pm \delta,$$

where ϕ is the approximate latitude, $z = 90^\circ - h$ where h is the approximate meridian altitude, and δ is the sun's declination.

With this approximate latitude an approximate correction of the chronometer to apparent solar time was calculated.

With this approximate chronometer correction the observed circum-meridian altitudes of the sun can be reduced to what they would have been had the observation been taken at the instant the sun was on the meridian. The reduction formula used was

$$\partial h = \frac{2\sin^2 \frac{t}{2}}{\sin 1''} \frac{\cos\phi \cos\delta}{\cos h},$$

where t is the interval from apparent noon, ϕ is the latitude, δ the sun's declination; t was obtained by applying the approximate chronometer correction to the times of observation finding the difference between this and twelve hours. The suitable correction ∂h is applied to each observation and the final latitude calculated from the formula

$$\phi = z \pm \delta,$$

where ϕ is the true latitude, z the zenith distance corrected for refraction and δ the sun's declination.

With this final latitude it is possible to calculate a final chronometer correction. The computed time of an observation was obtained from the formula

$$\sin\frac{\lambda}{2} = \sqrt{\frac{\sin\frac{1}{2}\{z + (\phi - \delta)\} \sin\frac{1}{2}\{z - (\phi - \delta)\}}{\cos\phi \cos\delta}},$$

where λ was the computed time of an observation, z the zenith distance corrected for refraction, ϕ the latitude of the place of observation, δ the sun's declination.

From the λ thus obtained and expressed in time the observed chronometer time was subtracted, the result was the chronometer correction ΔK made up of the chronometer error, the equation of time, and the difference of longitude from the standard meridian.

The first of these was known from an exchange of signals with the Royal Observatory of the Cape of Good Hope, or with the Government Observatory in Natal, the second is given in the nautical almanac, and the longitude with respect to the standard meridian could then be calculated.

Determination of azimuth. In order to determine the true bearing of one line whose circle reading on the magnetometer was known a theodolite was used instead of the mirror attachment to the Kew unifilar. The theodolite was set up at a suitable distance from the magnetometer—usually between 100 and 150 yards—and in such a position that its plumb line could be seen through the telescope of the magnetometer. The necessary observations were made by taking transits of the sun's limbs over the vertical wires in the theodolite and noting first the readings on the horizontal circle of the theodolite and second the corresponding chronometer times. The sun's limbs were taken in the same order as in the longitude observations, four sets being taken in all. The horizontal circle reading on the theodolite of the vertical axis of the magnetometer was read before observing the sun and again immediately after. The sun observations were combined in the same way as those for longitude and the proper level correction applied. The azimuth angle, A, was finally calculated from the formulae

$$\tan\frac{A-B}{2}=\frac{\sin\dfrac{\phi'-\delta'}{2}}{\sin\dfrac{\phi'+\delta'}{2}}\cot\frac{\lambda}{2},$$

$$\tan\frac{A+B}{2}=\frac{\cos\dfrac{\phi'-\delta'}{2}}{\cos\dfrac{\phi'+\delta'}{2}}\cot\frac{\lambda}{2},$$

where ϕ' is the colatitude, λ the apparent hour angle or the apparent time of an observation obtained from the observed chronometer time by applying the calculated correction to apparent time, and $\delta'=\delta+90^\circ$, where δ is the sun's declination.

The circle reading of the geographical meridian could now be obtained, and when this was known the true bearing of the reference line joining the centres of the magnetometer and the theodolite followed without difficulty.

A copy of the reduction of the observations taken at Ginginhlovu, Zululand, is given in appendix A.

Probable error of an azimuth determination. It will be seen from the description of the method employed to determine the azimuth of the reference line that there were two independent determinations of this azimuth. The mean of the two—or more if more determinations had been made—was assumed to be the correct azimuth, and the difference of any one determination from the mean taken irrespective of sign was called the error of that determination. The mean of all the errors was taken as the mean error of a determination. Its value was $\pm 0'{\cdot}3$.

The mean error of a determination of latitude or of longitude was not found.

Determination of magnetic declination. In the Kew unifilar the usual method of determining the true bearing of a reference line whose circle reading is known is to employ the transit mirror attached to the instrument. This method was not employed here. It was found at an early date in the survey that the drawbacks to accurate work were wind, sun, and dust; to mitigate these evils the magnetic instruments were always used in a tent, so made that it afforded by its double lining some protection from the sun, and so placed as to give the greatest protection from the wind; it was never possible by any arrangement to overcome the dust trouble at its worst. The most convenient method in these circumstances for determining the true bearing of a reference line was found in the use of a theodolite. The theodolite was placed so that its plumb line could be observed from the tent through the telescope of the magnetometer, this latter being placed well inside the tent. The reference line was that joining the centre of the two instruments. After setting up and adjusting the two instruments the circle reading of the reference line was taken on the magnetometer, the silk suspension of the magnetometer was then freed from torsion. The magnet was suspended and the instrument moved until the vertical cross wire of the telescope was coincident or approximately coincident with the zero of the scale affixed to the magnet. In the latter case the exact scale reading was noted and at the end a correction to the zero was applied. The magnet was next reversed and the scale reading taken without changing the position of the instrument. These two observations were repeated in the reverse order. The position of no torsion was determined at the end and the necessary correction calculated. The mean of the four circle readings corrected to the zero of the scale was corrected for torsion and the circle reading of the magnetic meridian thus was known. When the true bearing of the reference line (magnetometer—theodolite) was found by the method described in Appendix A the declination was calculated.

The results for a typical station are given in appendix B.

Errors of declination observations. The accidental error of an individual determination of the declination in so far as it depends on the magnetic part of the work was calculated by the method Thorpe and Rücker used in their survey of the British Isles*.

The mean angle between the magnetic and the geometric axis of the magnet was found by taking the mean of the half difference between the reading with the magnet erect and with it inverted. The difference between this mean angle and the angle determined by one set of observations gave its accidental error. The mean error has been determined by dividing the arithmetical sum of the individual errors by the number of individual observations.

If instead of the accidental error of one set of observations the error for one station be found, the latter differs slightly from the former because in the majority of cases more than one set of observations was made at each station. To determine the station

* *Phil. Trans.* A, Vol. 181, p. 70.

error the accidental errors of each set at a station were determined in the manner described above, and the algebraic sum taken; this gave the station error. The mean error for a station was determined by taking the arithmetical sum of the station errors and dividing this by the number of stations. The values of the accidental errors were:

Instrument	Error of an observation	Error of a station
31	± 0'·33	± 0'·18
73	± 0'·38	± 0'·29

The effective error of a declination observation. This is obtained by combining the accidental error of an azimuth determination with the error as obtained in the preceding paragraph. The values of the effective error are:

Instrument	Effective error of an observation	Effective error of a station
31	± 0'·63	± 0'·48
73	± 0'·68	± 0'·59

Determination of dip. In the dip observations the position of the magnetic meridian on the horizontal circle of the instrument was determined by the usual method of noting when the needle stood vertical and then turning through 90° from that position. The needles to be used were always magnetised at the beginning of the meridian determination by means of the permanent magnets supplied with the instrument, one needle only being used for this determination. The observations for dip were then made with this needle by taking four sets of readings in the positions necessary for the elimination of the errors of construction of the needle and the instrument; a second needle was then taken and the same set of observations made in the same order. The magnetism of each needle was next reversed and the four sets of readings taken again with each needle but in the reverse order. By this method of observing the mean time of a dip observation was the same for each needle. The time required for an observation with two needles was about half an hour.

The observations for a typical station are given in appendix C.

The accidental error of an observation was obtained according to the method described by Thorpe and Rücker*. It was found—except in a dust storm—that the needles used had as a rule a bias so that one read higher than the others.

The differences between the readings of any two needles at the same place and at the same time were obtained for all the observations taken during a given period; the algebraic mean of these was taken as the mean difference between the two for that period. Half the difference between this mean and the actual difference at any place is the accidental error of that particular observation. The values obtained in this way are as follows:

Instrument	Needles	Accidental error
9	1 and 2	± 0'·48
142	2 and 1	± 0'·45
142	4 and 1	± 0'·44

* *Phil. Trans.* A, Vol. 181, 1891, p. 79.

The effective error of a dip observation. The above method of considering the error in a dip observation is of interest in so far that it gives an idea of the accuracy of the observational work. Another way of considering the error is to assume that the mean of the dips obtained with different needles is the true dip, and to take the difference between this mean and the actual dip with any one needle as the error of the observation with that needle. The mean of all these errors irrespective of sign may then be called the effective error of an observation. The value of the error calculated in this way is given in the following table.

Instrument	Number of needles used	Effective error
9	2	$\pm$ 0′·68
142	2	$\pm$ 0′·62
142	3	$\pm$ 0′·55

Determination of the horizontal intensity. The method employed was to determine the ratio $\frac{M}{H}$ by Lamont's method of deflection, and the product MH by determining the period of vibration and then eliminating M, the moment of the magnet. The magnetometer was first set up for a determination of the period of vibration. After this had been obtained, the deflections at two distances, usually 30 cm. and 40 cm., were observed, and the end of the complete observation was reached with a second determination of the period of vibration, following as a rule immediately after the deflection. The deflections were observed in such a way that the mean time of the deflection at the different distances was the same. For example, if the deflection at 40 cm. with the deflecting magnet on the west side of the magnetometer, and its north end pointing west, was first observed, it was followed by the deflection at the same distance on the same side of the magnetometer with the south end of the magnet pointing west; the magnet was then placed—still on the same side of the magnetometer—at 30 cm. with its south end pointing west, and lastly the magnet was at this second distance placed with its north end west. The same four observations were finally made with the magnet on the east side of the magnetometer at the same distances beginning at 30 cm. and reversing the order employed in the first set.

The period of vibration was determined by recording the time of every fifth passage up to the sixty-fifth. The time of the hundredth was then calculated and every fifth passage starting from the hundredth again recorded till the one hundred and sixty-fifth. The method of observing was that known as the eye and ear method. In this way two sets of fourteen independent determinations of the time taken for one hundred vibrations were obtained, one before the deflections which we shall denote by set I, and one after—set II. The first three and the last three of I were combined and marked I A, the fifth to the tenth were combined and marked I B. In the same way II A and II B were found. The means were now taken of I A and II B and of II A and I B. The first mean divided by 100 was used with the deflection at

30 cm., and the second—also divided by 100—with that at 40 cm. In Rhodesia it was thought necessary to observe also at 25 cm. and 35 cm. for deflection. In such cases the first mean was used with 25 cm. and the second with 35 cm. It will be seen that by this method the mean times of the deflections and the corresponding vibrations were the same, and it was possible to get two values of the moment of the deflecting magnet and of the horizontal component of the earth's field all for the same mean time.

The formulae used for the reduction were

$$\frac{M}{H} = \tfrac{1}{2} r^3 \sin\theta \left\{1 + \frac{2\mu}{r^3} + qt + q_1 t^2\right\}\left(1 - \frac{P}{r^2}\right),$$

$$T^2 = T_1^2 \left\{1 + \frac{G}{F} - qt - q_1 t^2 + \mu \frac{2}{r^3 \sin\theta}\right\},$$

$$MH = \frac{\pi^2 K}{T^2},$$

where M is the moment of the deflecting magnet (used also for vibration).

H is the horizontal intensity of the earth's field.

r is the distance of the centre of the deflecting from the centre of the suspended magnet.

θ is the angle in degrees through which the deflected magnet is turned.

μ is the coefficient of induction.

q, q_1, are constants in the formula giving the effect of temperature on M.

T_1 is the period of a vibration corrected when necessary for rate of chronometer, and for infinitely small arc of vibration.

T is the period of a vibration corrected finally.

K is the moment of inertia of the magnet of magnetic moment M.

$\frac{G}{F}$ is the ratio of the torsion couple to the magnetic directive couple.

P is a constant depending on the distribution of magnetism in the deflecting and the suspended magnets, and is given by the equation

$$P = \frac{A - A_1}{\dfrac{A}{r^2} - \dfrac{A_1}{r_1^2}};$$

$$\text{where } A = \tfrac{1}{2} r^3 \sin\theta \left\{1 + \frac{2\mu}{r^3} + qt + q_1 t^2\right\},$$

$$A_1 = \tfrac{1}{2} r_1^3 \sin\theta_1 \left\{1 + \frac{2\mu}{r_1^3} + qt + q_1 t^2\right\},$$

r and r_1 representing here the two deflection distances.

t is the temperature in degrees centigrade reckoned from 0° C.

In appendix D will be found a copy of the observations taken at a typical station.

Values of P for different instruments.

Year	Magnetometer	Distances: 30 and 40 cm.	Distances: 25 and 35 cm.
1901–2	31	− 2·17	
1903	31	− 1·90	
1903	73 (Rhodesia)	+ 7·09	+ 6·45
	73 (Transvaal, Natal and Orange River Colony)	+ 7·33	
1904–5	73	+ 7·19	
1906	31	− 1·91	

Error of a determination of the horizontal intensity.

The error was determined by taking half the difference of the two values of H obtained in the manner explained above: the arithmetic mean of the errors gives the mean error of a determination. Its value for each instrument was $\pm 2\gamma$, where γ is ·00001 of a c.g.s. unit.

CHAPTER III

DAILY AND SECULAR VARIATIONS

Daily variation of declination. At the present time there is no permanent magnetic observatory in South Africa. This has made it impossible to apply in this survey a correction for daily variation—assuming it to be a function of local time—by taking the disturbances given on the continuous records of the instruments in such an observatory and applying them to correct the actual values of the magnetic elements as found by the field observations. It was therefore desirable to obtain some other method whose application could be justified. The known daily variations available for South African observations are those obtained at Cape Town, St Helena, and Mauritius. The Cape Town results are for the years 1841 to 1846, those for St Helena for the years 1840 to 1849, and the Mauritius results are taken from the 'Mauritius magnetical reductions' edited by C. Claxton, p. 17, table 17. The whole of these results were reduced so as to show the deviations of the declination at each hour of the day—time being Greenwich mean time—from the mean declination of the month. The mean of these deviations was then taken for four periods, viz. the winter, the summer, and the equinoctial months. Table 2 gives the daily variations so prepared and arranged for these three places.

The results given in this table were combined by taking the means of the latitudes, the longitudes, and the deviations of declination and obtaining thereby a mean station whose latitude was 23° 20′ S. and longitude 23° 27′ E., and where the declination deviations were at any hour the mean of those at the three stations at the same hour. The following formula was then applied,

$$\Delta D = \Delta D_m + x \sin \phi_r + y \sin \lambda_r,$$

where ΔD is the declination deviation at a particular hour at one of the stations,

ΔD_m is the declination deviation at the same hour at the mean station,

ϕ_r is the latitude of the station concerned with respect to the mean station,

λ_r is the longitude of the same station with respect to the mean,

x, y constants to be determined.

The values of ΔD_m, x, and y are given in table 3.

TABLE 2.

Solar diurnal variation of declination in minutes of arc. + denotes that the magnet at that time was East, and − that it was West, of the mean position. The time is Greenwich mean time.

CAPE OF GOOD HOPE. Latitude 33° 56′ S.; Longitude 18° 29′ E.

	noon	1 p.m.	2 p.m.	3 p.m.	4 p.m.	5 p.m.	6 p.m.	7 p.m.	8 p.m.	9 p.m.	10 p.m.	11 p.m.	midn't	1 a.m.	2 a.m.	3 a.m.	4 a.m.	5 a.m.	6 a.m.	7 a.m.	8 a.m.	9 a.m.	10 a.m.	11 a.m.
	′	′	′	′	′	′	′	′	′	′	′	′	′	′	′	′	′	′	′	′	′	′	′	′
Nov.—Feb.	+1·48	+2·09	+1·90	+1·35	+0·89	+0·67	+0·82	+1·01	+1·04	+1·04	+0·95	+0·85	+0·69	+0·51	+0·28	0·00	−0·26	−1·12	−2·43	−3·07	−3·64	−2·94	−1·41	+0·20
Mar.—Apr.	+1·26	+1·78	+1·61	+1·09	+0·64	+0·41	+0·65	+0·60	+0·49	+0·44	+0·46	+0·44	+0·51	+0·60	+0·58	+0·56	+0·57	+0·40	−0·54	−2·41	−3·87	−3·74	−2·13	−0·10
May—Aug.	−0·67	−0·02	+0·38	+0·20	−0·26	−0·39	−0·32	−0·20	−0·08	−0·02	0·00	+0·17	+0·31	+0·41	+0·54	+0·67	+0·75	+1·09	+1·59	+1·01	−0·40	−1·59	−1·90	−1·39
Sept.—Oct.	+0·70	+1·20	+1·15	+0·68	+0·18	+0·12	+0·30	+0·41	+0·42	+0·42	+0·39	+0·38	+0·44	+0·40	+0·35	+0·29	+0·23	+1·33	+0·53	−0·62	−2·12	−2·55	−1·82	−0·39

ST HELENA. Latitude 15° 57′ S.; Longitude 5° 40′ W.

	noon	1 p.m.	2 p.m.	3 p.m.	4 p.m.	5 p.m.	6 p.m.	7 p.m.	8 p.m.	9 p.m.	10 p.m.	11 p.m.	midn't	1 a.m.	2 a.m.	3 a.m.	4 a.m.	5 a.m.	6 a.m.	7 a.m.	8 a.m.	9 a.m.	10 a.m.	11 a.m.
	′	′	′	′	′	′	′	′	′	′	′	′	′	′	′	′	′	′	′	′	′	′	′	′
Nov.—Feb.	+1·10	+1·39	+1·07	+0·50	−0·02	−0·27	−0·17	+0·17	+0·53	+0·72	+0·78	+0·74	+0·58	+0·37	+0·18	+0·20	−0·19	−0·32	−0·47	−1·33	−2·22	−2·11	−1·18	+0·10
Mar.—Apr.	+1·07	+1·39	+0·92	+0·27	−0·21	−0·36	−0·23	−0·02	+0·04	+0·07	+0·18	+0·26	+0·28	+0·26	+0·20	+0·20	+0·28	+0·33	+0·54	+0·12	−1·26	−2·20	−1·80	−0·29
May—Aug.	−0·49	−0·55	−0·46	−0·25	−0·35	−0·73	−0·87	−0·66	−0·46	−0·33	−0·21	−0·09	+0·01	+0·06	+0·11	+0·19	+0·28	+0·49	+0·97	+1·82	+1·67	+0·61	−0·26	−0·53
Sept.—Oct.	+1·01	+1·23	+0·92	+0·25	−0·35	−0·55	−0·42	−0·21	−0·07	+0·03	+0·12	+0·14	+0·05	−0·07	−0·14	−0·17	−0·31	−0·19	+0·44	+0·16	−0·65	−0·97	−0·67	−0·10

MAURITIUS. Latitude 20° 6′ S.; Longitude 57° 33′ E.

	noon	1 p.m.	2 p.m.	3 p.m.	4 p.m.	5 p.m.	6 p.m.	7 p.m.	8 p.m.	9 p.m.	10 p.m.	11 p.m.	midn't	1 a.m.	2 a.m.	3 a.m.	4 a.m.	5 a.m.	6 a.m.	7 a.m.	8 a.m.	9 a.m.	10 a.m.	11 a.m.
	′	′	′	′	′	′	′	′	′	′	′	′	′	′	′	′	′	′	′	′	′	′	′	′
Nov.—Feb.	+1·9	+1·0	+0·5	+0·3	+0·25	+0·20	+0·1	0·0	−0·1	−0·1	−0·2	−0·3	−0·55	−0·9	−1·65	−2·6	−2·2	−2·35	−1·40	−0·2	+1·3	+2·25	+2·6	+2·55
Mar.—Apr.	+1·8	+0·8	+0·2	−0·1	−0·1	−0·15	−0·1	−0·1	−0·1	−0·1	−0·1	−0·15	−0·15	−0·2	−0·25	−0·65	−1·95	−2·85	−2·5	−1·2	+0·7	+2·0	+2·6	+2·4
May—Aug.	+1·7	+0·6	−0·3	−0·55	−0·55	−0·50	−0·35	−0·25	−0·1	0·0	+0·1	+0·15	+0·2	+0·3	+0·5	+1·20	+0·8	−0·6	−1·8	−1·9	−1·2	−0·2	+1·0	+1·8
Sept.—Oct.	+2·1	+1·15	+0·3	+0·1	+0·1	0·0	−0·1	−0·1	−0·1	−0·1	−0·1	−0·1	−0·1	−0·2	+0·15	−0·1	−0·9	−1·95	−2·55	−2·3	−0·9	+0·8	+2·25	+2·6

TABLE 3.

Values of ΔD_m, x, and y, in the formula $\Delta D = \Delta D_m + x \sin$ (relative latitude) $+ y \sin$ (relative longitude).

		noon	1 p.m.	2 p.m.	3 p.m.	4 p.m.	5 p.m.	6 p.m.	7 p.m.	8 p.m.	9 p.m.	10 p.m.	11 p.m.	midn't	1 a.m.	2 a.m.	3 a.m.	4 a.m.	5 a.m.	6 a.m.	7 a.m.	8 a.m.	9 a.m.	10 a.m.	11 a.m.
		′	′	′	′	′	′	′	′	′	′	′	′	′	′	′	′	′	′	′	′	′	′	′	′
Nov.	ΔD_m	+1·49	+1·49	+1·16	+0·72	+0·37	+0·10	+0·25	+0·39	+0·49	+0·55	+0·51	+0·43	+0·24	−0·01	−0·40	−0·80	−0·88	−1·26	−1·45	−0·53	− 1·52	− 0·93	0·00	+0·95
to	x	−3·44	−2·99	−3·65	−3·23	−2·80	−2·67	−3·12	−3·19	−2·63	−2·23	+0·89	−1·77	−1·89	−2·19	−2·79	−3·46	−2·40	+0·14	+5·20	+7·60	+ 9·64	+ 8·72	+5·79	+2·92
Feb.	y	+9·75	+0·58	−0·80	−0·42	+0·05	+0·27	+0·04	−0·39	−0·78	−0·92	−1·07	−1·12	−1·20	−1·36	−1·94	−2·89	−2·08	−1·93	−0·50	+1·60	+ 4·00	+ 4·71	+4·01	+2·57
Mar.	ΔD_m	+1·35	+1·32	+0·94	+0·42	+0·11	−0·03	+0·11	+0·16	+0·14	+0·14	+0·18	+0·18	+0·19	+0·22	+0·18	+0·04	−0·37	−0·71	−0·83	−1·16	− 1·48	− 1·31	−0·44	+0·67
to	x	+0·32	−2·17	−3·64	−3·60	−2·83	−2·41	−2·90	−2·29	−1·80	−1·50	−1·35	−1·28	−1·49	−1·80	−1·91	−2·37	−3·99	−4·47	−0·21	+7·18	+11·74	+10·98	+7·00	+2·89
Apr.	y	+0·72	−0·70	−0·94	−0·58	−0·09	+0·04	−0·08	−0·24	−0·23	−0·27	−0·36	−0·29	−0·54	−0·57	−0·56	−0·98	−2·38	−3·35	−2·92	−0·77	+ 2·68	+ 4·77	+4·69	+2·77
May	ΔD_m	+0·18	+0·01	−0·13	−0·20	−0·39	−0·54	−0·51	−0·37	−0·21	−0·12	−0·04	+0·08	+0·17	+0·26	+0·38	+0·69	+0·61	+0·33	+0·25	+0·31	+ 0·02	− 0·39	−0·39	−0·04
to	x	+3·53	−0·34	−2·77	−1·98	−0·60	−0·89	−1·23	−1·08	−0·84	−0·68	−0·35	−0·58	−0·82	−0·90	−1·03	−0·33	−0·96	−3·53	−5·86	−2·08	+ 3·45	+ 6·67	+7·43	+6·11
Aug.	y	+2·34	+1·08	−0·04	−0·42	−0·23	−0·16	+0·42	+0·32	+0·29	+0·27	+0·27	+0·19	+0·12	+0·16	+0·32	+0·94	+0·41	−1·28	−3·05	−3·70	− 2·50	− 0·31	+1·71	+2·65
Sept.	ΔD_m	+1·27	+1·19	+0·79	+0·34	−0·02	−0·14	−0·07	+0·03	+0·08	+0·12	+0·14	+0·14	+0·13	+0·04	+0·12	+0·01	−0·33	−0·27	−0·53	−0·92	− 1·22	− 0·91	−0·08	+0·70
to	x	+2·53	−0·02	−1·63	−1·73	−1·25	−1·60	−2·02	−2·05	−1·77	−1·52	−1·22	−1·16	−1·57	−1·83	−1·17	−1·51	−2·69	−7·67	−4·28	−0·52	+ 4·85	+ 7·88	+7·87	+4·60
Oct.	y	+1·21	−0·08	−0·70	−0·25	+0·35	+0·42	+0·16	−0·04	−0·15	−0·23	−0·30	−0·31	−0·25	−0·25	−0·18	−0·04	−0·75	−2·20	−3·15	−2·39	+ 0·10	+ 2·23	+3·32	+2·85

TABLE 4.

Observations at MATJESFONTEIN. Lat. 33° 14′·2 S.; Relative lat. −9° 54′·2. Long. 20° 36′·0 E.; Relative long. −2° 51′·0.

1900 April

W. 28° + Declination observed

Date	8 a.m.	9 a.m.	10 a.m.	11 a.m.	noon	1 p.m.	2 p.m.	3 p.m.	4 p.m.
	′	′	′	′	′	′	′	′	′
8th	—	—	47·8	—	—	—	—	—	45·1
9th	49·4	50·2	48·9	45·1	43·3	42·2	41·8	—	43·6
10th	47·7	47·6	45·6	44·1	43·9	—	44·0	—	45·4
11th	47·6	48·5	47·0	45·3	44·3	43·8	44·0	44·1	44·8
12th	49·7	48·4	46·0	44·1	43·5	43·3	43·3	43·1	—
13th	53·6	53·9	51·7	49·8	47·4	45·3	45·4	47·0	48·1
14th	52·0	52·4	47·2	44·3	42·2	41·6	42·7	—	—
Means	50·0	50·2	47·8	45·5	44·1	43·4	42·7	43·8	45·4

Greatest difference in means 28° 50′·2 − 28° 42′·7 = 7′·5

W. 28° + Declination corrected by formula $\Delta D = \Delta D_m + x \sin \phi_r + y \sin \lambda_r$

Date	8 a.m.	9 a.m.	10 a.m.	11 a.m.	noon	1 p.m.	2 p.m.	3 p.m.	4 p.m.
	′	′	′	′	′	′	′	′	′
8th	—	—	46·3	—	—	—	—	—	45·6
9th	46·2	47·1	47·2	45·1	44·3	44·0	43·4	—	44·2
10th	44·6	44·2	43·8	44·1	45·0	—	45·9	—	46·0
11th	44·4	45·0	45·2	45·3	45·3	45·2	45·6	45·1	45·4
12th	46·7	45·0	44·4	44·1	44·6	44·8	44·9	44·3	—
13th	50·4	50·5	49·9	49·8	48·4	47·1	46·9	47·6	48·7
14th	48·7	49·0	45·5	44·3	43·3	43·2	44·5	—	—
	46·8	46·8	46·1	45·5	45·2	45·0	45·3	45·6	46·4

28° 46′·8 − 28° 45′·0 = 1′·8

DURBAN.

W. 25° +

		6 a.m.	10 a.m.	1 p.m.	7 p.m.
		′	′	′	′
1893	July—Aug.	27·9	33·5	28·2	28·6
	Sept.—Oct.	29·4	29·4	24·3	26·6
	Nov.—Dec.	28·7	19·8	18·8	22·2
	Means	28·7	27·6	23·8	25·8
1894	Jan., Feb., Nov., Dec.	25·7	17·3	14·7	19·1
	Mar.—Apr.	24·8	21·2	17·8	—
	May—Aug.	18·7	22·6	17·4	—
	Sept.—Oct.	20·0	19·9	14·8	17·8
	Means	22·3	20·3	16·2	—
	Means for 1893–1894	25·5	24·0	20·0	—

Greatest difference in means 25° 25′·5 − 25° 20′·0 = 5′·5

W. 25° +

		6 a.m.	10 a.m.	1 p.m.	7 p.m.
		′	′	′	′
1893	July—Aug.	27·1	34·2	28·3	28·3
	Sept.—Oct.	28·0	30·2	25·5	26·4
	Nov.—Dec.	27·8	21·0	20·0	21·7
		27·6	28·5	24·6	25·5
1894	Jan.—Dec.	24·8	18·5	15·9	18·6
	Mar.—Apr.	23·6	22·2	18·8	—
	May—Aug.	17·9	23·3	17·5	—
	Sept.—Oct.	18·6	18·7	16·0	17·6
		21·2	20·7	17·1	—
		24·4	24·6	20·9	—

25° 24′·6 − 25° 20′·9 = 3′·7

The formula with known constants can now be used to calculate the declination deviation at some other station of known latitude and longitude where observations had been taken. The corrections when applied to values of the declination at the same place found at different times of the day should—were there no disturbance and were the means of obtaining the correction satisfactory—give the same value in each case for the declination.

It is to be noted that this formula has been taken as the simplest convenient one giving the proper value for the mean station on the assumption that there is no secular variation of the daily variation of declination, an assumption which so far as the writer is aware has no experimental foundation. It would be possible from a similar formula for the northerly and the westerly components to calculate the magnetic intensity round any closed curve and from that the electric current necessary to account for the daily variations. This has not been done here, the problem not being germane to the present purpose.

The only test of the validity of the formula lies in its application to results obtained independently of those which have been used in deriving it. At Matjesfontein observations of the declination were made throughout the day for about a week in April 1900. The relative latitude for it is 9° 54′ S. and the relative longitude 2° 51′ W. These values were used in the above formula, and the Δ_m, x, and y were taken for the corresponding time from table 2—a linear interpolation being used when the time was not an exact hour. In this way the declination deviation as given by the above formula was found. The same procedure was gone through for the results obtained over a number of years at the Natal Government observatory by Mr E. N. Nevill, F.R.S., the government astronomer. In table 4 the observed and the calculated values are shown for these two places.

It will be seen from table 4 that after the application of the correction the difference between the highest and the lowest mean hourly reading of the declination has been reduced in the case of Matjesfontein from 7′·3 to 1′·8 ; in the case of Durban from 5′·5 to 3′·7.

Observations have been taken at other places at different hours of the same day. The correction according to the above formula has been calculated and applied to the separate results. In the following table a list of the places is given where this has been done, the first column of figures gives the greatest difference in the observed declination, the second the greatest difference after the correction had been applied.

TABLE 5.

CAPE COLONY RESULTS—

Place	Greatest difference in observed declination ′	Greatest difference in corrected declination ′	Place	Greatest difference in observed declination ′	Greatest difference in corrected declination ′
Matjesfontein	7·5	1·8	Berg River Mouth	2·1	3·1
Stellenbosch	1·1	0·2	Malmesbury	0·3	0·4
Rosmead Junction	1·3	0·4	Naauwpoort	9·5	6·7
Bethesda Road	5·8	4·7	Signal Hill	4·2	1·1
Zuurpoort	8·8	7·1	Aliwal North	7·9	4·7
Aberdeen	2·1	1·3	Queenstown	2·5	1·7
Klipplast	3·2	2·3	Indwe	2·1	1·3
Cape Town	7·1	5·9	Tulbagh Road	0·3	4·5
Simonstown	10·0	5·7	Nelspoort	4·7	1·2
Sir Lowry's Pass	2·3	1·3	Worcester	1·6	0·0
Howhoek	8·0	4·0	Colesberg	5·6	4·4
Villiersdorp	7·4	3·5	Grahamstown	9·9	9·6
Hermanus	4·4	0·9	Port Alfred	5·0	3·1
Stanford	7·0	3·8	Middleton	0·6	0·9
L'Agulhas	2·9	1·3	Coerney	1·9	2·6
Bredasdorp	3·3	0·7	Witmoss	1·4	0·4
Hutchinson	2·1	2·2	Fish River	1·8	1·4
De Aar	9·6	4·7	Prince Albert Road	2·9	0·8
Elim	0·9	0·7	Prince Albert	3·9	2·0
Swellendam	6·8	2·9	Zeekoegat	1·0	0·7
Robertson	7·4	4·4	Oudtshoorn	4·7	1·6
Strandfontein	5·5	1·7	Cango	2·8	7·5
Darling	1·0	0·6	Ladismith, C. C.	7·1	3·1
Hoetjes Bay	1·8	1·0	Avontuur	4·8	0·8
Van Wyk's Farm	0·3	5·8	Buffelsklip	3·5	2·1
Riversdale	6·8	3·0	Vondeling	6·7	9·2
Still Bay	6·6	1·7	Potfontein	4·4	2·5
Tygerfontein	6·4	1·5	Honey Nest Kloof	13·8	9·9
Mossel Bay	2·9	1·0	Warrenton	9·8	7·4
Plettenberg Bay	5·1	5·4	Mafeking	6·9	5·1
Gamtoos River Bridge	2·4	1·0			

BECHUANALAND AND RHODESIAN RESULTS—

Place	Greatest difference in observed declination ′	Greatest difference in corrected declination ′	Place	Greatest difference in observed declination ′	Greatest difference in corrected declination ′
Gwaai	4·9	4·9	Salisbury	1·8	1·5
Shoshong Road	5·7	3·5	Marandellas	4·9	4·8
Malinde	1·2	1·0	Umtali	0·6	0·7
Magalapye	3·0	1·6	Malenje	7·1	5·8
Palapye	5·0	3·6	Beira	4·7	3·6
Shangani	2·1	1·3	Ayrshire Mine	8·5	8·2
Lochard	1·3	0·4	Forty-one mile Siding	2·0	1·8
Hartley	4·6	3·2	Makwiro Siding	10·0	9·7
Battlefields	3·1	2·6			

ORANGE RIVER COLONY RESULTS—

Place	Greatest difference in observed declination ′	Greatest difference in corrected declination ′	Place	Greatest difference in observed declination ′	Greatest difference in corrected declination ′
Vredefort	5·3	2·6	Karree	0·1	0·3
Springfontein	4·1	4·0	Smaldeel	3·6	3·5
Bethulie	3·4	3·5	Boschrand	1·0	1·7
Groenplaats	11·0	7·3	Honing Spruit	1·1	1·1
Krugers	4·4	4·3	Abelsdam	4·2	0·5
Bethany	2·2	2·1	Heilbron	0·3	0·4
Ferreira	1·2	0·9	Saxony	8·4	5·8
Thaba 'Nchu	7·5	7·2	De Jager's Farm	6·3	2·8
Winkeldrift	6·7	3·9			

TABLE 5 (*continued*).

TRANSVAAL AND NATAL RESULTS—

Place	Greatest difference in observed declination	Greatest difference in corrected declination	Place	Greatest difference in observed declination	Greatest difference in corrected declination
	′	′		′	′
Elsburg	0·2	0·0	Vlaklaagte	8·7	7·6
Randfontein	1·1	1·4	Kraal	10·2	9·1
Welverdiend	1·1	1·3	Charlestown	15·2	12·5
Nylstroom	9·9	9·6	Newcastle	6·5	5·6
Pietersburg	2·4	1·8	Dannhauser	5·1	2·5
Piet Potgietersrust	6·3	6·2	Waschbank	5·1	3·6
Kalkbank	2·0	0·6	Modderspruit	4·7	3·1
Klipfontein	0·7	0·7	Albert Falls	0·6	0·0
Bethal	8·1	7·3	Krantzkloof	5·5	3·5
Lydenburg	3·5	3·1	Stanger	9·0	6·9
Nelspruit	4·7	2·0	Hlabisa	6·0	3·3
Barberton	6·5	3·8	Kwambonambi	2·5	2·0
Uitkyk	5·5	5·1	Ginginhlovu	6·4	4·2
Naboomspruit	8·2	7·2	Colenso	3·3	2·8
Aberfeldy	10·2	8·5	Underberg	12·5	10·3
Balmoral	6·2	5·1	Indowane	6·7	3·2
Hamaans Kraal	5·1	4·4	Ibisi Bridge	6·4	3·0
Greylingstad	3·8	2·7			

On the whole it may be said that the correction improves the results. It is as a rule too small; this is particularly the case in winter. The further discussion of the matter can hardly be said to be justified with the data available. The two points which require elucidation are first the presence or otherwise of a secular variation of the daily variation and secondly the possibly greater value of the daily variation in the winter time.

A correction for daily variation obtained in the above manner has been applied to all the declination observations.

Daily variation of the horizontal intensity. No satisfactory method was found for applying a correction for the daily variation of this element. The method employed in the case of the declination was first tried but the corrections were too small. A second method was the following. It will be remembered that the horizontal intensity observations were first a vibration experiment, second a deflection at two distances, third another vibration. The values of the magnetic moment were obtained by the process described in Chapter II, and were plotted against time, a straight line was then drawn through the points; from it the probable value of the magnetic moment may be found at any time. The value obtained in this way at the time when a given vibration was made was then used with the corresponding period of vibration to determine the horizontal component at that time. To find the rate of daily variation from these values the difference of the two H values at the different times of the vibration experiments was taken and divided by the interval of time in hours; this was called the rate of change at the mean time of the two. After this had been done for a number of stations, the algebraic mean of all the rates between

8.31 a.m. and 9.29 a.m. (say) was found and called the rate of change at 9 a.m. The values obtained were as follows :—

Between	7 a.m.	and	8 a.m. G.M.T.	a decrease of	2·4 γ
..	8	..	9	..	4·8 γ
..	9	..	10	..	4·8 γ
..	10	..	11	..	5·6 γ
..	11	..	12 noon	..	7·2 γ
..	12 noon	..	1 p.m.	..	5·6 γ

These were derived from the data obtained in the field at places north of the Orange River.

The effect of applying the corrections obtained in this way to stations not used in deriving them is seen in the following table.

TABLE 6.

Place	Difference in H values as observed (unit 1γ)	Difference after correction (unit 1γ)	Place	Difference in H values as observed (unit 1γ)	Difference after correction (unit 1γ)
Kimberley	22	3	Grahamstown	10	0
Stellenbosch	10	10	Mossel Bay	47	34
Beaufort West	7	6	Barrington	29	22
Prince Albert	13	5	Knysna	19	11
Zuurpoort	13	15	Blaauwkrantz	17	1
Willowmore	7	7	Storms River	0	13
Malmesbury	13	13	Middelberg	21	8
Signal Hill	20	0	Gamtoos River Bridge	28	48
East London	15	6	Assegai Bosch	7	3
Cape Town	8	21	Buffelsklip	11	3
Cape Town	45	62			

The correction in most cases improves the results; it has not been applied to the values of the horizontal intensity given later.

Daily variation of dip. The field observations gave no sufficient means of testing whether or not a correction improved the results and none has been applied. In the case of this element—as well as the others—the time of observation has been given in the tabulated results of appendix E.

Secular variation. To determine the secular variation the three magnetic elements were observed at a number of places distributed as widely as possible and at intervals of one or more years.

Secular variation of declination. In the following table the secular variation is given for a number of stations. It has been determined for any particular year or years by taking the difference of the declination corrected for daily variation at the end and the beginning of the time interval and dividing this by the number of years. The minus sign denotes a decrease of westerly declination.

TABLE 7.

Place	Period	Secular variation ′	Remarks
Aberdeen Road	1900–1904	−10·1	Observations taken in July in both years
Beaufort West	1900–1905	− 4·4	.. .. January in both years
Beira	1892–1903	−10·9	
Cape Town, R.O.	1900–1902	− 3·8	.. .. Dec. 1900, Jan. 1901, and in March, July and Nov. 1902
Colesberg	1902–1904	− 7·7	.. .. July in both years
Cookhouse	1900–1904	− 9·2	.. .. January 1900 and July 1904
Hutchinson	1900–1904	− 8·4	.. .. January 1900 and June 1904
Mafeking	1903–1906	−10·3	.. .. March 1903 and January 1906
Matjesfontein	1900–1904	− 7·3	.. .. April 1900, Dec. 1902, Dec. 1903, Sept. and Oct. 1904
Miller Siding	1900–1904	−11·0	.. .. July of both years
Prince Albert Road	1900–1903	− 5·2	.. .. January of both years
Reuben Point (Delagoa Bay)	1894–1899	−10·0	
Rosmead Junction	1900–1902	−10·8	.. .. July of both years
Walfisch Bay	1892–1904	−10·6	
Wepener	1886–1904	−11·5	
Willowmore	1900–1903	−10·3	.. .. July 1900 and Feb. 1903
Worcester	1875–1902	− 2·6	
Picene	1903–1906	−10·5	.. .. Sept. 1903 and July 1906
Tulbagh	1870–1902	− 4·2	

Secular variation of declination at Durban.

Period	6 a.m. ′	7 a.m. ′	10 a.m. ′	1 p.m. ′	7 p.m. ′	Mean ′
1893–4	− 8·7		− 8·2	− 9·0	− 8·1	− 8·5
1894–5	− 11·1		− 6·5	− 7·0	− 12·2	− 9·2
1895–6	− 11·1	− 10·5	− 9·3			− 10·3
1896–1904	− 11·4	− 12·5	− 12·4			− 12·1

In Durban secular variations have been obtained by subtracting the mean value at a given hour from the mean value at the same hour in some succeeding year and dividing by the number of years. The results show that the secular variation at Durban has increased in value since 1893.

Observations have been taken at approximately the same spot in the grounds of the Royal Observatory, Cape Town, since 1841. The subjoined table gives the value of the secular variation at that place from 1841 to 1900.

Period	Secular variation ′	Remarks
1841–1846	+ 1·8	Observations taken hourly
1846–1852	+ 2·3	.. .. five times daily
1852–1860	+ 2·4	
1860–1869	+ 2·2	.. .. twice daily
1873–1890	− 1·8	Challenger results in 1873, Preston's results in 1890,
1890–1900	− 3·8	Beattie and Morrison's results in 1900

The accompanying map gives the geographical positions of the repeat stations with the values of the secular variation of declination at them.

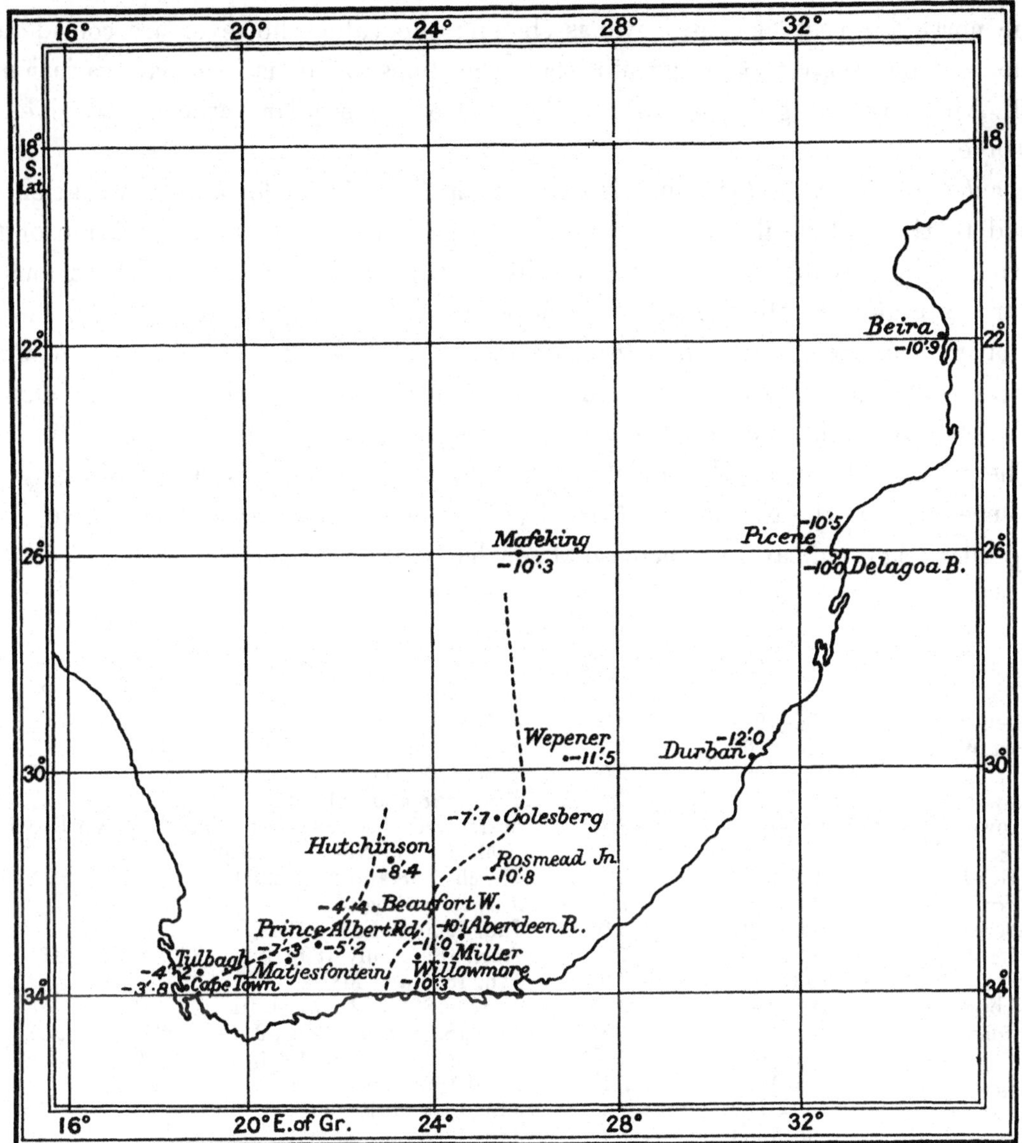

Map 1. Showing the secular variation of declination in minutes of arc per year. The minus sign denotes a decrease of westerly declination.

It will be seen from this map that the value of the secular variation increases towards the east and the north. A line cutting the south coast of the Cape Colony at Knysna and passing in a north-north-easterly direction to about Burghersdorp separates a region on the east with a secular variation of 10′ per year from a westerly region where it is less. This same line after passing Burghersdorp runs almost due north to Mafeking and then north-west by west to Walfisch Bay. A second line starting from Cape Town runs north-east to Beaufort West and then north to

Hutchinson. Between these two lines the variation is from 4′ to 10′ per year. This map should be compared with Neumayer's maps* giving the secular variation for the period 1870–1890, and the period 1890–1900. In both of the latter maps the decrease to the north and east of Cape Town is shown; the values however are considerably smaller. The comparison suggests that the lines of equal secular variation in South Africa are roughly parallel to the line of no secular variation as given by Neumayer.

Secular variation of the dip. No correction for daily or for annual variation was applied to the field results used for the determination of the secular variation of this element. The precaution was taken in this as well as in the other observations for secular observation to observe at the same place and at the same time of day, thereby eliminating the daily variation. In some instances it was possible to observe at the same place in different years at the same season of the year—spring, summer, autumn or winter—and at these places the annual variation was eliminated.

The results given in the following table have been derived by dividing the difference between the dips at two times by the interval in years between these times. The positive sign indicates an increase of southerly dip.

TABLE 8.

Place	Period	Secular variation	Observations taken in
		′	
Aberdeen Road	1900–1904	+ 8·5	July in both years
Beaufort West	1900–1905	+ 8·3	Jan. in both years
Beira	1903–1906	+ 6·0	Sept. 1903 and July 1906
Bulawayo	1898–1904	+ 7·2	Jan. 1898 and July 1904
Cape Town	1898–1904	+ 8·7	March and May 1898, Oct. 1899, Aug. and Dec. 1900, March and Nov. 1902, April 1904
Ceres Road	1898–1900	+ 8·0	March 1898 and May 1900
Colesberg	1900–1904	+ 7·0	July in both years
Cookhouse	1900–1904	+ 8·0	Jan. 1900 and July 1904
Hutchinson	1898–1904	+ 7·9	April 1898, July 1899, Jan. 1900, June 1904
Mafeking	1898–1903	+ 7·6	Jan. 1898 and March 1903
Malmesbury	1898–1901	+ 9·0	March 1898 and Aug. 1901
Matjesfontein	1899–1905	+ 8·3	July 1899, April, May, July 1900, Dec. 1902, Dec.
	1904–1906	+ 6·2	1903, Jan., Sept. 1904, Feb. 1905
Naauwpoort	1898–1901	+ 7·1	April 1898, Dec. 1901
Orange River	1898–1902	+ 8·4	Jan. 1898, July 1902
Palapye	1898–1903	+ 7·7	Jan. 1898, March 1903
Picene	1903–1906	+ 3·7	Sept. 1903, July 1906
Prince Albert Road	1899–1903	+ 8·8	July 1899, Jan. 1900, Jan. 1903
Rosmead Junction	1900–1904	+ 7·9	July in both years
Sir Lowry's Pass	1898–1901	+ 9·2	May 1898 and Jan. 1901
Miller Siding	1900–1904	+ 7·8	July in both years
Uitenhage	1900–1904	+ 8·0	July in both years
Willowmore	1900–1903	+ 8·8	July 1900, Feb. 1903
Worcester	1899–1902	+10·0	Oct. 1899, April 1902

Of the above stations the only one where observations have been taken for any length of time at approximately the same position is in Cape Town in the grounds of the Royal Observatory of the Cape of Good Hope.

* "Atlas des Erdmagnetismus von Neumayer," *Terrestrial Magnetism*, Vol. VI. p. 62.

The values of the secular variation for this station since 1841 are as follows:

Period	Secular variation	Remarks
1841–1846	+ 5·5	Observations taken twice weekly
1843–1854	+ 5·1	.. .. twice weekly
1854–1873	+ 2·2	.. .. twice weekly, 1854, *Challenger* results, 1873
1873–1890	+ 4·9	Preston's results in 1890
1890–1900	+ 7·8	Beattie and Morrison's results in 1900

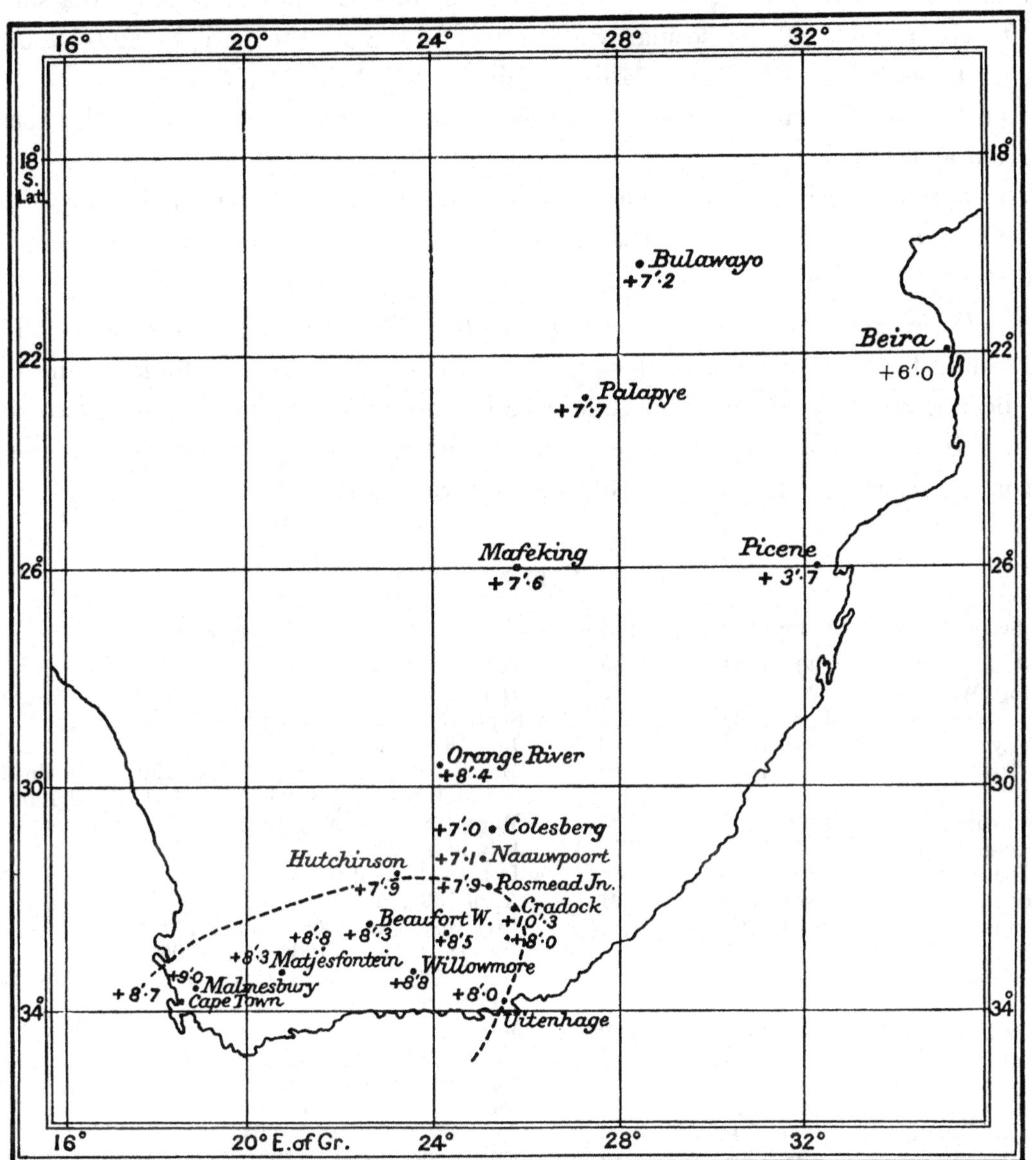

Map 2. Showing the secular variation of dip in minutes of arc per year. The positive sign denotes an increase of southerly dip.

The secular change in dip at Cape Town reached a minimum at the time that of the declination was zero. Since then it has increased in value and the results found in connection with this survey show that the acceleration continued till about 1902.

This is seen if instead of taking the mean change between 1898 and 1904 we take those for different parts of that period; it is then found that the yearly change between 1898 and 1900 was $+7'{\cdot}3$, between 1900 and 1902 $+10'{\cdot}4$, and between 1902 and 1904 $+9'{\cdot}0$. The other station in the above list at which a number of determinations of dip has been made and where therefore it is possible to get values of the secular change from year to year is Matjesfontein. The values found there —extending however only from 1899 to 1906—show the same effect only to a smaller extent, viz. a value of the secular variation of $+7'{\cdot}8$ in the period 1899–1900, increasing to $+8'{\cdot}8$ in 1902–3, and falling again to $+7'{\cdot}8$ in 1904–6.

Map 2 gives the distribution of stations, and the values adopted for the secular variation at them.

The dotted line is a line of equal secular variation, the value on it being $+8'{\cdot}0$. The effect as one goes north or east from this line is a decrease in the value of the variation.

Secular variation of the horizontal intensity. No corrections were made on the field results for daily or for annual variation. The results given in the following table have been derived by dividing the difference between the two intensities at two times by the interval in years between the times. The negative sign indicates a decrease of horizontal intensity. The values are given in terms of γ.

TABLE 9.

Station	Period	Variation	Observations taken in
Aberdeen Road	1900–1904	– 83	July in both years
Beaufort West	1900–1905	– 88	Jan. in both years
Beira	1903–1906	– 70	Sept. 1903, and July 1906
Bulawayo	1898–1904	– 60	Jan. 1898, and July 1904
Cape Town	1898–1905	– 96	March 1898, Dec. 1900, Jan. 1901, March 1902, March and May 1903, May 1904, April 1905
Ceres Road	1898–1900	– 69	March 1898 and May 1900
Colesberg	1902–1904	– 87	July in both years
Cookhouse	1900–1904	– 88	Jan. 1900 and July 1904
Cradock	1898–1900	– 91	April 1898, and Jan. 1900
Hutchinson	1899–1904	– 83	July 1899, Jan. 1900, June 1904
Mafeking	1898–1906	– 70	Jan. 1898, March 1903, Jan. 1906
Matjesfontein	1899–1904	– 85	July 1899, May, July 1900, Dec. 1903, Jan., Sept. Oct. 1904
Malmesbury	1898–1901	– 80	March 1898, Aug. 1901
Naauwpoort	1898–1901	– 96	April 1898, Jan. 1901
Palapye	1898–1903	– 60	Jan. 1898, March 1903
Picene	1903–1906	– 77	Sept. 1903, July 1906
Rosmead Junction	1900–1904	– 86	July 1900, 1902, and 1904
Stellenbosch	1898–1906	– 84	April 1898, Sept. 1899, June 1900, Aug. 1901, March 1906
Miller Siding	1900–1904	– 88	July in both years
Uitenhage	1900–1904	– 95	July in both years
Willowmore	1900–1903	– 90	July 1900, Feb. 1903
Orange River	1898–1902	– 72	Jan. 1898, July 1902

The only station where there is a number of observations extending over a considerable period of time is that at the Royal Observatory, already referred to.

The mean annual secular variation of this element there since 1843 is given in the following table:

Period	Secular variation
	γ
1843–1855	– 35
1855–1873	– 33
1873–1890	– 45
1890–1897	– 43
1898–1905	– 96

There seems to be no doubt that the value of the horizontal intensity secular variation was increasing at the time of this survey.

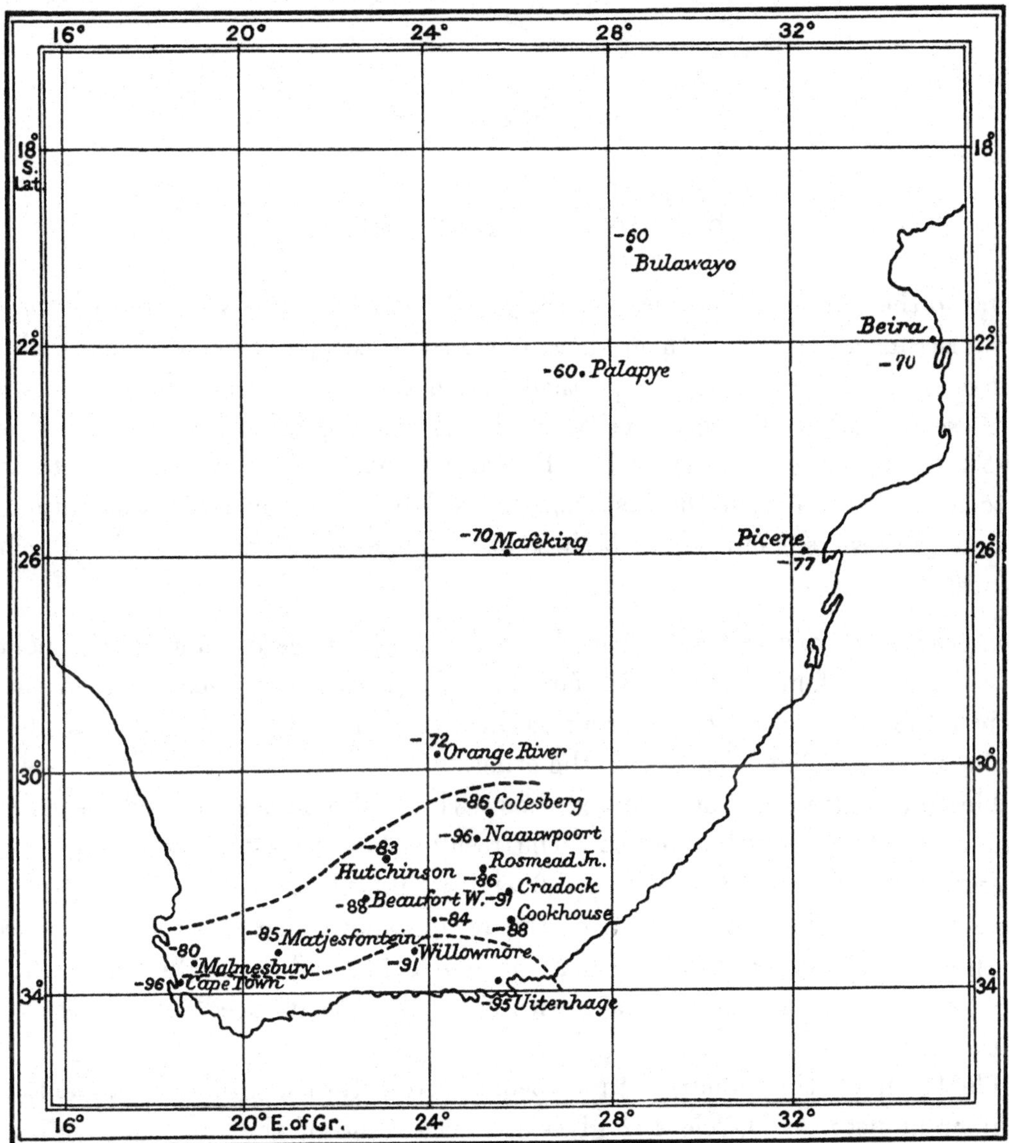

Map 3. Showing secular variation of the horizontal intensity per year in terms of γ. The minus sign denotes a decrease in the value of the intensity.

Map 3 gives the distribution of the stations and the value of the secular variation in terms of γ for each station. The two dotted lines are lines of equal secular variation of H, the lower for 90γ, the more northerly for 80γ. The line of maximum decrease in the value of secular variation is almost due north.

CHAPTER IV

COMPARISON OF INSTRUMENTS

DURING the course of the survey it was found possible not only to compare the two sets of instruments amongst themselves, but also to carry out comparisons with the instruments of other observers, in particular with the "Discovery" instruments, and with those of Major Chaves, Director of the Meteorological service of the Azores, who—thanks to the generosity of the Prince of Monaco—was able to devote some time to a comparison with the instruments used here and by later observations to show how the results compared with those of the magnetic observatory of Val Joyeux in France.

Comparison of dip circles 142 *and* 9. Comparisons were made in the field at Cape Town, Strandfontein and Matjesfontein. The instruments were used in such a way that the mean time of observation at one place with one instrument was approximately the same as that with the other.

Table 10 contains a list of observed dips with the dip circles 142 and 9 each with its pair of needles marked 1 and 2. The results show that the average difference between the dip observed with circle 142 and with circle 9 is 3′·4, the former being the greater. A further examination of the results shows that this total difference is made up of two parts, an excess of 0′·9 with circle 142 and its needle 1 over circle 9 and its needle 1, and an excess of 5′·9 with circle 142 and its needle 2 over that of 9 with its No. 2.

Table 11 shows the values of the dip obtained with the same instruments but with different needles. Circle 142 gives a lower value of the dip than circle 9—when each is used with its own No. 3 and No. 4 needles—by 1′·25. The total difference is practically identical with the separate differences. The second part of this table gives a comparison of the dip obtained with needles No. 3 and 4 of circle 142 with that of needles No. 1 and 2 of circle 9.

TABLE 10.

CAPE TOWN. Observer, Beattie

Date	G.M.T.	Instrument	Needle	Dip	Mean	Needle	Dip	Mean	Diff. 142 − 9
	h m			° ′	° ′		° ′	° ′	′
1900 Dec. 26th	8 7 a.m.	142	1	58 43·0		2	58 44·5		
Dec. 27th	10 0 a.m.	..	..	58 40·2		..	58 42·4		
					58 41·6			58 43·4	
Dec. 26th	10 0 a.m.	9	1	58 40·3		2	58 38·9		
Dec. 27th	8 0 a.m.	..	..	58 41·1		..	58 38·5		
					58 40·7			58 38·7	
				Differences	+ 0·9			+ 4·7	+ 2·8
STRANDFONTEIN									
1901 Feb. 21st	9 46 a.m.	142	1	58 54·6		2	58 57·4		
	3 22 p.m.	..	..	58 53·6		..	58 57·3		
Feb. 22nd	11 42 a.m.	..	..	58 53·7		..	58 57·3		
		..	..	58 51·1		..	58 54·8		
					58 53·3			58 56·7	
Feb. 21st	11 42 a.m.	9	1	58 52·9		2	58 50·3		
	1 46 p.m.	..	..	58 54·9		..	58 49·9		
Feb. 22nd	9 15 a.m.	..	..	58 51·8		..	58 50·2		
	12 42 p.m.	..	..	58 52·6		..	58 48·5		
					58 53·1			58 49·7	
				Differences	+ 0·2			+ 7·0	+ 3·6
MATJESFONTEIN									
1902 Dec. 25th	6 10 a.m.	142	1	59 47·3		2	59 51·2		
Dec. 26th	8 58 a.m.	..	..	59 49·9		..	59 49·6		
					59 48·6			59 50·4	
Dec. 25th	9 52 a.m.	9	1	59 45·4		2	59 46·4		
Dec. 26th	5 39 a.m.	..	..	59 47·2		..	59 44·6		
					59 46·3			59 45·5	
				Differences	+ 2·3			+ 4·9	+ 3·6
				Means of differences	+ 0·9			+ 5·9	+ 3·4

TABLE 11.

MATJESFONTEIN. Observer, Beattie

Date	G.M.T.	Instrument	Needle	Dip	Mean	Needle	Dip	Mean	Diff. 142 − 9
	h m			° ′	° ′		° ′	° ′	′
1902 Dec. 25th	6 13 a.m.	142	3	59 46·1		4	59 47·2		
Dec. 26th	8 58 a.m.	..	..	59 45·9		..	59 45·6		
					59 46·0			59 46·4	
Dec. 25th	9 52 a.m.	9	3	59 47·9		4	59 48·8		
Dec. 26th	5 37 a.m.	..	..	59 45·5		..	59 47·7		
					59 46·7			59 48·2	
				Differences	− 0·7			− 1·8	− 1·25
1906 June 28th	9 22 a.m.	142	3	60 13·0		4	60 14·5		
..	10 24 a.m.	..	..	60 11·7		..	60 15·4		
June 30th	1 13 p.m.	..	..	60 10·6		..	60 12·8		
					60 11·8			60 14·2	
Observer, Morrison									
June 28th	9 26 a.m.	9	1	60 12·1		2	60 10·7		
..	10 30 a.m.	..	..	60 12·2		..	60 11·1		
June 30th	1 8 p.m.	..	..	60 11·3		..	60 10·1		
					60 11·9			60 10·6	
				Differences	− 0·1			+ 3·6	+ 1·75

The first is the greater by 1′·75. This is composed of two differences, the first due to needle 3 of instrument 142, which gives a lower value of 0′·1, and the second due to needle 4 of instrument 142, which gives a higher value by 3′·6.

Table 12 gives the result of a comparison for the purpose of discovering to what extent the differences are due to peculiarities of the dip circles themselves. To elucidate this point a number of observations was made with each instrument at approximately the same time and with the same needles on both. The mean difference is 0′·2.

It appears that the differences are due almost entirely to peculiarities in the needles. It will be seen also that the comparison of one circle with another has a different value according to the needles used, and how to get a correction to a circle such that the corrected result gives the true dip is a matter of some difficulty.

TABLE 12.

CAPE TOWN. Observer, Beattie

	Date	Instrument	Needles	Dip	Instrument	Needles	Dip	Diff. 142 − 9
				° ′			° ′	′
1904	April 24th	142	1_{142} 4_9	59 15·5	9	1_{142} 4_9	59 14·3	+ 1·2
MATJESFONTEIN								
	Sept. 21st	..	1_{142} 4_9	59 58·7	..	1_{142} 4_9	59 58·7	− 0·0
	..	..	1_{142} 4_9	59 59·4	..	1_{142} 4_9	59 57·8	+ 1·6
	Sept. 22nd	..	1_{142} 4_9 3_9	59 59·8	..	1_{142} 4_9 3_9	59 59·6	+ 0·2
	..	..	1_{142} 4_9 3_9	59 58·9	..	1_{142} 4_9 3_9	59 59·4	− 0·5
	Sept. 23rd	..	4_9 3_9	60 0·6	..	4_9 3_9	60 1·4	− 0·8
	..	..	4_9 3_9	59 59·7	..	4_9 3_9	59 59·8	− 0·1
							Mean difference	+ 0·2

NOTE. 1_{142} means needle No. 1 of dip circle 142. When no index is attached to the number of a needle it is to be understood that the needle is used on the dip circle to which it belongs.

If we suppose that the circles give the same dip with the same needles—an assumption made from the results in table 12 by neglecting the small difference 0′·2—and if further we take into account the mean differences between the needles as found from the field observations, viz. for

circle 9 No. 1 > No. 2 by + 1′·7
,, 142 No. 2 > No. 1 by + 2′·1
,, ,, No. 3 > No. 4 by − 0′·1
,, ,, No. 4 > No. 1 by − 0′·7
,, ,, No. 1 > No. 4_9 by + 0′·5

We have the following table giving the correction to be applied to each needle reading to reduce it to No. 4 of circle 9 used on circles 142 or 9.

TABLE 13.

Circle	Needle	Correction	Circle	Needle	Correction
		′			′
9	1	+0·4	142	1	−0·5
..	2	+2·1	..	2	−2·6
..	3	−0·2	..	3	+0·3
..	4	0·0	..	4	+0·2
			..	3_9	−0·2
			..	4_9	0·0
		—— +0′·6			—— −0′·5

A rough mean correction to be applied to the results of the separate circles irrespective of what needle is used is obtained by taking the mean of the above separate corrections. The dip therefore as given by circle 142 is 1′·1 greater than that given by circle 9.

If the results of needles marked 2 in each instrument be neglected, the dip as given by instrument 142 is higher than that by instrument 9 by a quantity which can certainly be neglected.

In actual field observations No. 2 of circle 142 was rejected at an early date, a spot of rust having appeared on the axle.

Comparison of magnetometers 31 *and* 73. *Horizontal intensity.* For this comparison observations were taken at Matjesfontein and at Cape Town. When only one observer was available the instruments were used in such a manner as to make the mean time of observation with one of them the same as that of the other. When two observers were available the instruments were used simultaneously, each in its own tent. After each determination, consisting of a vibration, a deflection, and a second vibration, the instruments and the observers exchanged positions, in this way eliminating any effect due to the position of observation. An attempt was also made in Cape Town to see if the difference could be allocated to any definite part of the instruments; unfortunately it was found that all the parts were not interchangeable.

TABLE 14.

ROYAL OBSERVATORY GROUNDS, CAPE TOWN

Date	G.M.T.	Value of H with instrument		Mean difference	
	h m	73	31	73 − 31	Observer
1903 March 9th	8 6 a.m.	·18318			Beattie
..	1 29 p.m.	·18276			..
..	1 56 p.m.		·18315		..
March 10th	10 34 a.m.	·18287			..
..	10 12 a.m.		·18304		..
May 19th	10 31 a.m.		·18276		..
..	1 3 p.m.		·18290		..
..	1 3 p.m.	·18268			..
May 20th	11 3 a.m.	·18280			..
..	12 16 p.m.		·18320		..
		——	——	−15 γ	

TABLE 14 (*continued*).

MATJESFONTEIN

Date	G.M.T. h m	Value of H with instrument 73	31	Mean difference 73 − 31	Observer
1904 Sept. 21st	1 17 p.m.		·17914		Beattie
..	2 37 p.m.	·17895			..
Sept. 20th	9 40 a.m.	·17932			..
..	9 56 a.m.		·17965		..
..	12 50 p.m.		·17899		..
..	2 1 p.m.	·17892			..
Sept. 22nd	12 46 p.m.	·17908			..
..	2 5 p.m.		·17904		..
Sept. 23rd	1 13 p.m.		·17908		..
..	2 43 p.m.	·17907			..
Sept. 24th	8 40 a.m.	·17950			..
..	10 35 a.m.		·17928		..
		———	———	− 6 γ	
1904 Oct. 29th	10 22 a.m.		·17903		Morrison
..	10 7 a.m.	·17905			Beattie
Oct. 30th	10 36 a.m.		·17891		Morrison
..	10 14 a.m.	·17903			Beattie
		———	———	+ 7 γ	
1906 June 29th	8 51 a.m.		·17818		Morrison
..	9 30 a.m.	·17796			Beattie
July 2nd	9 10 a.m.		·17805		Morrison
..	9 8 a.m.	·17791			Beattie
..	10 33 a.m.		·17798		Morrison
..	10 30 a.m.	·17791			Beattie
..	1 21 p.m.		·17788		Morrison
..	1 19 p.m.	·17767			Beattie
..	2 24 p.m.		·17794		Morrison
..	2 30 p.m.	·17774			Beattie
		———	———	− 17 γ	

In taking the mean of the means the four have been weighted and the resultant difference is 12γ—the horizontal intensity determined with instrument 31 being the greater.

Comparison of magnetometers 31 *and* 73. *Declination.* The observations for comparison were made at Matjesfontein. Of the results only those have been taken which were obtained at the same mean time. When a single observer only was available the procedure was to obtain the declination with one instrument first, then with the other, and finally with the one again. In the observations of a magnetic azimuth made in 1906 with the two instruments, two tents were used, and the comparisons arranged in such a way that the same number of observations was made with each instrument in each tent.

TABLE 15.

MATJESFONTEIN

Date	G.M.T.	Declination West with 73	Declination West with 31	Difference in D. 73 − 31	Observer
	h m	° ′	° ′	′	
1902 Dec. 27th	2 11 p.m.	28 25·7			
..	1 22 p.m.		28 24·8	+ 1·3	Beattie
..	2 52 p.m.		28 24·0		
Dec. 29th	2 20 p.m.	28 27·3			
..	2 58 p.m.		28 27·3	+ 1·0	..
..	1 3 p.m.		28 25·3		
1904 Oct. 29th	6 9 a.m.	28 12·0		− 1·0	..
..	6 9 a.m.		28 13·0		Morrison
Oct. 30th	8 27 a.m.	28 14·6			Beattie
..	9 24 a.m.	28 13·6		+ 2·0	
..	8 25 a.m.		28 12·6	+ 1·5	Morrison
..	9 23 a.m.		28 12·1		

Date	G.M.T.	Magnetic Azimuth with 73	Magnetic Azimuth with 31	diff. in Az. 73 − 31	Difference in D. 73 − 31	Observer
		° ′	° ′	′		
1906 July 1st	1 33 p.m.	52 11·1		− 1·5	+ 1·5	Beattie with 73
..	1 34 p.m.		52 12·6			Morrison .. 31
..	1 47 p.m.	52 10·4		− 2·2	+ 2·2	..
..	1 43 p.m.		52 12·6			
..	2 9 p.m.	52 10·3		− 2·3	+ 2·3	..
..	2 21 p.m.		52 12·6			
..	2 28 p.m.	52 9·9		− 1·3	+ 1·3	..
..	2 37 p.m.		52 11·2			
..	2 38 p.m.	52 9·0		− 1·4	+ 1·4	..
..	2 45 p.m.		52 10·4			
..	2 48 p.m.	52 8·6		− 1·5	+ 1·5	..
..	2 54 p.m.		52 10·1			
..	3 20 p.m.	52 7·0	52 9·7	− 2·7	+ 2·7	..
..	3 32 p.m.	52 6·2		− 2·6	+ 2·6	..
..	3 27 p.m.		52 8·8			

The average difference in the value of the same declination determined by these instruments is 1′·6, the declination as determined by magnetometer 73 being greater than that by 31.

Comparison with instruments of the "Discovery." This comparison was carried out on the Rifle Range, Simonstown. The Rear Admiral Commanding in Chief, A. W. Moore, had tents pitched and the instruments conveyed there. The result of the comparison was as follows for dip, and declination ; no satisfactory comparison of the horizontal intensity was obtained.

TABLE 16.

	Instrument	Reference Instrument	Difference
			′
Dip	Dip circle 27 (of "Discovery")	Dip circle 142 (of Survey)	1·7 (142 − 27)
	.. 26 ..	.. 142 ..	3·2 (142 − 26)
Declination	Magnetometer 25 (of "Discovery")	Magnetometer 31 (of Survey)	0·8 (31 − 25)
	.. 36 ..	.. 31 ..	−0·8 (31 − 36)

Comparison with the instruments of Major Chaves and through them with the standards of the Central Magnetic Observatory of France, Val Joyeux. The comparison of the survey instruments with those of Major Chaves was carried out at Matjesfontein, at a place where observations had been made in connection with the survey for several years. The observers were Major Chaves, Professor Morrison and the writer. Major Chaves's instruments were Dip Circle No. 15 Abadie Brunner, made by Chasselon, Paris, and theodolite declinometer No. 28 Mascart Brunner, also made by Chasselon. Professor Morrison worked with dip circle 9 and magnetometer 31; the writer with dip circle 142 and magnetometer 73. Two tents were used sufficiently far apart to preclude any action of one instrument on another. Preliminary experiments were made on the 28th and on the 29th in the case of the declination. A satisfactory comparison was made on the 1st July, when the magnetic azimuth of a given line was determined simultaneously by the three observers.

The following table gives the results of the comparison.

TABLE 17.

Date	Instrument	G.M.T.		Magnetic azimuth of line	Means	Observer
		h m		° ′	° ′	
1906 July 1st	28	8 36 a.m.	Bar 1	52 18·9		Chaves
..	..	9 14 ..	Bar 2	52 20·4		..
					52 19·6	
..	..	9 31 ..	Bar 1	52 20·7		..
..	..	10 14 ..	Bar 2	52 21·3		..
					52 21·0	
					52 20·3	
..	31	8 41 a.m.		52 22·8		Morrison
..	..	8 51 ..		52 21·8		..
..	..	9 1 ..		52 21·8		..
..	..	9 9 ..		52 21·2		..
..	..	9 41 ..		52 22·7		..
..	..	9 49 ..		52 22·0		..
..	..	10 0 ..		52 22·1		..
..	..	10 8 ..		52 22·4		..
					52 22·1	
..	73	8 40 a.m.		52 19·2		Beattie
..	..	8 57 ..		52 19·4		..
..	..	9 38 ..		52 19·9		..
..	..	9 48 ..		52 19·6		..
..	..	9 59 ..		52 19·7		..
..	..	10 8 ..		52 19·8		..
					52 19·6	

The corrections to reduce to Chaves's instrument are as follows:

Instrument	Magnetic azimuth of line	Declination
	′	′
31	−1·8	+1·8
73	+0·7	−0·7

That is to say the declination given by magnetometer 31 is 1′·8 less and that by 73 is 0′·7 greater than it would be had it been determined by Chaves's No. 28.

Comparison of values of horizontal intensity as determined by instruments 28, 31, *and* 73.

The value of H was calculated by Chaves from the formula

$$H = C\frac{1}{t\sqrt{\sin\theta}},$$

where C is the constant of the magnet employed,
t is the period of vibration,
and θ is the angle of deflection.

The constants were calculated by Chaves at Ponta Delgada in May of 1906*.

TABLE 18.

	Date	Instrument	G.M.T.	Horizontal intensity		Means	Observer
			h m				
1906	June 29th	28	8 47 a.m.	Bar 1	·17810		Chaves
				Bar 2	·17803		..
						·17807	
	June 30th	..	9 34 ..	Bar 1	·17830		..
				Bar 2	·17832		..
						·17831	
	June 29th	31	8 54 ..	H_{30}	·17819		Morrison
				H_{40}	·17817		..
						·17818	
	June 30th	73	9 30 ..	H_{30}	·17797		Beattie
				H_{40}	·17797		..
						·17797	

The following table gives the correction to be applied to instruments 31 and 73 to reduce the values of H as determined by them to the value of Chaves's instrument.

Instrument	Correction to reduce to Chaves's instrument
31	-11γ
73	$+34\gamma$

Comparison of dip circles 15, 9, *and* 142.

TABLE 19.

	Date	Instrument	G.M.T.	Needle	Dip	Means	Observer
			h m		° ′	° ′	
1906	June 28th	15	10 31 a.m.	1	60 12·2		Chaves
				2	60 8·3		..
						60 10·3	
	June 30th	..	12 41 p.m.	1	60 9·8		..
			1 39 ..	2	60 7·1		..
						60 8·5	
	June 28th	9	10 30 a.m.	1	60 11·1		Morrison
				2	60 12·1		..
						60 11·6	
	June 30th	..	1 8 p.m.	1	60 10·1		..
				2	60 11·3		..
						60 10·7	
	June 28th	142	10 25 a.m.	3	60 11·7		Beattie
			9 24 ..	4	60 15·4		..
						60 13·5	
	June 30th	..	1 13 p.m.	3	60 10·6		..
				4	60 12·8		..
						60 11·7	

* Contribution aux études de magnétisme terrestre en Afrique par F. A. Chaves. *Bull. de l'Institute Oceanographique,* No. 120.

The following table gives the differences with respect to No. 15 Chaves's instrument.

Instrument	Correction to reduce to Chaves's instrument
	′
9	−1·8
142	−3·2

The dip therefore as given by Chaves's instrument is 1′·8 less than that given by circle 9 and 3′·2 less than that given by circle 142.

Comparison with Val Joyeux standard instruments. Major Chaves was good enough at the end of his African tour to compare his instruments with those of Val Joyeux and to communicate to the writer the results given in the following table.

Instruments	D	H	θ
	° ′		° ′
Val Joyeux	14 55·6	·19739	64 47·6
Chaves	14 55·0	·19745	64 47·8
Correction to reduce to Chaves	−0·6	+6 γ	+0·2

With these comparisons it is possible to refer to the Val Joyeux standard instruments the results as they have been obtained in the survey here. For this purpose it must be remembered that in addition to the comparisons of the two sets of survey instruments with those of Chaves there have been numerous comparisons of the two sets of survey instruments with each other, the results of which have been given in the earlier part of this chapter. It will be seen that the result of all the comparisons of magnetometers 31 and 73 is in good agreement with the special comparison made during Chaves's visit; the whole comparisons show—page 39—that 73 gives a declination greater than that with 31 by 1′·6, whereas in the particular case here considered the difference was 2′·5 in the same direction. The result in the case of the horizontal intensity is not so satisfactory. The difference from the whole of the comparisons shows—page 38—that the result from 31 exceeds that from 73 by 12γ, whereas by the indirect comparison through Chaves's instrument the difference—in the same direction—is 45γ. From a consideration of the results it has been thought best to take the comparison with 73 as the correct one and to connect 31 to Val Joyeux through the whole comparisons with 73 directly. The reasons for this are that in the field book the day is marked as particularly suitable for observational work, and the results obtained on that date by Chaves with his two magnets are in better accord than on the other day.

In the comparison with the dip circles the results obtained in June, 1906, give a difference between the two survey circles in close agreement with those obtained with the same instruments and needles on other occasions. The following table gives the correction to the various survey instruments to reduce them to the Val Joyeux standards.

TABLE 20.

Instrument	Difference (Val Joyeux value – instrument value)		
	D	H	θ
	′		′
31	+ 2·4	+ 16 γ	
73	– 0·1	+ 28 γ	
9			– 2·0
142			– 3·4

CHAPTER V

EPOCH OF THE SURVEY AND SUMMARY OF OBSERVED RESULTS REDUCED TO THAT EPOCH; TABLES BEARING ON THE MAGNETIC STATE OF THE SURVEYED COUNTRY, METHODS OF DRAWING MAPS AND CONSIDERATION OF MAGNETIC DISTURBANCES

Epoch. The epoch is 1st July, 1903. This date was chosen because the greatest number of stations was occupied between Dec. 1902, and Feb. 1904. Further these stations were situated in Rhodesia, the Transvaal, Natal and the Orange River Colony, where a knowledge of the secular variation had to be derived from few and widely distributed stations. On the other hand the observations in the Cape Colony were taken chiefly before Dec. 1902 and after Feb. 1904; the repeat stations however were numerous and well distributed. A secular correction was calculated from the results obtained at the repeat stations.

The various stations are arranged in alphabetical order. The number of each station is given in the first and the last columns. The name of the station is given in the second column, the latitude, the longitude, the declination, D, the dip, θ, and the horizontal intensity, H, are given in the third, fourth, fifth, sixth and seventh columns respectively.

The declination results have been obtained from the field results by applying corrections for daily and for secular variation.

The dip and the horizontal intensity results have been obtained from the field results by applying corrections for the secular variations only.

The actual results of observation are given in appendix E.

TABLE 21.

Summary of Declinations, Dips and Horizontal Intensities reduced to the Epoch 1st July, 1903.

No.	Station Name	Lat.	Long.	Declination (D)	Dip (θ)	Horizontal Intensity (H)	Station No.
		° ′	° ′	° ′	° ′		
1	Abelsdam	27 37·1 S.	26 29·6 E.	23 38·1 W.	59 2·5 S.	·18790	1
2	Aberdeen, C. C.	32 29·1 ..	24 3·0 ..	27 14·3 ..	60 24·7 ..	·17991	2
3	Aberdeen (Transvaal)	26 3·8 ..	29 33·0 ..	21 15·6 ..	58 43·3 ..	·19217	3
4	Aberdeen Road	32 46·0 ..	24 20·0 ..	27 6·2 ..	60 20·8 ..	·18192	4
5	Aberfeldy	25 45·7 ..	28 34·5 ..	22 14·4 ..	58 3·5 ..	·19425	5
6	Adelaide	32 43·0 ..	26 18·0 ..	26 21·5 ..	61 22·0 ..	·17743	6

TABLE 21 (*continued*).

No.	Station Name	Lat. ° ′	Long. ° ′	Declination (D) ° ′	Dip (θ) ° ′	Horizontal Intensity (H)	Station No.
7	Albert Falls	29 26·0 S.	30 29·0 E.	23 46·5 W.	60 54·7 S.	·18338	7
8	Alicedale	33 18·9 ..	26 2·5 ..	27 27·2 ..	61 36·5 ..	·17546	8
9	Aliwal North	30 41·7 ..	26 42·0 ..	25 21·0 ..	60 22·1 ..	·18205	9
10	Alma	27 7·6 ..	31 5·5 ..	22 37·1 ..	59 21·8 ..	·18857	10
11	Amabele Junction	32 43·1 ..	27 19·2 ..	26 2·9 ..	61 22·1 ..	·17815	11
12	Amaranja	31 14·7 ..	29 30·0 ..		61 0·2 ..	·18073	12
13	Amatongas	19 11·2 ..	33 45·0 ..	16 4·0 ..	53 38·0 ..	·21839	13
14	Ashton	33 50·0 ..	20 4·0 ..		59 40·8 ..	·18135	14
15	Assegai Bosch	33 56·7 ..	24 20·5 ..	27 33·4 ..		·17746	15
16	Avontuur	33 44·2 ..	23 13·0 ..	27 48·7 ..	60 42·1 ..	·17842	16
17	Ayrshire Mine	17 11·5 ..	30 23·0 ..	16 5·0 ..	50 54·9 ..	·22805	17
17A	Balmoral	25 51·3 ..	28 58·0 ..	22 7·0 ..	58 9·9 ..	·19362	17A
18	Bamboo Creek	19 16·5 ..	34 12·0 ..	15 5·1 ..	53 46·0 ..	·21930	18
19	Bankpan	26 18·0 ..	29 35·0 ..		59 43·2 ..	·18485	19
20	Barberton	25 47·3 ..	31 0·0 ..	21 12·8 ..	58 36·2 ..	·19287	20
21	Barrington	33 55·2 ..	22 52·0 ..	26 37·7 ..	60 7·7 ..	·18308	21
22	Battlefields	18 36·4 ..	30 0·0 ..	17 23·0 ..	52 31·0 ..	·21872	22
23	Bavaria	27 30·7 ..	29 8·4 ..	22 29·3 ..	59 15·6 ..	·18843	23
24	Baviaanskrantz	27 19·0 ..	26 55·0 ..	26 49·4 ..	58 41·5 ..	·18903	24
24A	Beaconsfield	28 45·0 ..	24 44·0 ..		58 39·8 ..		24A
25	Beaufort West	32 20·9 ..	22 34·1 ..	27 8·9 ..	59 43·0 ..	·18205	25
26	Beira	19 49·2 ..	34 50·0 ..	15 55·9 ..	54 34·4 ..	·21673	26
27	Belleville	33 49·0 ..	18 39·0 ..		59 11·3 ..	·18262	27
28	Berg River Mouth	32 46·5 ..	18 10·0 ..	28 18·3 ..	58 29·4 ..	·18435	28
29	Bethal	26 28·1 ..	29 27·5 ..	22 44·5 ..	58 13·0 ..	·19416	29
30	Bethany	29 37·0 ..	26 2·0 ..	24 50·0 ..	59 42·4 ..	·18443	30
31	Bethesda Road	31 55·3 ..	24 38·0 ..	26 21·4 ..	60 28·9 ..	·18062	31
32	Bethlehem	28 13·8 ..	28 17·3 ..	23 38·9 ..	59 17·0 ..	·18931	32
33	Bethulie	30 30·5 ..	25 59·0 ..	25 30·1 ..	59 50·8 ..	·18400	33
34	Biesjespoort	31 43·8 ..	23 12·0 ..	26 1·5 ..	59 37·5 ..	·18445	34
35	Birthday	23 19·5 ..	30 46·0 ..	16 7·5 ..	59 7·0 ..	·18566	35
36	Blaauwbosch	30 38·9 ..	22 14·1 ..	26 33·9 ..	58 52·2 ..	·18496	36
37	Blaauwkrantz	33 57·0 ..	23 35·0 ..	27 43·9 ..	60 57·8 ..	·17813	37
38	The Bluff	29 52·5 ..	31 4·0 ..	23 40·1 ..	61 22·1 ..	·17924	38
39	Boschkopjes	23 11·5 ..	29 55·0 ..	19 50·9 ..	56 49·2 ..	·19957	39
40	Boschrand	27 45·8 ..	27 12·0 ..	23 38·7 ..	59 2·7 ..	·18842	40
41	Boston	29 41·0 ..	30 1·0 ..	23 44·9 ..	60 54·1 ..	·18242	41
42	Botha's Berg	25 25·0 ..	29 49·0 ..	22 23·5 ..	57 57·4 ..	·20347	42
43	Brak River	22 52·2 ..	29 13·0 ..	19 31·6 ..	55 55·5 ..	·20452	43
44	Brandboontjes	23 28·0 ..	30 16·0 ..	19 52·3 ..	56 40·2 ..	·20065	44
45	Bredasdorp	34 32·2 ..	20 3·0 ..	28 39·2 ..	59 56·3 ..	·18015	45
46	Breekkerrie	30 6·7 ..	21 35·0 ..	27 3·1 ..	58 20·9 ..	·18531	46
47	Britstown	30 35·0 ..	23 33·0 ..	26 28·7 ..	59 6·3 ..	·18097	47
48	Buffelsberg	23 36·7 ..	30 1·0 ..	20 55·3 ..	57 11·4 ..	·19484	48
49	Buffelshoek	23 8·3 ..	28 55·0 ..	20 7·2 ..	57 5·4 ..	·19809	49
50	Buffelsklip	33 31·7 ..	22 52·5 ..	27 46·6 ..	60 38·1 ..	·17867	50
51	Bulawayo	20 9·1 ..	28 36·3 ..	18 37·3 ..	53 27·9 ..	·21461	51
52	Bult and Baatjes	26 8·0 ..	30 16·0 ..		59 12·4 ..	·19223	52
52A	Bulwer	29 48·4 ..	29 41·0 ..	24 21·5 ..		·18080	52A
53	Burghersdorp	31 0·0 ..	26 18·0 ..	25 42·2 ..	60 17·7 ..	·18212	53
54	Bushmanskop	32 20·8 ..	22 14·5 ..	27 16·8 ..	59 40·7 ..	·18231	54
55	Butterworth	32 21·3 ..	28 4·0 ..	25 51·1 ..	61 20·1 ..	·17841	55
56	Caledon River	30 16·8 ..	26 41·7 ..	25 43·8 ..	60 22·9 ..	·18089	56
57	Calitzdorp	33 32·1 ..	21 41·0 ..	28 17·9 ..	60 18·8 ..	·17918	57
58	Camperdown	29 44·0 ..	30 37·0 ..		61 28·8 ..	·17833	58
59	Cango	33 24·8 ..	22 14·5 ..	28 18·6 ..	60 43·7 ..	·17780	59
60	Cape Town (Royal Ob.)	33 56·1 ..	18 28·7 ..	28 43·9 ..	59 6·0 ..	·18271	60
61	Cathcart	32 18·0 ..	27 9·0 ..		61 46·0 ..	·17477	61
62	Ceres Road	33 25·6 ..	19 19·0 ..	28 31·7 ..	59 10·7 ..	·18284	62

TABLE 21 (*continued*).

No.	Station Name	Lat. ° ′	Long. ° ′	Declination (D) ° ′	Dip (θ) ° ′	Horizontal Intensity (H)	Station No.
63	Charlestown	27 24·9 S.	29 54·0 E.	21 33·4 W.	59 23·7 S.	·18861	63
64	Clarkson	34 1·0 ..	24 10·0 ..		61 8·6 ..	·17800	64
65	Coerney	33 27·6 ..	25 44·0 ..	27 17·5 ..	61 30·2 ..	·17626	65
66	Colenso	28 44·0 ..	29 50·0 ..	22 48·3 ..	60 9·3 ..	·18548	66
67	Colesberg	30 42·8 ..	25 8·0 ..	25 36·5 ..	60 0·6 ..	·18183	67
68	Connan's Farm	28 58·4 ..	21 19·3 ..	24 54·8 ..	56 16·2 ..	·19602	68
69	Cookhouse	32 44·0 ..	25 48·0 ..	26 43·0 ..	61 14·4 ..	·17732	69
70	Cotswold Hotel	30 42·7 ..	29 53·4 ..	24 23·2 ..	61 27·7 ..	·17769	70
71	Cradock	32 9·6 ..	25 38·0 ..	26 27·2 ..	60 49·1 ..	·17981	71
72	Cream of Tartarfontein	22 35·3 ..	29 1·0 ..	19 41·6 ..	55 44·5 ..	·20406	72
73	Crocodile Pools	24 46·7 ..	25 50·0 ..	21 37·6 ..	56 14·3 ..	·19995	73
74	Dalton	29 16·0 ..	30 42·0 ..		61 23·6 ..	·17974	74
75	Dambiesfontein	31 24·2 ..	21 17·6 ..	27 7·4 ..	59 16·1 ..	·18355	75
76	Dannhauser	28 1·2 ..	30 5·0 ..	22 29·8 ..	59 43·7 ..	·18597	76
77	Dargle Road	29 29·1 ..	30 11·0 ..	24 5·7 ..	60 38·8 ..	·18468	77
78	Darling	33 22·1 ..	18 22·0 ..	28 30·6 ..	58 55·5 ..	·18351	78
79	De Aar	30 40·0 ..	24 2·0 ..	25 42·0 ..	59 52·0 ..	·18447	79
80	De Doorns	33 29·0 ..	19 36·0 ..		59 20·4 ..	·18205	80
81	Deelfontein Farm	30 5·8 ..	26 31·7 ..	25 22·0 ..	60 6·4 ..	·18243	81
82	Deelfontein	28 20·0 ..	27 48·0 ..		59 25·0 ..	·18706	82
83	De Jager's Farm	28 15·9 ..	28 57·9 ..	23 4·4 ..	59 33·7 ..	·18799	83
84	Dewetsdorp	29 26·1 ..	26 39·3 ..	25 43·3 ..	59 37·1 ..	·18528	84
85	Draghoender	29 22·3 ..	22 7·4 ..	27 14·4 ..	58 16·1 ..	·18723	85
86	Drew	33 59·5 ..	20 13·0 ..	28 25·1 ..	59 51·1 ..	·18130	86
87	Driefontein	26 29·4 ..	29 13·0 ..	22 23·4 ..	59 6·0 ..	·18788	87
88	Driehoek	27 11·6 ..	30 41·0 ..	22 18·1 ..	59 25·4 ..	·18898	88
89	East London	33 0·0 ..	27 56·0 ..	26 31·4 ..	61 48·3 ..	·17600	89
90	Elandshoek	25 30·0 ..	30 41·0 ..		59 52·0 ..	·18867	90
91	Elandskloof Farm	28 0·0 ..	26 24·0 ..		58 56·6 ..	·18830	91
92	Elim	34 35·8 ..	19 46·0 ..	28 42·0 ..	59 55·1 ..	·18065	92
93	Ellerton	23 19·0 ..	30 30·0 ..		56 53·5 ..	·19819	93
94	Elliot	31 18·0 ..	27 54·0 ..	24 54·7 ..	61 4·7 ..	·17998	94
95	Elsburg	26 15·0 ..	28 11·0 ..	22 28·8 ..	58 26·1 ..	·19140	95
96	Emmasheim	28 17·2 ..	28 7·3 ..	23 55·2 ..	59 28·7 ..	·18697	96
97	Estcourt	29 0·9 ..	29 54·0 ..	23 17·2 ..	60 12·0 ..	·18376	97
98	Ferreira	29 12·0 ..	26 11·0 ..	24 47·3 ..	59 30·7 ..	·18480	98
99	Fish River	31 55·3 ..	25 27·0 ..	26 18·5 ..	60 44·0 ..	·18018	99
100	Forty-one mile Siding	17 43·0 ..	30 33·0 ..	15 55·2 ..	51 47·3 ..	·22465	100
101	Fountain Hall	29 15·8 ..	29 59·0 ..	23 20·4 ..	60 19·6 ..	·18500	101
102	Francistown	21 4·0 ..	27 32·0 ..		56 29·8 ..	·20380	102
102A	,, (10 miles S. of)				54 6·0 ..		102A
103	Fraserburg	31 55·2 ..	21 31·3 ..	27 33·1 ..	59 15·5 ..	·18290	103
104	Fraserburg Road	32 46·0 ..	22 0·0 ..		59 59·0 ..	·18251	104
105	Gamtoos River Bridge	33 55·2 ..	25 2·5 ..	27 25·4 ..	61 14·4 ..	·17736	105
106	George Town	33 57·0 ..	22 29·0 ..		60 29·0 ..	·17970	106
107	Gemsbokfontein	31 22·8 ..	22 57·5 ..	27 47·3 ..	59 27·1 ..	·18326	107
108	Ginginhlovu	29 1·7 ..	31 35·0 ..	23 6·5 ..	61 3·1 ..	·18251	108
109	Glenallen	29 39·0 ..	22 36·0 ..	47 59·5 ..	41 53·0 ..	·23457	109
110	Glenconnor	33 25·0 ..	25 10·0 ..	27 28·8 ..	61 19·0 ..	·17637	110
111	Globe and Phoenix	18 56·0 ..	29 48·0 ..	16 53·5 ..	52 43·6 ..	·21945	111
112	Goedgedacht	26 38·9 ..	29 37·0 ..	22 13·1 ..	58 44·2 ..	·19138	112
113	Gordon's Bay	34 8·0 ..	18 55·0 ..		59 23·7 ..	·18100	113
114	Graaff Reinet	32 16·9 ..	24 36·0 ..	26 57·5 ..	60 34·6 ..	·17931	114
115	Grahamstown	33 19·7 ..	26 32·0 ..	27 1·1 ..	61 43·7 ..	·17589	115
116	Grange	29 37·9 ..	30 23·0 ..		61 6·2 ..	·18097	116
117	Graskop	27 15·0 ..	29 53·0 ..	22 34·6 ..	59 18·4 ..	·18865	117
118	Greylingstad	26 44·6 ..	28 45·5 ..	22 23·5 ..	58 56·7 ..	·18969	118
119	Greytown	29 4·9 ..	30 38·0 ..	23 53·6 ..	60 28·9 ..	·18690	119
120	Grobler's Bridge	25 53·5 ..	30 13·0 ..	20 55·2 ..	58 25·8 ..	·19310	120

TABLE 21 (*continued*).

No.	Station Name	Lat.	Long.	Declination (D)	Dip (θ)	Horizontal Intensity (H)	Station No.
		° ′	° ′	° ′	° ′		
121	Groenkloof	29 28·4 S.	27 11·4 E.	24 31·0 W.	59 56·6 S.	·18451	121
122	Groenplaats	27 16·0 ..	28 33·8 ..	22 24·9 ..	59 1·3 ..	·19076	122
123	Grootfontein	33 7·6 ..	21 15·0 ..	28 16·2 ..	60 1·5 ..	·18044	123
124	Gwaai	19 17·5 ..	27 42·2 ..	18 17·0 ..	52 27·2 ..	·21749	124
125	Gwelo	19 28·2 ..	29 47·0 ..	17 56·9 ..	54 23·1 ..	·22034	125
126	Hamaan's Kraal	25 24·3 ..	28 17·0 ..	22 5·6 ..	57 29·9 ..	·19710	126
127	Hankey	33 52·0 ..	24 53·0 ..	27 22·6 ..	61 12·9 ..	·17773	127
128	Hartley	18 8·3 ..	30 8·0 ..	16 40·7 ..	51 55·7 ..	·22273	128
129	Hector Spruit	25 26·2 ..	31 40·5 ..	20 34·8 ..	58 10·5 ..	·19673	129
130	Heidelberg, C. C.	34 5·3 ..	20 58·0 ..	28 22·3 ..	60 6·8 ..	·18041	130
131	Heilbron	27 18·2 ..	27 58·0 ..	23 17·5 ..	58 50·9 ..	·18998	131
132	Helvetia	29 52·1 ..	26 33·0 ..	25 15·5 ..	59 58·9 ..	·18336	132
133	Hermanus	34 25·3 ..	19 16·0 ..	28 46·3 ..	59 38·2 ..	·18138	133
134	Hermon	33 26·7 ..	18 58·0 ..	28 29·6 ..	59 5·4 ..	·18309	134
135	Highlands	27 16·0 ..	31 23·0 ..	21 52·8 ..	59 55·8 ..	·18742	135
136	Hlabisa	28 18·5 ..	32 6·0 ..	22 9·1 ..	60 53·8 ..	·18262	136
137	Hluti	27 11·6 ..	31 35·0 ..	21 45·3 ..	59 44·1 ..	·18760	137
138	Hoetjes Bay	33 1·0 ..	17 57·0 ..	28 26·8 ..	58 35·0 ..	·18407	138
139	Holfontein	29 14·9 ..	27 22·5 ..	24 37·6 ..	59 50·1 ..	·18502	139
140	Honey Nest Kloof	29 12·2 ..	24 33·0 ..	25 36·3 ..	59 1·9 ..	·18623	140
141	Honing Spruit	27 27·0 ..	27 25·0 ..	23 15·0 ..	58 42·1 ..	·19005	141
142	Hopefield	33 14·4 ..	18 21·0 ..	28 21·5 ..	58 46·5 ..	·18348	142
143	Howhoek	34 12·7 ..	19 10·0 ..	28 38·3 ..	59 31·8 ..	·18128	143
144	Huguenot	33 45·3 ..	19 0·0 ..	28 29·8 ..	59 13·6 ..	·18286	144
145	Humansdorp	34 2·0 ..	24 38·5 ..	26 58·4 ..	61 16·0 ..	·17702	145
146	Hutchinson	31 29·6 ..	23 15·0 ..	26 47·0 ..	59 34·1 ..	·18316	146
147	Ibisi Bridge	30 24·4 ..	29 54·5 ..	24 26·7 ..	61 2·0 ..	·18170	147
148	Idutywa	32 0·8 ..	28 20·4 ..	25 55·8 ..	61 44·1 ..	·17559	148
149	Igusi	19 40·8 ..	28 6·0 ..	17 19·3 ..	53 1·0 ..	·21633	149
150	Illovo River	30 6·1 ..	30 51·0 ..	23 49·1 ..	61 19·1 ..	·18001	150
151	Imvani	32 2·0 ..	27 5·0 ..	26 3·7 ..	61 28·0 ..	·17781	151
152	Indowane	29 57·5 ..	29 26·7 ..	24 13·7 ..	61 0·5 ..	·18074	152
153	Indwe	31 27·8 ..	27 21·0 ..	25 32·4 ..	60 56·1 ..	·18186	153
154	Inoculation	20 49·7 ..	27 38·0 ..	19 11·6 ..	54 0·2 ..	·21121	154
155	Inyantué	18 32·5 ..	26 41·8 ..	18 27·4 ..	51 8·8 ..	·22341	155
156	Kaalfontein	26 0·5 ..	28 16·5 ..		58 13·1 ..	·19459	156
157	Kaalkop Farm	27 47·3 ..	28 58·3 ..	22 54·6 ..	59 26·8 ..	·18770	157
158	Kaapmuiden	25 31·7 ..	31 19·0 ..	20 57·9 ..	58 29·4 ..	·19310	158
159	Kalkbank	23 31·5 ..	29 20·0 ..	20 8·7 ..	56 22·4 ..	·20283	159
160	Kaloombies	22 39·3 ..	29 14·0 ..	20 2·1 ..	56 35·4 ..	·20150	160
161	Karree	28 52·5 ..	26 21·0 ..	23 27·4 ..	59 19·5 ..	·18548	161
162	Kathoek	34 23·3 ..	20 20·0 ..	28 37·1 ..	60 2·7 ..	·18012	162
163	Kenhardt	29 18·0 ..	21 9·0 ..	26 22·4 ..	58 1·4 ..	·18925	163
163A	Kenilworth (nr Kimb.)	28 42·0 ..	24 27·0 ..		57 39·2 ..		163A
164	Kimberley	28 43·0 ..	24 46·0 ..		58 48·5 ..	·18727	164
165	King William's Town	32 52·5 ..	27 25·0 ..	26 19·1 ..	61 39·0 ..	·17726	165
166	Klaarstroom	33 20·0 ..	22 32·5 ..	27 56·0 ..	60 28·3 ..	·17906	166
167	Klerksdorp	26 52·3 ..	26 38·0 ..	23 41·9 ..	58 12·2 ..	·19123	167
168	Klipfontein, C. C.	30 42·1 ..	22 23·5 ..	23 6·3 ..	59 16·1 ..	·18340	168
169	,, (Spelonken)	23 5·7 ..	30 10·0 ..	19 31·3 ..	56 21·1 ..	·20344	169
170	Klipplaat	33 2·0 ..	24 26·0 ..	27 27·0 ..	61 11·1 ..	·17745	170
171	Knysna	34 1·7 ..	23 3·0 ..	27 52·8 ..	60 50·7 ..	·17840	171
172	Kokstad	30 32·8 ..	29 28·0 ..	24 40·1 ..	61 11·9 ..	·18016	172
173	Komati Poort	25 26·0 ..	31 54·0 ..		59 45·3 ..	·18869	173
174	Komgha	32 35·6 ..	27 54·5 ..	26 12·0 ..	61 35·0 ..	·17782	174
175	Kraal	26 25·1 ..	28 26·0 ..	22 5·7 ..	58 38·6 ..	·19068	175
176	Krantz Kloof	29 48·0 ..	30 54·0 ..	23 31·8 ..	61 20·8 ..	·18001	176
177	Krantz Kop	30 48·8 ..	20 45·4 ..	26 54·7 ..	58 26·9 ..	·18677	177
178	Kromm River	27 19·0 ..	28 18·8 ..	24 5·8 ..	58 55·9 ..	·18940	178

TABLE 21 (*continued*).

No.	Station Name	Lat. ° ′	Long. ° ′	Declination (D) ° ′	Dip (θ) ° ′	Horizontal Intensity (H)	Station No.
179	Krugers	29 57·1 S.	25 50·0 E.	25 0·2 W.	59 52·5 S.	·18490	179
180	Kruispad	32 56·8 ..	20 33·3 ..	28 10·4 ..	59 45·7 ..	·18146	180
181	Kwambonambi	28 36·2 ..	32 5·0 ..	22 57·7 ..	60 37·5 ..	·18454	181
182	Laat Rivier	29 38·2 ..	21 19·3 ..	26 25·7 ..	58 13·3 ..	·18793	182
183	Ladismith, C. C.	33 29·0 ..	21 17·0 ..	28 14·7 ..	60 6·0 ..	·17974	183
184	L'Agulhas	34 50·0 ..	20 0·0 ..	28 41·4 ..	60 4·7 ..	·17970	184
185	Laingsburg	33 12·0 ..	20 52·0 ..	28 16·9 ..	59 55·8 ..	·18022	185
186	Lake Banagher	26 22·0 ..	30 19·0 ..	21 38·8 ..	58 54·6 ..	·19056	186
187	Langlaagte	26 11·8 ..	28 1·0 ..	23 4·6 ..	58 56·2 ..	·18907	187
188	Letjesbosch	32 34·0 ..	22 18·0 ..		59 39·4 ..	·18346	188
189	Libode	31 32·1 ..	29 1·5 ..	25 13·0 ..	61 31·1 ..	·18006	189
190	Lobatsi	25 13·8 ..	25 40·0 ..	22 27·8 ..	57 0·4 ..	·19542	190
191	Lochard	19 55·3 ..	29 3·0 ..	18 17·0 ..	53 33·6 ..	·21506	191
192	Lydenburg	25 5·8 ..	30 26·0 ..	20 53·5 ..	58 5·3 ..	·19483	192
193	Machadodorp	25 39·9 ..	30 15·0 ..	20 50·1 ..	58 43·9 ..	·19073	193
194	Macheke	18 8·3 ..	31 51·0 ..	16 14·8 ..	52 8·9 ..	·22424	194
195	Mafeking	25 52·0 ..	25 39·0 ..	23 4·8 ..	57 11·9 ..	·19487	195
196	Magalapye	23 6·8 ..	26 50·0 ..	20 23·5 ..	55 34·0 ..	·20496	196
197	Magnet Heights	24 44·8 ..	29 58·0 ..	19 44·9 ..	57 58·6 ..	·19272	197
198	Makwiro Siding	17 57·3 ..	30 25·0 ..	16 34·4 ..	51 36·3 ..	·22447	198
199	Malagas	34 18·5 ..	20 36·0 ..	28 40·9 ..	60 5·0 ..	·18066	199
200	Malenje Siding	18 55·2 ..	32 15·0 ..	16 7·0 ..	53 3·4 ..	·22040	200
201	Malinde	18 45·0 ..	27 1·3 ..	18 1·6 ..	51 50·3 ..	·21859	201
202	Malmesbury	33 28·0 ..	18 43·0 ..	28 28·9 ..	59 5·7 ..	·18220	202
203	Mandegos	19 7·0 ..	33 28·0 ..	15 58·1 ..	53 25·6 ..	·21930	203
204	Mapani Loep	22 17·5 ..	29 3·0 ..	19 4·2 ..	55 47·8 ..	·20532	204
205	Mara	23 8·0 ..	29 21·0 ..		57 16·1 ..	·20181	205
206	Marandellas	18 11·3 ..	31 32·9 ..	15 55·2 ..	52 54·6 ..	·22211	206
207	Maribogo	26 25·1 ..	25 15·0 ..	23 35·4 ..	57 17·7 ..	·19366	207
208	Matetsi	18 12·5 ..	26 1·5 ..	17 57·9 ..	51 41·1 ..	·21696	208
209	Matjesfontein	33 14·2 ..	20 36·0 ..	28 22·9 ..	59 50·5 ..	·18028	209
210	Meyerton	26 33·2 ..	28 1·0 ..	22 42·4 ..	58 30·3 ..	·19148	210
211	Middelberg (Tzitzikama)	34 0·0 ..	24 9·0 ..	27 43·8 ..	61 2·3 ..	·17779	211
212	Middlepost	31 54·2 ..	20 14·0 ..	27 29·0 ..	58 49·5 ..	·18455	212
213	Middleton	32 57·8 ..	25 51·0 ..	26 56·1 ..	61 22·0 ..	·17742	213
214	Mill River	33 36·0 ..	22 55·0 ..		60 38·4 ..	·17839	214
215	Miller Siding	33 5·4 ..	24 8·0 ..	27 31·3 ..	61 0·2 ..	·17794	215
216	Miller's Point	34 14·2 ..	18 26·0 ..	28 41·1 ..	59 17·2 ..	·18259	217
217	Misgund	33 45·5 ..	23 32·0 ..	27 53·4 ..	60 48·6 ..	·17825	217
218	Mission Station	23 12·7 ..	30 27·0 ..	19 31·8 ..	57 5·9 ..	·19910	218
219	Modder Spruit	28 28·9 ..	29 53·0 ..	23 9·8 ..	60 0·3 ..	·18560	219
220	Molteno	31 24·0 ..	26 21·0 ..	25 58·2 ..	60 35·9 ..		220
221	Mossel Bay	34 10·8 ..	22 9·5 ..	28 12·1 ..	60 33·3 ..	·17930	221
222	Mount Ayliff (near)	30 48·2 ..	29 31·5 ..	24 33·1 ..	61 16·4 ..	·17890	222
223	Mount Frere	30 53·5 ..	28 59·0 ..	24 33·6 ..	61 1·6 ..	·18067	223
224	Mount Moreland	29 38·4 ..	31 11·0 ..	23 12·1 ..	61 19·6 ..	·18064	224
224A	Movene	25 34·0 ..	32 7·0 ..	17 42·5 ..	59 6·5 ..	·19573	224A
225	M'Phatele's Location	24 19·8 ..	29 41·0 ..	20 58·8 ..	59 11·7 ..	·19535	225
226	Naauwpoort	31 14·0 ..	24 55·0 ..	26 1·6 ..	60 1·8 ..	·18166	226
227	Naboomspruit	24 31·3 ..	28 43·0 ..	21 3·4 ..	57 4·9 ..	·19889	227
228	Nelspoort	32 7·7 ..	23 1·0 ..	27 35·2 ..	60 18·1 ..	·17795	228
229	Nelspruit	25 28·1 ..	30 58·5 ..	20 59·5 ..	58 28·4 ..	·19290	229
230	Newcastle	27 45·3 ..	29 58·0 ..	22 33·1 ..	59 38·6 ..	·18845	230
231	,, (Transvaal)	26 32·1 ..	30 27·0 ..	21 58·1 ..	58 56·6 ..	·18982	231
232	Nooitgedacht	25 38·1 ..	30 31·0 ..	19 0·7 ..	57 22·1 ..	·19204	232
233	Norval's Pont	30 39·0 ..	25 27·0 ..		59 50·1 ..	·18320	233
234	'Nqutu Road	28 5·0 ..	30 26·0 ..		60 13·0 ..	·18584	234
235	Nylstroom	24 42·4 ..	28 26·0 ..	21 11·8 ..	57 28·9 ..	·19843	235
235A	Orange River	29 38·0 ..	24 16·0 ..	25 24·0 ..	59 7·3 ..	·18559	235A

TABLE 21 (*continued*).

No.	Station Name	Lat. ° ′	Long. ° ′	Declination (D) ° ′	Dip (θ) ° ′	Horizontal Intensity (H)	Station No.
236	Orjida	33 26·0 S.	23 19·0 E.	27 48·7 W.	60 45·1 S.	·17836	236
237	Oudemuur	31 5·8 ..	20 19·1 ..	28 19·4 ..	58 42·2 ..	·18549	237
238	Oudtshoorn	33 35·2 ..	22 12·5 ..	28 11·4 ..	60 24·5 ..	·17944	238
239	Paardevlei	30 36·1 ..	21 54·0 ..	27 0·7 ..	58 43·6 ..	·18588	239
240	Paarl	33 45·0 ..	18 57·0 ..		59 13·5 ..		240
241	Palapye	22 33·4 ..	27 7·0 ..	20 33·4 ..	55 7·7 ..	·20639	241
242	Pampoenpoort	31 3·5 ..	22 39·1 ..	26 47·7 ..	59 19·0 ..	·18324	242
243	Payne's Farm	30 36·9 ..	29 47·5 ..	24 36·7 ..	61 34·4 ..	·17826	243
244	Picene	25 40·8 ..	32 18·5 ..	20 48·1 ..	59 0·0 ..	·19308	244
245	Pienaar's River	25 12·7 ..	28 19·0 ..	23 21·5 ..	57 48·8 ..	·19729	245
246	Pietersburg	23 50·3 ..	29 27·0 ..	20 12·0 ..	56 57·0 ..	·19855	246
247	Piet Potgietersrust	24 11·2 ..	29 1·0 ..	20 40·2 ..	56 58·9 ..	·19831	247
248	Piet Retief	27 0·5 ..	30 48·5 ..	22 14·5 ..	59 32·6 ..	·18831	248
249	Pilgrim's Rest	24 56·8 ..	30 45·0 ..	20 59·2 ..	57 49·8 ..	·19612	249
250	Piquetberg	32 55·0 ..	18 43·0 ..	28 10·7 ..	58 56·4 ..	·18340	250
251	Pivaan's Poort	27 33·8 ..	30 28·0 ..	22 39·2 ..	59 43·0 ..	·18796	251
252	Platrand	27 6·4 ..	29 29·0 ..	22 36·4 ..	59 12·2 ..	·18907	252
253	Plettenberg Bay	34 2·2 ..	23 21·0 ..	27 48·1 ..	60 54·9 ..	·17812	253
254	Plumtree	20 30 ..	27 50 ..		53 46·9 ..	·21219	254
255	Pokwani (Transvaal)	24 52·2 ..	29 46·0 ..	21 17·9 ..	57 41·3 ..	·19753	255
256	Port Alfred	33 35·8 ..	26 54·0 ..	26 54·8 ..	61 42·0 ..	·17595	256
257	Port Beaufort	34 23·8 ..	20 49·0 ..	28 27·3 ..	60 12·4 ..	·17990	257
258	Port Elizabeth	33 58·0 ..	25 37·0 ..	27 39·4 ..	61 35·4 ..	·17640	258
259	Port Shepstone	30 43·7 ..	30 27·0 ..	24 16·2 ..	61 53·3 ..	·17801	259
260	Port St Johns	31 37·8 ..	29 33·0 ..	25 18·2 ..	62 1·5 ..	·17547	260
261	Potchefstroom	26 42·8 ..	27 5·0 ..	23 0·4 ..	58 21·7 ..	·19079	261
262	Potfontein	30 12·2 ..	24 7·0 ..	25 38·5 ..	59 21·4 ..	·18433	262
263	Pretoria	25 45·3 ..	28 12·0 ..	22 16·4 ..	58 2·7 ..	·19377	263
264	Prince Albert	33 13·2 ..	22 3·0 ..	28 4·3 ..	60 20·5 ..		264
265	Prince Albert Road	32 58·7 ..	21 42·0 ..	27 51·2 ..	60 13·7 ..	·18010	265
266	Queenstown	31 54·0 ..	26 52·0 ..	26 16·0 ..	61 1·6 ..	·18049	266
267	Randfontein	26 10·7 ..	27 42·0 ..	22 46·2 ..	58 19·2 ..	·19151	267
268	Rateldraai	28 45·7 ..	21 17·9 ..	24 53·5 ..	57 45·5 ..	·19194	268
269	Rateldrift	31 31·6 ..	20 17·6 ..	27 32·3 ..	58 50·5 ..	·18439	269
270	Richmond (Natal)	29 54·0 ..	30 20·0 ..		61 5·4 ..	·18091	270
271	Richmond Road	31 13·0 ..	23 38·0 ..	27 11·5 ..	59 54·4 ..	·18231	271
272	Rietkuil Farm	30 14·4 ..	29 22·0 ..	24 12·2 ..	61 5·3 ..	·17978	272
273	Rietpoort	31 4·4 ..	20 55·1 ..	26 53·5 ..	58 54·5 ..	·18523	273
274	Rietvlei	24 35·0 ..	30 40·0 ..	20 24·2 ..	57 30·2 ..	·19793	274
275	Rietvlei, C. C.	33 32·0 ..	22 29·0 ..	27 58·8 ..	60 29·0 ..		275
276	Riversdale	34 5·0 ..	21 16·0 ..	28 17·2 ..	60 11·0 ..	·17997	276
277	Rivierplaats	32 8·5 ..	20 24·0 ..	27 17·1 ..	58 49·1 ..	·18456	277
278	Roadside	30 44·3 ..	20 25·5 ..	26 51·5 ..	58 17·5 ..	·18734	278
279	Robertson	33 48·8 ..	19 53·0 ..	28 24·2 ..	59 40·7 ..	·18145	279
280	Rodekrantz	24 38·0 ..	30 35·0 ..	20 52·9 ..	57 43·4 ..	·19734	280
281	Roodepoort	30 13·0 ..	23 22·0 ..	25 14·1 ..	57 58·3 ..	·18534	281
282	Rooidam	29 50·7 ..	23 11·8 ..	24 59·6 ..	58 40·5 ..	·18657	282
283	Revué	18 59·0 ..	33 3·0 ..	16 5·8 ..		·21995	283
284	Rooipüts	29 17·4 ..	21 38·6 ..	27 0·2 ..	58 16·4 ..	·18545	284
285	Rooival	32 12·0 ..	21 58·3 ..	27 19·0 ..	59 27·0 ..	·18274	285
286	Rosmead Junction	31 39·6 ..	25 5·0 ..	26 13·7 ..	60 21·0 ..	·18170	286
287	Rouxville	30 31·6 ..	26 47·3 ..	25 6·8 ..	60 12·5 ..	·18286	287
288	Rusapi	18 32·0 ..	32 8·0 ..	16 3·4 ..	52 35·2 ..	·22179	288
289	Rustplaats	24 50·6 ..	30 38·0 ..	20 52·7 ..	57 50·0 ..	·19639	289
290	Ruyterbosch	33 55·7 ..	22 2·0 ..	28 0·3 ..	60 19·3 ..	·18004	290
291	Sabie River	25 6·1 ..	30 45·0 ..	21 4·0 ..	58 16·6 ..	·19694	291
292	Salisbury	17 50·3 ..	31 3·0 ..	16 7·5 ..	51 42·7 ..	·22132	292
293	Saxony	28 44·1 ..	27 44·4 ..	24 11·9 ..	59 47·1 ..	·18562	293
294	Schietfontein	32 41·7 ..	20 46·6 ..	27 52·5 ..	59 32·0 ..	·18314	294

TABLE 21 (*continued*).

No.	Station Name	Lat. ° ′	Long. ° ′	Declination (D) ° ′	Dip (θ) ° ′	Horizontal Intensity (H)	Station No.
295	Schikhoek	27 24·6 S.	30 34·0 E.	23 16·0 W.	59 48·5 S.	·18656	295
296	Schoemanshoek	25 27·9 ..	30 21·0 ..	21 8·9 ..	58 29·6 ..	·19293	296
297	Schuilplaats	26 54·2 ..	29 47·0 ..	22 16·1 ..	59 7·5 ..	·18922	297
298	Secocoeni's Stad	24 28·3 ..	29 52·0 ..	19 17·9 ..	56 27·3 ..	·20070	298
299	Seruli	21 55·7 ..	27 19·0 ..	20 26·2 ..	54 55·7 ..	·20658	299
300	Shangani	19 45·8 ..	29 24·0 ..	18 37·9 ..	54 41·1 ..	·22431	300
301	Shashi	21 23·2 ..	27 27·0 ..	19 28·6 ..	57 0·4 ..	·21195	301
302	Shela River	26 51·0 ..	30 43·0 ..	21 43·8 ..	59 10·1 ..	·18896	302
303	Shoshong Road	23 34·8 ..	26 34·0 ..	21 54·3 ..	56 40·0 ..	·19750	303
304	Signal Hill	33 55·0 ..	18 24·3 ..	28 50·3 ..	59 7·4 ..	·18271	304
305	Simonstown (Rifle Range)	34 12·0 ..	18 26·0 ..	28 41·9 ..	59 15·8 ..	·18209	305
305A	Simonstown (Glencairn)	34 10·8 ..	18 26·0 ..	28 37·2 ..	59 11·7 ..	·18252	305A
306	Sir Lowry's Pass	34 7·3 ..	18 55·0 ..	28 58·0 ..	59 25·0 ..	·18108	306
307	Smaldeel	28 24·3 ..	26 44·0 ..	24 21·4 ..	59 7·0 ..	·18680	307
308	Spitzkopje	25 18·2 ..	30 49·0 ..	20 36·4 ..	58 48·0 ..	·19224	308
309	Springfontein	30 16·7 ..	25 44·0 ..	25 57·3 ..	60 15·0 ..	·18102	309
310	Springs	26 13·0 ..	28 27·0 ..	22 19·7 ..	58 21·7 ..	·19116	310
311	Stanford	34 26·7 ..	19 28·0 ..	28 43·0 ..	59 41·2 ..	·18127	311
312	Stanger	29 21·1 ..	31 15·0 ..	22 59·3 ..	61 16·3 ..	·18011	312
313	Steenkampspoort	32 6·3 ..	21 44·1 ..	27 23·4 ..	59 23·6 ..	·18335	313
314	Stellenbosch	33 56·0 ..	18 50·0 ..	28 44·9 ..	59 20·4 ..	·18192	314
315	Sterkstroom	31 34·5 ..	26 33·0 ..	25 58·2 ..	60 49·3 ..	·18095	315
316	Steynsburg	31 18·5 ..	25 48·0 ..	26 6·1 ..	60 15·1 ..	·18258	316
317	Still Bay	34 22·0 ..	21 25·0 ..	28 21·7 ..	60 18·3 ..	·17967	317
318	Stormberg Junction	31 17·5 ..	26 16·0 ..	25 58·5 ..		·18138	318
319	Storms River	33 58·0 ..	23 49·5 ..	27 41·5 ..	61 0·8 ..	·17790	319
320	Strandfontein	34 5·3 ..	18 34·0 ..	28 52·5 ..	59 14·2 ..		320
321	Sutherland	32 25·0 ..	20 39·3 ..	27 31·1 ..	59 15·8 ..	·18495	321
322	Swellendam	34 2·0 ..	20 27·0 ..	28 24·7 ..	59 53·0 ..	·18107	322
323	Taungs	27 34·8 ..	24 45·0 ..	24 5·3 ..	58 21·8 ..	·18887	323
324	Thaba 'Nchu	29 10·7 ..	26 49·0 ..	24 27·3 ..	59 36·6 ..	·18548	324
325	Tinfontein	30 24·0 ..	26 54·8 ..	24 45·0 ..	60 33·0 ..	·18077	325
326	Toise River	32 27·3 ..	27 28·7 ..	26 28·2 ..	61 46·7 ..	·17559	326
327	Touws River	33 21·0 ..	20 3·0 ..	28 49·6 ..	59 31·9 ..	·18151	327
328	Tsolo	31 18·2 ..	28 45·6 ..	24 49·6 ..	60 57·3 ..	·18002	328
329	Tugela	29 12·0 ..	31 25·0 ..		61 9·0 ..	·18139	329
330	Tulbagh Road	33 19·3 ..	19 10·0 ..	28 29·1 ..	59 2·4 ..	·18313	330
331	Tweepoort	26 36·7 ..	30 43·0 ..	22 3·4 ..	58 50·9 ..	·19260	331
332	Twee Rivieren	33 50·3 ..	23 56·5 ..	27 40·4 ..	60 59·3 ..	·17783	332
333	Twelfelhoek	27 27·4 ..	29 20·4 ..	22 1·0 ..	59 28·6 ..	·18796	333
334	Tygerfontein	34 10·0 ..	21 35·5 ..	28 13·0 ..	60 18·4 ..	·17967	334
335	Tygerkloof Drift	28 10·8 ..	28 35·2 ..	22 56·8 ..	59 21·5 ..	·18836	335
336	Thirtyfirst	25 40·6 ..	29 37·6 ..	22 0·9 ..	58 25·0 ..	·19197	336
337	Uitenhage	33 47·0 ..	25 24·0 ..	27 22·6 ..	61 24·6 ..	·17673	337
338	Uitkyk	25 49·5 ..	29 25·0 ..	21 24·0 ..	58 10·0 ..	·19456	338
339	Uitspan Farm	31 41·2 ..	21 27·2 ..	27 38·5 ..	59 34·3 ..	·18157	339
340	Umhlatuzi	28 51·7 ..	31 54·0 ..	22 28·9 ..	60 38·4 ..	·18602	340
341	Umhlengana Pass	31 36·0 ..	29 19·6 ..	25 10·5 ..	61 50·2 ..	·17787	341
342	Umtali	18 59·2 ..	32 39·0 ..	15 50·1 ..	53 0·0 ..	·22005	342
343	Umtata	31 35·9 ..	28 47·1 ..	25 12·3 ..	61 35·4 ..	·17879	343
344	Umtwalumi	30 28·0 ..	30 40·0 ..		61 45·4 ..	·17735	344
345	Umzinto	30 19·4 ..	30 39·0 ..	23 40·1 ..	61 22·2 ..	·18102	345
346	Underberg Hotel	29 47·9 ..	29 30·5 ..	24 39·2 ..	61 1·8 ..	·18068	346
347	Upington	28 27·7 ..	21 14·9 ..	27 3·5 ..	57 42·2 ..	·19183	347
348	Utrecht (W. of)	27 39·9 ..	30 16·0 ..	22 41·4 ..	59 33·8 ..	·18780	348
349	Van Reenen	28 22·2 ..	29 24·5 ..	23 5·1 ..		·18596	349
350	Van Wyk's Farm	33 49·4 ..	21 12·0 ..	28 11·4 ..	60 3·0 ..	·18025	350
351	Van Wyk's Vlei	30 22·3 ..	21 50·0 ..	27 4·8 ..	58 46·6 ..	·18461	351
352	Victoria Falls	17 55·6 ..	25 51·0 ..	17 52·5 ..	51 24·4 ..	·22070	352

TABLE 21 (*continued*).

No.	Station Name	Lat. ° ′	Long. ° ′	Declination (D) ° ′	Dip (θ) ° ′	Horizontal Intensity (H)	Station No.
353	Villiersdorp	33 59·5 S.	19 19·0 E.	28 37·9 W.	59 30·8 S.	·18118	353
354	Virginia	28 7·5 ..	26 55·0 ..	24 6·4 ..	59 2·4 ..	·18749	354
355	Vlaklaagte	26 50·6 ..	29 5·0 ..	22 23·4 ..	58 54·3 ..	·19059	355
356	Vogelvlei	29 8·3 ..	27 31·1 ..	24 4·9 ..	60 0·1 ..	·18422	356
357	Vondeling	33 19·8 ..	23 4·0 ..	27 46·3 ..	60 34·0 ..	·17858	357
358	Vredefort	27 1·2 ..	27 22·9 ..	22 35·6 ..	57 37·4 ..	·19276	358
359	Vredefort Road	27 7·0 ..	27 45·0 ..		59 3·9 ..	·17948	359
360	Vryburg	26 57·1 ..	24 43·0 ..	22 57·5 ..	58 2·3 ..	·19179	360
361	Wakkerstroom	27 21·5 ..	30 9·0 ..	22 27·0 ..	59 24·0 ..	·18874	361
362	Wankie	18 22·3 ..	26 28·5 ..	16 6·1 ..	51 5·6 ..	·22192	362
363	Warmbad (Waterberg)	24 53·0 ..	28 20·0 ..	21 21·1 ..	57 11·0 ..	·19691	363
364	Warmbad(Zoutpansberg)	22 24·9 ..	29 12·0 ..	21 16·7 ..	57 17·2 ..	·19223	364
365	Warrenton	28 6·9 ..	24 52·0 ..	24 51·6 ..	58 48·9 ..	·18798	365
366	Waschbank	28 18·8 ..	30 8·0 ..	23 16·0 ..	59 59·0 ..	·18608	366
367	Waterworks	29 4·5 ..	26 28·0 ..	24 29·6 ..	59 37·2 ..	·18500	367
368	Welverdiend	26 22·7 ..	27 17·0 ..	22 59·5 ..	58 43·7 ..	·18985	368
369	Wepener	29 43·6 ..	27 3·7 ..	25 23·1 ..	60 9·9 ..	·18303	369
370	Williston	31 20·4 ..	20 55·2 ..	26 39·6 ..	59 2·9 ..	·18451	370
371	Willowmore	33 9·4 ..	23 30·0 ..	27 51·0 ..	60 48·8 ..	·17794	371
372	Winburg	28 31·2 ..	27 3·0 ..	24 12·7 ..	59 17·4 ..	·18646	372
373	Winkeldrift	27 10·6 ..	27 7·6 ..	24 15·3 ..	58 43·5 ..	·18988	373
374	Witklip	23 16·5 ..	29 17·0 ..	21 58·0 ..	56 30·9 ..	·19958	374
375	Witmoss	32 33·0 ..	25 45·0 ..	26 26·8 ..	61 14·6 ..	·17852	375
376	Wolvefontein	33 19·0 ..	24 55·0 ..		61 19·2 ..	·17600	376
377	Wolvehoek	26 54·9 ..	27 50·0 ..	23 1·4 ..	58 36·2 ..	·19040	377
378	North of Limpopo	22 7·2 ..	29 10·0 ..	19 59·5 ..	55 52·0 ..	·20291	378
379	Wonderfontein	25 48·3 ..	29 53·0 ..	23 35·4 ..	58 0·0 ..	·19847	379
380	Woodville	33 56·3 ..	22 41·0 ..	27 56·8 ..	60 37·3 ..	·17881	380
381	Worcester	33 39·0 ..	19 26·0 ..	28 34·0 ..	59 23·5 ..	·18199	381
382	Zak Rivier	30 30·9 ..	20 31·0 ..	27 0·9 ..	58 17·2 ..	·18707	382
383	Zand River	23 3·8 ..	29 34·0 ..	21 13·0 ..	56 20·7 ..	·20234	383
384	Zeekoegat	33 3·0 ..	22 31·0 ..	27 57·2 ..	60 38·4 ..	·17639	384
385	Zuurbraak	34 0·3 ..	20 39·0 ..	28 20·2 ..	59 58·9 ..	·18092	385
386	Zuurfontein	32 51·0 ..	18 35·0 ..	28 16·6 ..	58 44·4 ..	·18426	386
387	Zuurpoort	32 2·9 ..	24 8·0 ..	26 53·0 ..	60 23·4 ..	·18072	387
388		23 42·7 ..	29 44·0 ..	20 47·7 ..	56 35·2 ..	·20114	388
389		24 8·0 ..	29 28·0 ..	20 23·9 ..	56 55·4 ..	·19908	389
390		25 9·8 ..	29 44·8 ..	21 33·7 ..	58 14·0 ..	·19557	390
391		25 47·5 ..	29 36·0 ..	21 10·0 ..	58 29·5 ..	·19228	391
392		24 47·0 ..	30 40·0 ..		58 9·0 ..	·19547	392
393		28 54·6 ..	27 44·1 ..	24 5·5 ..	59 54·2 ..	·18476	393
394		28 31·6 ..	27 42·3 ..	23 58·1 ..	59 31·9 ..	·18697	394
395		28 6·7 ..	29 3·1 ..	22 57·9 ..	59 39·3 ..	·18634	395
396		27 22·7 ..	29 0·0 ..	22 32·3 ..	59 8·2 ..	·18892	396
397		27 30·0 ..	26 38·3 ..		58 57·5 ..	·18777	397
398	East of Komgha	32 32·6 ..	27 58·3 ..	26 7·2 ..	61 32·6 ..	·17730	398
399	Outspan	32 13·4 ..	28 10·4 ..	25 56·9 ..	61 18·7 ..	·17803	399
400	Bashee	31 42·0 ..	28 30·0 ..	25 58·4 ..	61 35·1 ..	·17786	400
401		31 26·0 ..	29 31·5 ..		61 2·1 ..	·18449	401
402		31 0·6 ..	29 30·5 ..	24 17·9 ..	61 6·8 ..	·17973	402
403		30 49·7 ..	29 15·5 ..	24 14·4 ..	61 17·8 ..	·17953	403
404		31 6·0 ..	28 52·0 ..		61 19·4 ..	·17793	404
405	Ugie (on road to)	31 8·3 ..	28 26·2 ..	25 7·3 ..	61 12·0 ..	·17971	405

To obtain some idea of the magnetic state of the regions surveyed from the results given in the previous table the country was divided into a number of overlapping districts each containing about thirty stations. For each station the vertical, the total, the westerly and the northerly components of the intensity were calculated according to the formulae given in the introductory chapter. For each district the latitudes and the longitudes of all the stations in it were added together and divided by the number of stations, in this way the latitude and the longitude of a mean station was obtained; the magnetic elements at it were assumed to be those found by taking the mean of these elements at the various stations in that district. Table 22 gives the latitudes, ϕ_m, the longitudes, λ_m, the declinations, D_m, the dips, θ_m, the horizontal intensities, H_m, the vertical intensities, Z_m, the total intensities, T_m, the northerly intensities, X_m, and the westerly intensities, Y_m, for the different mean stations. The last four elements were obtained by calculating the values for each station and then proceeding in the manner explained above.

TABLE 22.

District	ϕ_m ° ′	λ_m ° ′	D_m ° ′	θ_m ° ′	H_m	Z_m	T_m	X_m	Y_m
I	33 46 S.	18 57 E.	28 33 W.	59 16 S.	·18234	·30669	·35681	·16034	·08727
II	33 52 ..	20 49 ..	28 23 ..	60 3 ..	·18037	·31266	·36096	·15861	·08573
III	33 36 ..	23 0 ..	27 52 ..	60 41 ..	·17877	·31845	·36530	·15771	·08347
IV	33 25 ..	25 1 ..	27 14 ..	61 14 ..	·17736	·32298	·36848	·15775	·08132
V	31 39 ..	28 32 ..	25 23 ..	61 26 ..	·17840	·32758	·37302	·16117	·07630
VI	29 43 ..	30 31 ..	23 45 ..	61 6 ..	·18127	·32838	·37510	·16611	·07305
VII	27 42 ..	30 3 ..	22 35 ..	59 37 ..	·18753	·32008	·37100	·17313	·07213
VIII	27 57 ..	27 49 ..	23 33 ..	59 14 ..	·18788	·31570	·36736	·17258	·07515
IX	25 59 ..	30 14 ..	21 29 ..	58 35 ..	·19289	·31572	·36999	·17946	·07065
X	26 35 ..	27 56 ..	22 52 ..	58 31 ..	·19134	·31252	·36645	·17618	·07429
XI	29 38 ..	26 12 ..	25 0 ..	59 44 ..	·18431	·31587	·36573	·16698	·07748
XII	31 38 ..	25 19 ..	26 17 ..	60 28 ..	·18087	·31927	·36696	·16217	·08009
XIII	31 58 ..	21 31 ..	27 21 ..	59 24 ..	·18311	·30962	·35973	·16255	·08409
XIV	39 26 ..	22 0 ..	26 36 ..	58 47 ..	·18546	·30608	·35792	·16620	·08310
XV	23 53 ..	29 44 ..	20 26 ..	57 3 ..	·19916	·30722	·36615	·18671	·06947

TABLE 22A.

District	Z_c	T_c	X_c	Y_c	$Z_m - Z_c$	$T_m - T_c$	$X_m - X_c$	$Y_m - Y_c$
I	·30668	·35680	·16017	·08715	1 γ	1 γ	17 γ	12 γ
II	·31303	·36129	·15869	·08574	– 37 ..	– 33 ..	– 8 ..	– 1 ..
III	·31836	·36511	·15804	·08348	9 ..	19 ..	– 33 ..	– 1 ..
IV	·32307	·36855	·15770	·08116	– 9 ..	– 7 ..	5 ..	16 ..
V	·32766	·37308	·16118	·07632	– 8 ..	– 6 ..	– 1 ..	– 2 ..
VI	·32837	·37508	·16592	·07301	1 ..	2 ..	19 ..	4 ..
VII	·31985	·37077	·17314	·07202	23 ..	23 ..	– 1 ..	11 ..
VIII	·31558	·36728	·17223	·07507	12 ..	8 ..	35 ..	8 ..
IX	·31581	·37004	·17949	·07064	– 9 ..	– 5 ..	– 3 ..	1 ..
X	·31244	·36638	·17630	·07435	8 ..	7 ..	– 12 ..	– 6 ..
XI	·31583	·36567	·16704	·07789	4 ..	6 ..	– 6 ..	– 41 ..
XII	·31925	·36693	·16217	·08009	2 ..	3 ..	0 ..	0 ..
XIII	·30962	·35972	·16264	·08413	0 ..	1 ..	– 9 ..	– 4 ..
XIV	·30603	·35784	·16583	·08304	5 ..	8 ..	37 ..	6 ..
XV	·30726	·36616	·18663	·06953	– 4 ..	– 1 ..	8 ..	– 6 ..

Table 22A gives the values of the vertical component (Z_c), the total intensity (T_c), the northerly (X_c), and the westerly (Y_c) components calculated from the mean values of the declination (D_m), the dip (θ_m), and the horizontal intensity (H_m) as given in the previous table. The latter part gives the differences between the values of the four elements calculated in this way and those tabulated in Table 22. The intensities are in c.g.s. units and the differences in γ.

A set of equations can now be formed for each district,—the number of equations being equal to the number of its stations,—of the form

$x(\phi_m - \phi) + y(\lambda_m - \lambda) =$ value of the magnetic element at the mean station (as given in Table 22) minus that at the station considered; where x and y are constants to be determined, ϕ_m, λ_m are the latitude and the longitude of the mean station of the district, ϕ, λ the latitude and the longitude of some station where the magnetic elements are known from observation. The x and the y in each set of equations were determined by making (1) the coefficients of every x positive and by addition obtaining a single equation in x and in y, and (2) the coefficient of every y positive thereby obtaining a second equation in x and y. The values of x and y found in this way give for the various districts mean values of the change of the magnetic element concerned with latitude and longitude respectively.

In Table 23 the change per degree for each district of the values of the declination, D, the dip, θ, and the horizontal intensity, H, is given.

TABLE 23.

District	Change per degree of latitude			Change per degree of longitude		
	$\frac{\partial D}{\partial \phi}$ ′	$\frac{\partial \theta}{\partial \phi}$ ′	$\frac{\partial H}{\partial \phi}$	$\frac{\partial D}{\partial \lambda}$ ′	$\frac{\partial \theta}{\partial \lambda}$ ′	$\frac{\partial H}{\partial \lambda}$
I	12·0	28·6	−156 γ	− 7·8	15·0	− 60 γ
II	6·0	21·6	− 48 ..	−13·2	26·4	− 90 ..
III	6·0	6·6	10 ..	−13·2	15·6	− 50 ..
IV	22·2	13·2	− 76 ..	− 9·6	19·2	− 88 ..
V	52·2	34·2	−318 ..	−10·2	22·8	−150 ..
VI	33·0	42·0	−432 ..	−24·6	25·8	−161 ..
VII	34·8	32·4	−222 ..	0·0	16·2	− 39 ..
VIII	32·4	28·2	−210 ..	−25·2	7·8	17 ..
IX	34·8	30·6	−327 ..	−32·4	19·8	−120 ..
X	18·6	30·0	−210 ..	−28·8	7·8	66 ..
XI	37·2	30·6	−210 ..	−18·0	17·4	− 72 ..
XII	31·8	34·2	−222 ..	−12·0	13·2	− 8 ..
XIII	31·8	28·2	−178 ..	− 6·6	15·6	− 62 ..
XIV	31·8	32·4	−214 ..	− 5·4	17·4	−108 ..
XV	26·4	34·8	−222 ..	−21·0	15·6	−126 ..

Table 24 gives the change per degree for each district of the values of the vertical intensity Z, the total intensity, T, the northerly intensity, X, and the westerly intensity, Y.

TABLE 24.

District	Change per degree of latitude				Change per degree of longitude			
	$\frac{\partial Z}{\partial \phi}$	$\frac{\partial T}{\partial \phi}$	$\frac{\partial X}{\partial \phi}$	$\frac{\partial Y}{\partial \phi}$	$\frac{\partial Z}{\partial \lambda}$	$\frac{\partial T}{\partial \lambda}$	$\frac{\partial X}{\partial \lambda}$	$\frac{\partial Y}{\partial \lambda}$
I	312 γ	192 γ	− 43 γ	121 γ	294 γ	234 γ	−133 γ	−133 γ
II	288 ..	222 ..	− 29 ..	16 ..	330 ..	222 ..	− 40 ..	− 99 ..
III	108 ..	96 ..	36 ..	−52 ..	223 ..	152 ..	− 18 ..	− 93 ..
IV	156 ..	78 ..	−168 ..	37 ..	228 ..	163 ..	− 78 ..	− 96 ..
V	60 ..	−90 ..	−360 ..	78 ..	119 ..	42 ..	− 92 ..	−121 ..
VI	198 ..	−10 ..	−480 ..	−21 ..	252 ..	183 ..	− 68 ..	−172 ..
VII	348 ..	192 ..	−277 ..	121 ..	246 ..	204 ..	− 38 ..	7 ..
VIII	240 ..	102 ..	−270 ..	64 ..	246 ..	216 ..	39 ..	−122 ..
IX	171 ..	4 ..	−470 ..	38 ..	143 ..	135 ..	− 19 ..	−255 ..
X	240 ..	83 ..	−198 ..	10 ..	240 ..	237 ..	126 ..	−126 ..
XI	264 ..	120 ..	−286 ..	81 ..	252 ..	216 ..	0 ..	− 92 ..
XII	372 ..	222 ..	−264 ..	53 ..	300 ..	246 ..	− 29 ..	− 87 ..
XIII	277 ..	148 ..	−276 ..	52 ..	187 ..	159 ..	− 71 ..	− 71 ..
XIV	360 ..	196 ..	−311 ..	38 ..	148 ..	78 ..	−145 ..	−102 ..
XV	312 ..	156 ..	−288 ..	53 ..	147 ..	34 ..	− 12 ..	−138 ..

From the three tables given above, viz. Tables 22, 23 and 24, it is now possible to calculate the value of any one of the elements at the intersection of each degree of latitude and of longitude. In the following tables the results of such a calculation are given; in the case where an intersection falls in two or more districts the algebraic mean is given of the value of the element so calculated.

Table 25.

Showing the values of the declination, D, at the intersection of degrees of latitude and of longitude. Epoch, 1st July, 1903.

The figures in minutes in heavy type in the intermediate columns are arranged so that, starting at the right of a row, the declination in a column plus the number of minutes on the left gives the declination in the next column on the left; starting from the top of a column the declination plus the minutes in heavy type in the intermediate line below gives that in the next lower line.

φ/λ	18° E.		19°		20°		21°		22°		23°		24°		25°		26°		27°		28°		29°		30°		31°		32°	λ/φ
	° ′	′	° ′	′	° ′	′	° ′	′	° ′	′	° ′	′	° ′	′	° ′	′	° ′	′	° ′	′	° ′	′	° ′	′	° ′	′	° ′	′	° ′	
22° S.	—		—		—		—		—		—		—		—		—		—		—		19 52		—		—		—	22° S.
																							26							
23°	—		—		—		—		—		—		—		—		—		—		—		20 18	**21**	19 57		—		—	23°
																									26					
24°	—		—		—		—		—		—		—		—		—		—		—		—		20 23		—		—	24°
																									33					
25°	—		—		—		—		—		—		—		—		—		—		—		—		20 56	**28**	20 28		—	25°
																									41		**37**			
26°	—		—		—		—		—		—		—		—		—		—		22 40	**29**	22 11	**34**	21 37	**32**	21 5		—	26°
																					18		**21**		**34**		**35**			
27°	—		—		—		—		—		—		—		—		—		23 26	**28**	22 58	**26**	22 32	**21**	22 11	**31**	21 40		—	27°
																			30		**33**		**24**		**34**					
28°	—		—		—		—		—		—		—		24 22		—		23 56	**25**	23 31	**35**	22 56	**11**	22 45		—		—	28°
															37				**27**						**42**					
29°	—		—		—		—		—		—		—		24 59	**18**	24 41	**18**	24 23		—		—		23 27	**18**	23 9	**25**	22 44	29°
															37		**37**		**37**						**40**		**33**			
30°	—		—		—		—		—		—		—		25 36	**18**	25 18	**18**	25 0		—		—		24 7	**25**	23 42		—	30°
															25		**34**		**37**						**30**					
31°	—		—		—		26 59	**5**	26 54		—		26 13	**12**	26 1	**9**	25 52	**15**	25 37		—		24 45	**8**	24 37		—		—	31°
							27		**25**				**32**		**32**		**29**		**26**				**52**							
32°	—		—		27 33	**7**	27 26	**7**	27 19		—		26 45	**12**	26 33	**12**	26 21	**18**	26 3	**16**	25 47	**10**	25 37		—		—		—	32°
							41		**37**				**41**		**33**		**36**				**52**									
33°	28 32	**8**	28 24		—		28 7	**11**	27 56	**8**	27 48	**22**	27 26	**20**	27 6	**9**	26 57		—		26 39		—		—		—		—	33°
	12		**12**				**15**		**10**		**6**		**12**		**23**		**22**													
34°	28 44	**8**	28 36	**5**	28 31	**9**	28 22	**16**	28 6	**12**	27 54	**16**	27 38	**9**	27 29	**10**	27 19		—		—		—		—		—		—	34°
φ/λ	18° E.		19°		20°		21°		22°		23°		24°		25°		26°		27°		28°		29°		30°		31°		32°	λ/φ

TABLE 26.

Showing the values of the dip, θ, at the intersection of degrees of latitude and of longitude. Epoch, 1st July, 1903.

The figures in minutes in heavy type in the intermediate columns are arranged so that, starting at the left of a row, the dip in the column plus the minutes in heavy type on the right gives the dip in the succeeding column; starting at the top of a column the dip plus the minutes in the intermediate line below gives the dip in the next lower line.

ϕ/λ	18° E.		19°		20°		21°		22°		23°		24°		25°		26°		27°		28°		29°		30°		31°		32°	λ/ϕ
	° ′	′	° ′	′	° ′	′	° ′	′	° ′	′	° ′	′	° ′	′	° ′	′	° ′	′	° ′	′	° ′	′	° ′	′	° ′	′	° ′	′	° ′	
23° S.	—		—		—		—		—		—		—		—		—		—		—		56 21	**15**	56 36		—		—	23° S.
																							35		**36**					
24°	—		—		—		—		—		—		—		—		—		—		—		56 56	**16**	57 12		—		—	24°
																									42					
25°	—		—		—		—		—		—		—		—		—		—		—		—		57 54	**26**	58 20		—	25°
																									37		**31**			
26°	—		—		—		—		—		—		—		—		—		—		58 14	**8**	58 22	**9**	58 31	**20**	58 51		—	26°
																					33		**33**		**36**		**30**			
27°	—		—		—		—		—		—		—		—		—		58 39	**8**	58 47	**8**	58 55	**12**	59 7	**14**	59 21		—	27°
																			30		**30**		**33**		**39**					
28°	—		—		—		—		—		—		—		58 33		—		59 9	**8**	59 17	**11**	59 28	**18**	59 46		—		—	28°
															31				**29**						**35**					
29°	—		—		—		57 43	**17**	58 0		—		—		59 4	**17**	59 21	**17**	59 38		—		—		60 21	**27**	60 48	**25**	61 13	29°
							32		**32**						**30**		**31**		**31**						**44**		**42**			
30°	—		—		—		58 15	**17**	58 32	**18**	58 50		—		59 34	**18**	59 52	**17**	60 9		—		—		61 5	**25**	61 30		—	30°
							33		**32**		**32**				**28**		**27**		**19**						**37**					
31°	—		—		—		58 48	**16**	59 4	**18**	59 22	**27**	59 49	**13**	60 2	**17**	60 19	**9**	60 28		—		61 14	**28**	61 42		—		—	31°
							29		**29**				**34**		**34**		**31**		**35**				**35**							
32°	—		—		59 1	**16**	59 17	**16**	59 33		—		60 23	**13**	60 36	**14**	60 50	**13**	61 3	**23**	61 26	**23**	61 49		—		—		—	32°
							30		**38**				**28**		**33**		**37**				**34**									
33°	58 40	**15**	58 55		—		59 47	**24**	60 11	**26**	60 37	**14**	60 51	**18**	61 9	**18**	61 27		—		62 0		—		—		—		—	33°
	28		**28**				**23**		**22**		**7**		**10**		**12**		**14**													
34°	59 8	**15**	59 23	**18**	59 41	**29**	60 10	**23**	60 33	**11**	60 44	**17**	61 1	**20**	61 21	**20**	61 41		—		—		—		—		—		—	34°
ϕ/λ	18° E.		19°		20°		21°		22°		23°		24°		25°		26°		27°		28°		29°		30°		31°		32°	λ/ϕ

TABLE 27.

Giving the values of the horizontal intensity, H, in terms of γ at the intersections of degrees of latitude and of longitude.

Epoch, 1st July, 1903.

The figures are arranged so that when starting from the bottom of a column the horizontal intensity plus the figures in heavy type in the intermediate line above it gives the value on the next line, and so that when starting from the right-hand side of a row the horizontal intensity plus the figures in heavy type in the intermediate column gives the next value.

φ/λ	18° E.		19°		20°		21°		22°		23°		24°		25°		26°		27°		28°		29°		30°		31°		32°	λ/φ
22° S.	—		—		—		—		—		—		—		—		—		—		—		20426		—		—		—	22° S.
																							222							
23°	—		—		—		—		—		—		—		—		—		—		—		20204	**126**	20078		—		—	23°
																							222		**222**					
24°	—		—		—		—		—		—		—		—		—		—		—		19982	**126**	19856		—		—	24°
																									219					
25°	—		—		—		—		—		—		—		—		—		—		—		—		19637	**118**	19519		—	25°
																									325		**327**			
26°	—		—		—		—		—		—		—		—		—		—		19264	**−66**	19330	**18**	19312	**120**	19192		—	26°
																					245		**269**		**365**		**338**			
27°	—		—		—		—		—		—		—		—		—		18978	**−41**	19019	**−42**	19061	**114**	18947	**93**	18854		—	27°
																			215		**238**		**299**		**259**					
28°	—		—		—		—		—		—		—		18870		—		18763	**−18**	18781	**19**	18762	**74**	18688		—		—	28°
															216				**231**						**195**					
29°	—		—		—		18964	**108**	18856		—		—		18654	**72**	18582	**50**	18532		—		—		18493	**134**	18359	**161**	18198	29°
							214		**214**						**216**		**216**		**238**						**405**		**432**			
30°	—		—		—		18750	**108**	18642	**108**	18534		—		18438	**72**	18366	**72**	18294		—		—		18088	**161**	17927		—	30°
							220		**212**		**214**				**208**		**180**		**80**						**347**					
31°	—		—		—		18530	**100**	18430	**110**	18320	**82**	18238	**8**	18230	**44**	18186	**−28**	18214		—		17977	**236**	17741		—		—	31°
							193		**155**				**222**		**222**		**186**		**239**				**318**							
32°	—		—		18398	**61**	18337	**62**	18275		—		18016	**8**	18008	**8**	18000	**25**	17975	**166**	17809	**150**	17659		—		—		—	32°
							224		**266**				**177**		**231**		**311**				**318**									
33°	18411	**60**	18351		—		18113	**104**	18009	**138**	17871	**32**	17839	**62**	17777	**88**	17689		—		17491		—		—		—		—	33°
	156		**156**				**98**		**81**		**−10**		**33**		**84**		**84**													
34°	18255	**60**	18195	**75**	18120	**105**	18015	**87**	17928	**47**	17881	**75**	17806	**113**	17693	**88**	17605		—		—		—		—		—		—	34°
φ/λ	18° E.		19°		20°		21°		22°		23°		24°		25°		26°		27°		28°		29°		30°		31°		32°	λ/φ

TABLE 28.

Giving the values of the total intensity, T, in terms of γ at the intersections of degrees of latitude and of longitude.
Epoch, 1st July, 1903.

The numbers are so arranged that starting from the top of a column the total intensity plus the figures in heavy type in the intermediate line below gives that on the adjacent lower line, and so that starting from the left-hand side of a row the vertical intensity plus the figures in heavy type of the intermediate column gives the next value.

ϕ/λ	18° E.		19°		20°		21°		22°		23°		24°		25°		26°		27°		28°		29°		30°		31°		32°	λ/ϕ
22° S.	—		—		—		—		—		—		—		—		—		—		—		36296		—		—		—	22° S.
																							156							
23°	—		—		—		—		—		—		—		—		—		—		—		36452	**34**	36486		—		—	23°
																							156		**156**					
24°	—		—		—		—		—		—		—		—		—		—		—		36608	**34**	36642		—		—	24°
																									239					
25°	—		—		—		—		—		—		—		—		—		—		—		—		36881	**218**	37099		—	25°
																									87		**4**			
26°	—		—		—		—		—		—		—		—		—		—		36611	**238**	36849	**119**	36968	**135**	37103		—	26°
																					77		**65**		**−5**		**3**			
27°	—		—		—		—		—		—		—		—		—		36461	**227**	36688	**226**	36914	**49**	36963	**143**	37106		—	27°
																			104		**93**		**56**		**184**					
28°	—		—		—		—		—		—		—		36118		—		36565	**216**	36781	**189**	36970	**177**	37147		—		—	28°
															120				**103**						**234**					
29°	—		—		—		35429	**78**	35507		—		—		36238	**214**	36452	**216**	36668		—		—		37381	**225**	37606	**183**	37789	29°
							197		**197**						**120**		**122**		**122**						**31**		**−10**			
30°	—		—		—		35626	**78**	35704	**78**	35782		—		36358	**216**	36574	**216**	36790		—		—		37412	**184**	37596		—	30°
							159		**200**		**196**				**119**		**135**		**180**						**0**					
31°	—		—		—		35785	**119**	35904	**74**	35978	**253**	36231	**246**	36477	**232**	36709	**261**	36970		—		37380	**32**	37412		—		—	31°
							111		**151**				**222**		**223**		**237**		**229**				**−90**							
32°	—		—		35737	**159**	35896	**159**	36055		—		36453	**247**	36700	**246**	36946	**253**	37199	**49**	37248	**42**	37290		—		—		—	32°
							98		**167**				**217**		**168**		**30**				**−90**									
33°	35311	**235**	35546		—		35994	**228**	36222	**243**	36465	**205**	36670	**198**	36868	**108**	36976		—		37158		—		—		—		—	33°
	193		**192**				**172**		**147**		**108**		**92**		**23**		**78**													
34°	35504	**234**	35738	**220**	35958	**208**	36166	**203**	36369	**204**	36573	**189**	36762	**129**	36891	**163**	37054		—		—		—		—		—		—	34°
ϕ/λ	18° E.		19°		20°		21°		22°		23°		24°		25°		26°		27°		28°		29°		30°		31°		32°	λ/ϕ

Table 29.

Giving the values of the vertical intensity, Z, in terms of γ at the intersections of degrees of latitude and of longitude. Epoch, 1st July, 1903.

The table is arranged so that starting from the upper value in a column the next is obtained by adding the figures in heavy type in the intermediate line below, and so that when starting from the left-hand side of a row the vertical intensity plus the figures in heavy type in the intermediate column gives the next value.

φ/λ	18° E.		19°		20°		21°		22°		23°		24°		25°		26°		27°		28°		29°		30°		31°		32°	λ/φ
22° S.	—		—		—		—		—		—		—		—		—		—		—		30027		—		—		—	22° S.
																							312							
23°	—		—		—		—		—		—		—		—		—		—		—		30339	**147**	30486		—		—	23°
																							312		**312**					
24°	—		—		—		—		—		—		—		—		—		—		—		30651	**147**	30798		—		—	24°
																									442					
25°	—		—		—		—		—		—		—		—		—		—		—		—		31240	**272**	31512		—	25°
																									301		**171**			
26°	—		—		—		—		—		—		—		—		—		—		31128	**240**	31368	**173**	31541	**142**	31683		—	26°
																					249		**253**		**191**		**172**			
27°	—		—		—		—		—		—		—		—		—		31135	**242**	31377	**244**	31621	**111**	31732	**123**	31855		—	27°
																			246		**250**		**243**		**368**					
28°	—		—		—		—		—		—		—		30854		—		31381	**246**	31627	**237**	31864	**236**	32100		—		—	28°
															264				**241**						**408**					
29°	—		—		—		29939	**147**	30086		—		—		31118	**252**	31370	**252**	31622		—		—		32508	**310**	32818	**252**	33070	29°
							359		**360**						**264**		**264**		**264**						**258**		**198**			
30°	—		—		—		30298	**148**	30446	**148**	30594		—		31382	**252**	31634	**252**	31886		—		—		32766	**250**	33016		—	30°
							330		**349**		**360**				**214**		**263**		**310**						**163**					
31°	—		—		—		30628	**167**	30795	**159**	30954	**342**	31296	**300**	31596	**301**	31897	**299**	32196		—		32774	**155**	32929		—		—	31°
							247		**266**				**372**		**372**		**371**		**387**				**60**							
32°	—		—		30688	**187**	30875	**186**	31061		—		31668	**300**	31968	**300**	32268	**315**	32583	**132**	32715	**119**	32834		—		—		—	32°
							240		**388**				**334**		**367**		**189**				**61**									
33°	30150	**294**	30444		—		31115	**334**	31449	**331**	31780	**222**	32002	**333**	32335	**122**	32457		—		32776		—		—		—		—	33°
	312		**313**				**250**		**232**		**108**		**131**		**50**		**156**													
34°	30462	**295**	30757	**289**	31046	**319**	31365	**316**	31681	**207**	31888	**245**	32133	**252**	32385	**228**	32613		—		—		—		—		—		—	34°
φ/λ	18° E.		19°		20°		21°		22°		23°		24°		25°		26°		27°		28°		29°		30°		31°		32°	λ/φ

TABLE 30.

Giving the values of the northerly intensity, X, in terms of γ at the intersections of degrees of latitude and of longitude. Epoch, 1st July, 1903.

The table is so arranged that starting from the bottom of a column the addition of the number in heavy type in the intermediate line above gives the next value, and so that starting on the right-hand side of a row and adding the number in heavy type in the intermediate column to the left the next value is obtained.

18° E.		19°		20°		21°		22°		23°		24°		25°		26°		27°		28°		29°		30°		31°		32°	λ/φ
—		—		—		—		—		—		—		—		—		—		—		19185		—		—		—	22° S.
																						266							
—		—		—		—		—		—		—		—		—		—		—		18919	**12**	18907		—		—	23°
																						282		**282**					
—		—		—		—		—		—		—		—		—		—		—		18637	**12**	18625		—		—	24°
																								247					
—		—		—		—		—		—		—		—		—		—		—		—		18378	**−15**	18393		—	25°
																								436		**470**			
—		—		—		—		—		—		—		—		—		—		17742	**−126**	17868	**−74**	17942	**19**	17923		—	26°
																				206		**250**		**451**		**470**			
—		—		—		—		—		—		—		—		—		17454	**−82**	17536	**−82**	17618	**127**	17491	**38**	17453		—	27°
																		235		**278**		**335**		**259**					
—		—		—		—		—		—		—		17164		—		17219	**−39**	17258	**−25**	17283	**51**	17232		—		—	28°
														286				**305**						**258**					
—		—		17202	**144**	17058		—		—		—		16878	**0**	16878	**−36**	16914		—		—		16974	**48**	16926	**68**	16858	29°
				312		**312**								**285**		**285**		**321**						**460**		**480**			
—		—		16890	**144**	16746	**144**	16602		—		—		16593	**0**	16593	**0**	16593		—		—		16514	**68**	16446		—	30°
				322		**286**		**312**						**198**		**256**		**266**						**384**					
—		—		16568	**108**	16460	**170**	16290		—		16425	**30**	16395	**58**	16337	**10**	16327		—		16318	**188**	16130		—		—	31°
				213		**177**		**79**				**265**		**264**		**235**		**229**				**360**							
—		—		16355	**72**	16283	**72**	16211		—		16160	**29**	16131	**29**	16102	**4**	16098	**58**	16040	**82**	15958		—		—		—	32°
						343		**361**				**334**		**275**		**333**				**361**									
16193	**133**	16060		—		15940	**90**	15850	**103**	15747	**−79**	15826	**−30**	15856	**87**	15769		—		15679		—		—		—		—	33°
43		**42**				**90**		**45**		**−35**		**66**		**178**		**168**													
16150	**132**	16018	**131**	15887	**37**	15850	**45**	15805	**23**	15782	**22**	15760	**−18**	15678	**77**	15601		—		—		—		—		—		—	34°
18° E.		19°		20°		21°		22°		23°		24°		25°		26°		27°		28°		29°		30°		31°		32°	λ/φ

TABLE 31.

Giving the values of the westerly intensity, Y, in terms of γ at the intersections of degrees of latitude and of longitude.

Epoch, 1st July, 1903.

The table is arranged so that by starting from the top of a column the value plus the number in heavy type in the intermediate line below gives the value on the next line, and so that by starting at the right of a row the value in a column plus the figures in heavy type in the intermediate column to the left gives the value in the next column on the left.

18°E.		19°		20°		21°		22°		23°		24°		25°		26°		27°		28°		29°		30°		31°		32°	λ/φ
—		—		—		—		—		—		—		—		—		—		—		6946		—		—		—	22°S.
																						54							
—		—		—		—		—		—		—		—		—		—		—		7000	**137**	6863		—		—	23°
																						54		**54**					
—		—		—		—		—		—		—		—		—		—		—		7054	**137**	6917		—		—	24°
																								112					
—		—		—		—		—		—		—		—		—		—		—		—		7029	**194**	6835		—	25°
																								96		**36**			
—		—		—		—		—		—		—		—		—		—		7414	**125**	7289	**164**	7125	**254**	6871		—	26°
																				14		**15**		**21**		**38**			
—		—		—		—		—		—		—		—		—		7552	**124**	7428	**124**	7304	**158**	7146	**237**	6909		—	27°
																		64		**66**		**3**		**103**					
—		—		—		—		—		—		—		7726		—		7616	**122**	7494	**187**	7307	**58**	7249		—		—	28°
														81				**36**						**141**					
—		—		8366	**102**	8264		—		—		—		7807	**92**	7715	**63**	7652		—		—		7390	**154**	7236	**174**	7062	29°
				37		**37**								**81**		**81**		**52**						**−1**		**−21**			
—		—		8403	**102**	8301	**102**	8199		—		—		7888	**92**	7796	**92**	7704		—		—		7389	**174**	7215		—	30°
				14		**29**		**37**						**114**		**100**		**124**						**−4**					
—		—		8417	**87**	8330	**94**	8236		—		8089	**87**	8002	**106**	7896	**68**	7828		—		7523	**138**	7385		—		—	31°
				103		**118**		**140**				**54**		**54**		**73**		**33**				**78**							
—		—		8520	**72**	8448	**72**	8376		—		8143	**87**	8056	**87**	7969	**108**	7861	**140**	7721	**120**	7601		—		—		—	32°
						73		**76**				**109**		**59**		**54**				**78**									
8761	**132**	8629		—		8521	**69**	8452	**71**	8381	**129**	8252	**137**	8115	**92**	8023		—		7799		—		—		—		—	33°
120		**120**				**36**		**−12**		**−52**		**−8**		**40**		**36**													
8881	**132**	8749	**113**	8636	**79**	8557	**117**	8440	**111**	8329	**85**	8244	**89**	8155	**96**	8059		—		—		—		—		—		—	34°
18°E.		19°		20°		21°		22°		23°		24°		25°		26°		27°		28°		29°		30°		31°		32°	λ/φ

In addition to presenting the results on maps by isomagnetic lines, much information is gained by calculating what we may conveniently by analogy call the magnetic anomaly for each element. In this instance the anomaly has been found in the following manner. The constants x and y of the equation on p. 53 were calculated for each element for each district in a manner already explained on that page.

The magnetic elements were then calculated from this equation for each station at which observations had been made by substituting the necessary quantities belonging to the mean station and to the one under consideration, and the known values of x and of y for the district. The difference between the calculated and the observed values is the anomaly for a given element for that particular station. In the case where the same station appeared in two or more districts, the anomaly for each district was obtained in the above manner and the algebraic mean of the several anomalies adopted as the correct value.

Table 32 contains a list of stations with their latitudes and longitudes, the observed and the calculated values of the declination, the northerly intensity, and the westerly intensity, together with the anomalies of these magnetic elements.

Table 33 contains a list of stations with the observed and the calculated values of the dip, the horizontal intensity, the vertical intensity, the total intensity, together with their anomalies.

Table 32.

Summary of observed and calculated values and anomalies of the declination, the northerly intensity, and the westerly intensity.

Station No.	Station Name	Lat.	Long.	Declination (D) Obs.	Declination (D) Calc.	Declination (D) Diff.	Northerly Comp. (X) Obs.	Northerly Comp. (X) Calc.	Northerly Comp. (X) Diff.	Westerly Comp. (Y) Obs.	Westerly Comp. (Y) Calc.	Westerly Comp. (Y) Diff.	Station No.
		° ′	° ′	° ′	° ′	′			(γ)			(γ)	
1	Abelsdam	27 37·1 S.	26 29·6 E.	23 38 W.	23 54 W.	16	·17214	·17268	54	·07533	·07637	104	1
2	Aberdeen, C. C.	32 29·1 ..	24 3·0 ..	27 14 ..	27 2 ..	−12	·15996	·15937	− 59	·08234	·08173	− 61	2
3	Aberdeen (Transvaal)	26 3·8 ..	29 33·0 ..	21 16 ..	21 54 ..	38	·17909	·17919	10	·06968	·07243	275	3
4	Aberdeen Road	32 46·0 ..	24 20·0 ..	27 6 ..	27 14 ..	8	·16193	·15867	−326	·08288	·08197	− 91	4
5	Aberfeldy	25 45·7 ..	28 34·5 ..	22 14 ..	22 18 ..	4	·17980	·17862	−118	·07352	·07339	− 13	5
6	Adelaide	32 43·0 ..	26 18·0 ..	26 22 ..	26 44 ..	22	·15898	·15848	− 50	·07878	·07983	105	6
7	Albert Falls	29 26·0 ..	30 29·0 ..	23 47 ..	23 36 ..	−11	·16782	·16753	− 29	·07393	·07317	− 76	7
8	Alicedale	33 18·9 ..	26 2·5 ..	27 27 ..	27 4 ..	−23	·15570	·15711	141	·08089	·08029	− 60	8
9	Aliwal North	30 41·7 ..	26 42·0 ..	25 21 ..	25 31 ..	10	·16452	·16409	− 43	·07794	·07814	20	9
10	Alma	27 7·6 ..	31 5·5 ..	22 37 ..	22 15 ..	−22	·17407	·17430	23	·07252	·07152	−100	10
11	Amabele Junction	32 43·1 ..	27 19·2 ..	26 3 ..	26 31 ..	28	·16140	·15843	−297	·07542	·07860	318	11
13	Amatongas	19 11·2 ..	33 45·0 ..	16 4 ..	15 16 ..	−48	·20986	·21585	599	·06044	·05814	−230	13
15	Assegai Bosch	33 56·7 ..	24 20·5 ..	27 33 ..	27 35 ..	2	·15733	·15748	15	·08210	·08205	− 5	15
16	Avontuur	33 44·2 ..	23 13·0 ..	27 49 ..	27 50 ..	1	·15781	·15769	− 12	·08325	·08325	0	16
17	Ayrshire Mine	17 11·5 ..	30 23·0 ..	16 5 ..	16 6 ..	1	·21921	·21776	−145	·06318	·06200	−118	17
17A	Balmoral	25 51·3 ..	28 58·0 ..	22 7 ..	22 9 ..	2	·17937	·17894	− 43	·07290	·07389	99	17A
18	Bamboo Creek	19 16·5 ..	34 12·0 ..	15 5 ..	15 3 ..	− 2	·21174	·21001	−173	·05707	·05737	30	18
20	Barberton	25 47·3 ..	31 0·0 ..	21 13 ..	20 57 ..	−16	·17980	·18026	46	·06979	·06863	−116	20
21	Barrington	33 55·2 ..	22 52·0 ..	26 38 ..	27 56 ..	78	·16366	·15783	−583	·08206	·08357	151	21
22	Battlefields	18 36·4 ..	30 0·0 ..	17 23 ..	17 12 ..	−11	·20873	·21177	304	·06535	·06459	− 76	22
23	Bavaria	27 30·7 ..	29 8·4 ..	22 29 ..	22 37 ..	8	·17410	·17416	6	·07207	·07255	48	23
24	Baviaanskrantz	27 23·0 ..	26 47·0 ..	26 49 ..	23 16 ..	−213	·16869	·17466	597	·08528	·07460	−1068	24
25	Beaufort West	32 20·9 ..	22 34·1 ..	27 9 ..	27 26 ..	17	·16183	·16077	−106	·08299	·08351	52	25
26	Beira	19 49·2 ..	34 50·0 ..	15 56 ..			·20841			·05949			26
28	Berg River Mouth	32 46·5 ..	18 10·0 ..	28 18 ..	28 28 ..	10	·16231	·16181	− 50	·08741	·08713	− 28	28
29	Bethal	26 28·1 ..	29 27·5 ..	22 45 ..	22 11 ..	−34	·17907	·17733	−174	·07506	·07280	−226	29
30	Bethany	29 37·0 ..	26 2·0 ..	24 50 ..	25 3 ..	13	·16738	·16702	− 36	·07746	·07762	16	30
31	Bethesda Road	31 55·3 ..	24 38·0 ..	26 21 ..	26 35 ..	14	·16185	·16164	− 21	·08019	·08084	65	31
32	Bethlehem	28 13·8 ..	28 17·3 ..	23 39 ..	23 31 ..	− 8	·17341	·17206	−135	·07594	·07475	−119	32
33	Bethulie	30 30·5 ..	25 59·0 ..	25 30 ..	25 35 ..	5	·16607	·16469	−138	·07922	·07865	− 57	33
34	Biesjespoort	31 43·8 ..	23 12·0 ..	26 2 ..	27 3 ..	61	·16577	·16198	−379	·08093	·08276	183	34
35	Birthday	23 19·5 ..	30 46·0 ..	16 8 ..	19 56 ..	228	·17836	·18808	972	·05156	·06820	1664	35
36	Blaauwbosch	30 38·9 ..	22 14·1 ..	26 34 ..	26 41 ..	7	·16543	·16515	− 28	·08272	·08300	28	36
37	Blaauwkrantz	33 57·0 ..	23 35·0 ..	27 44 ..	27 46 ..	2	·15767	·15771	4	·08289	·08305	16	37

TABLE 32 (*continued*).

Station No.	Station Name	Lat. ° ′	Long. ° ′	Declination (D) Obs. ° ′	Declination (D) Calc. ° ′	Declination (D) Diff. ′	Northerly Comp. (X) Obs.	Northerly Comp. (X) Calc.	Northerly Comp. (X) Diff. (γ)	Westerly Comp. (Y) Obs.	Westerly Comp. (Y) Calc.	Westerly Comp. (Y) Diff. (γ)	Station No.
38	The Bluff	29 52·5 S.	31 4·0 E.	23 40 W.	23 37 W.	– 3	·16416	·16497	81	·07195	·07206	11	38
39	Boschkopjes	23 11·5 ..	29 55·0 ..	19 51 ..	20 3 ..	12	·18771	·18852	81	·06776	·06887	111	39
40	Boschrand	27 45·8 ..	27 12·0 ..	23 39 ..	23 44 ..	5	·17260	·17290	30	·07557	·07577	20	40
41	Boston	29 41·0 ..	30 1·0 ..	23 45 ..	23 56 ..	11	·16697	·16665	– 32	·07347	·07392	45	41
42	Botha's Berg	25 25·0 ..	29 49·0 ..	22 24 ..	21 23 ..	–61	·18815	·18220	–595	·07751	·07150	–601	42
43	Brak River	22 52·2 ..	29 13·0 ..	19 32 ..	20 10 ..	38	·19276	·18955	–321	·06836	·06965	129	43
44	Brandboontjes	23 28·0 ..	30 16·0 ..	19 52 ..	20 3 ..	11	·18870	·18775	– 95	·06820	·06852	32	44
45	Bredasdorp	34 32·2 ..	20 3·0 ..	28 39 ..	28 37 ..	– 2	·15809	·15873	64	·08638	·08659	21	45
46	Breekkerrie	30 6·7 ..	21 35·0 ..	27 3 ..	26 28 ..	–35	·16504	·16770	266	·08428	·08348	– 80	46
47	Britstown	30 35·0 ..	23 33·0 ..	26 29 ..	26 5 ..	–24	·16199	·15871	–328	·08069	·08118	49	47
48	Buffelsberg	23 36·7 ..	30 1·0 ..	20 55 ..	20 12 ..	–43	·18199	·18734	535	·06958	·06895	– 63	48
49	Buffelshoek	23 8·3 ..	28 55·0 ..	20 7 ..	20 23 ..	16	·18600	·18883	283	·06814	·07021	207	49
50	Buffelsklip	33 31·7 ..	22 52·5 ..	27 47 ..	27 53 ..	6	·15808	·15768	– 40	·08327	·08370	43	50
51	Bulawayo	20 9·1 ..	28 36·3 ..	18 37 ..	18 36 ..	– 1	·20327	·20466	139	·06850	·06887	37	51
52A	Bulwer	29 48·4 ..	29 41·0 ..	24 22 ..	24 8 ..	–14	·16471	·16632	161	·07457	·07448	– 9	52A
53	Burghersdorp	31 0·0 ..	26 18·0 ..	25 42 ..	25 47 ..	5	·16410	·16332	– 78	·07899	·07870	– 29	53
54	Bushmanskop	32 20·8 ..	22 14·5 ..	27 17 ..	27 29 ..	12	·16203	·16097	–106	·08356	·08376	20	54
55	Butterworth	32 21·3 ..	28 4·0 ..	25 51 ..	26 4 ..	13	·16056	·15907	–149	·07779	·07741	– 38	55
56	Caledon River	30 16·8 ..	26 41·7 ..	25 44 ..	25 15 ..	–29	·16296	·16512	216	·07853	·07755	– 98	56
57	Calitzdorp	33 32·1 ..	21 41·0 ..	28 18 ..	28 27 ..	9	·15777	·15813	36	·08494	·08485	– 9	57
59	Cango	33 24·8 ..	22 14·5 ..	28 19 ..	28 3 ..	–16	·15653	·15796	143	·08432	·08428	– 4	59
60	Cape Town (Royal Ob.)	33 56·1 ..	18 28·7 ..	28 44 ..	28 39 ..	– 5	·16023	·16089	66	·08785	·08810	25	60
62	Ceres Road	33 25·6 ..	19 19·0 ..	28 32 ..	28 27 ..	– 5	·16066	·16000	– 66	·08733	·08639	– 94	62
63	Charlestown	27 24·9 ..	29 54·0 ..	21 33 ..	22 25 ..	52	·17542	·17397	–145	·06930	·07178	248	63
65	Coerney	33 27·6 ..	25 44·0 ..	27 18 ..	27 10 ..	– 8	·15664	·15710	46	·08082	·08065	– 17	65
66	Colenso	28 44·0 ..	29 50·0 ..	22 48 ..	23 11 ..	23	·17098	·17036	– 62	·07189	·07336	147	66
67	Colesberg	30 42·8 ..	25 8·0 ..	25 37 ..	25 55 ..	18	·16397	·16427	30	·07859	·07923	64	67
68	Connan's Farm	28 58·4 ..	21 19·3 ..	24 55 ..	25 57 ..	62	·17778	·17167	–611	·08257	·08334	77	68
69	Cookhouse	32 44·0 ..	25 48·0 ..	26 43 ..	26 50 ..	7	·15837	·15871	34	·07976	·08030	54	69
70	Cotswold Hotel	30 42·7 ..	29 53·4 ..	24 23 ..	24 27 ..	4	·16184	·16254	70	·07337	·07394	57	70
71	Cradock	32 9·6 ..	25 38·0 ..	26 27 ..	26 31 ..	4	·16098	·16069	– 29	·08010	·08011	1	71
72	Cream of Tartarfontein	22 35·3 ..	29 1·0 ..	19 42 ..	20 6 ..	24	·19212	·19037	–175	·06877	·06978	101	72
73	Crocodile Pools	24 46·7 ..	25 50·0 ..	21 38 ..			·18587			·07369			73
75	Dambiesfontein	31 24·2 ..	21 17·6 ..	27 7 ..	27 8 ..	1	·16337	·16418	81	·08368	·08409	41	75
76	Dannhauser	28 1·2 ..	30 5·0 ..	22 30 ..	22 46 ..	16	·17182	·17225	43	·07116	·07251	135	76
77	Dargle Road	29 29·1 ..	30 11·0 ..	24 6 ..	23 45 ..	–21	·16859	·16749	–110	·07540	·07409	–131	77
78	Darling	33 22·1 ..	18 22·0 ..	28 31 ..	28 33 ..	2	·16146	·16129	– 17	·08772	·08757	– 15	78
79	De Aar	30 40·0 ..	24 2·0 ..	25 42 ..	26 2 ..	20	·16622	·15907	–715	·08000	·08059	59	79
81	Deelfontein Farm	30 5·8 ..	26 31·7 ..	25 22 ..	25 9 ..	–13	·16484	·16564	80	·07816	·07740	– 76	81

83	De Jager's Farm	28 15·9 ..	28 57·9 ..	23 4 ..	23 22 ..	18	·17296	·17224	– 72	·07368	·07393	25	83
84	Dewetsdorp	29 26·1 ..	26 39·3 ..	25 43 ..	24 51 ..	–52	·16692	·16707	15	·08041	·07704	–337	84
85	Draghoender	29 22·3 ..	22 7·4 ..	27 14 ..	26 1 ..	–73	·16646	·16927	281	·08570	·08257	–313	85
86	Drew	33 59·5 ..	20 13·0 ..	28 25 ..	28 30 ..	5	·15945	·15868	– 77	·08628	·08611	– 17	86
87	Driefontein	26 29·4 ..	29 13·0 ..	22 23 ..	22 16 ..	– 7	·17372	·17764	392	·07159	·07307	148	87
88	Driehoek	27 11·6 ..	30 41·0 ..	22 18 ..	22 17 ..	– 1	·17484	·17427	– 57	·07172	·07157	– 15	88
89	East London	33 0·0 ..	27 56·0 ..	26 31 ..	26 40 ..	9	·15748	·15685	– 63	·07859	·07807	– 52	89
92	Elim	34 35·8 ..	19 46·0 ..	28 42 ..	28 40 ..	– 2	·15846	·15886	40	·08675	·08705	30	92
94	Elliot	31 18·0 ..	27 54·0 ..	24 55 ..	25 11 ..	16	·16233	·16300	67	·07580	·07679	99	94
95	Elsburg	26 15·0 ..	28 11·0 ..	22 29 ..	22 39 ..	10	·17686	·17717	31	·07308	·07394	86	95
96	Emmasheim	28 17·2 ..	28 7·3 ..	23 55 ..	23 37 ..	–18	·17091	·17186	95	·07581	·07497	– 84	96
97	Estcourt	29 0·9 ..	29 54·0 ..	23 17 ..	23 28 ..	11	·16881	·16974	93	·07263	·07399	136	97
98	Ferreira	29 12·0 ..	26 11·0 ..	24 47 ..	24 45 ..	– 2	·16777	·16821	44	·07748	·07714	– 34	98
99	Fish River	31 55·3 ..	25 27·0 ..	26 19 ..	26 25 ..	6	·16152	·16140	– 12	·07986	·08013	27	99
100	Forty-one mile Siding	17 43·0 ..	30 33·0 ..	15 55 ..	16 19 ..	24	·21604	·21563	– 41	·06162	·06353	191	100
101	Fountain Hall	29 15·8 ..	29 59·0 ..	23 20 ..	23 43 ..	23	·16986	·16867	–119	·07330	·07407	77	101
103	Fraserburg	31 55·2 ..	21 31·3 ..	27 33 ..	27 20 ..	–13	·16216	·16269	53	·08460	·08406	– 54	103
105	Gamtoos River Bridge	33 55·2 ..	25 2·5 ..	27 25 ..	27 27 ..	2	·15743	·15688	– 55	·08169	·08147	– 22	105
107	Gemsbokfontein	31 22·8 ..	22 57·5 ..	27 47 ..	26 57 ..	–50	·16214	·16244	30	·08543	·08263	–280	107
108	Ginginhlovu	29 1·7 ..	31 35·0 ..	23 7 ..	22 56 ..	–11	·16787	·16870	83	·07163	·07134	– 29	108
109	Glenallen	29 39·0 ..	22 36·0 ..	48 0 ..	26 7 ..	–1313	·15696	·16769	1073	·17432	·08228	–9204	109
110	Glenconnor	33 25·0 ..	25 10·0 ..	27 29 ..	27 15 ..	–14	·15647	·15763	116	·08139	·08117	– 22	110
111	Globe and Phœnix	18 56·0 ..	29 48·0 ..	16 54 ..	17 22 ..	28	·20998	·21072	74	·06376	·06526	150	111
112	Goedgedacht	26 38·9 ..	29 37·0 ..	22 13 ..	22 13 ..	0	·17717	·17644	– 73	·07237	·07249	12	112
114	Graaff Reinet	32 16·9 ..	24 36·0 ..	26 58 ..	26 47 ..	–11	·15983	·16068	85	·08129	·08107	– 22	114
115	Grahamstown	33 19·7 ..	26 32·0 ..	27 1 ..	27 0 ..	– 1	·15669	·15670	1	·07990	·07983	– 7	115
117	Graskop	27 15·0 ..	29 53·0 ..	22 35 ..	22 19 ..	–16	·17419	·17443	24	·07243	·07158	– 85	117
118	Greylingstad	26 44·6 ..	28 45·5 ..	22 24 ..	22 31 ..	7	·17539	·17689	150	·07226	·07324	98	118
119	Greytown	29 4·9 ..	30 38·0 ..	23 54 ..	23 22 ..	–32	·17088	·16910	–178	·07570	·07340	–230	119
120	Grobler's Bridge	25 53·5 ..	30 13·0 ..	20 55 ..	21 26 ..	31	·18037	·17993	– 44	·06895	·07066	171	120
121	Groenkloof	29 28·4 ..	27 11·4 ..	24 31 ..	24 36 ..	5	·16787	·16745	– 42	·07656	·07644	– 12	121
122	Groenplaats	27 16·0 ..	28 33·8 ..	22 25 ..	22 49 ..	24	·17635	·17523	–112	·07274	·07362	88	122
123	Grootfontein	33 7·6 ..	21 15·0 ..	28 16 ..	28 6 ..	–10	·15892	·15913	21	·08546	·08504	– 42	123
124	Gwaai	19 17·5 ..	27 42·2 ..	18 17 ..	18 19 ..	2	·20651	·20658	7	·06823	·06832	9	124
125	Gwelo	19 28·2 ..	29 47·0 ..	17 57 ..	17 45 ..	–12	·20953	·20857	– 96	·06791	·06578	–213	125
126	Hamaan's Kraal	25 24·3 ..	28 17·0 ..	22 6 ..	22 20 ..	14	·18263	·17897	–366	·07413	·07373	– 40	126
127	Hankey	33 52·0 ..	24 53·0 ..	27 23 ..	27 27 ..	4	·15781	·15710	– 71	·08173	·08161	– 12	127
128	Hartley	18 8·3 ..	30 8·0 ..	16 41 ..	16 50 ..	9	·21336	·21377	41	·06392	·06430	38	128
129	Hector Spruit	25 26·2 ..	31 40·5 ..	20 35 ..	20 23 ..	–12	·18418	·18178	–240	·06916	·06675	–241	129
130	Heidelberg, C. C.	34 5·3 ..	20 58·0 ..	28 22 ..	28 23 ..	1	·15874	·15849	– 25	·08573	·08561	– 12	130
131	Heilbron	27 18·2 ..	27 58·0 ..	23 18 ..	23 7 ..	–11	·17450	·17464	14	·07512	·07442	– 70	131
132	Helvetia	29 52·1 ..	26 33·0 ..	25 16 ..	25 3 ..	–13	·16583	·16631	48	·07824	·07735	– 89	132
133	Hermanus	34 25·3 ..	19 16·0 ..	28 46 ..	28 39 ..	– 7	·15899	·15963	64	·08730	·08764	34	133
134	Hermon	33 26·7 ..	18 58·0 ..	28 30 ..	28 30 ..	0	·16092	·16045	– 47	·08735	·08687	– 48	134
135	Highlands	27 16·0 ..	31 23·0 ..	21 53 ..	22 20 ..	27	·17392	·17382	– 10	·06984	·07170	186	135

Table 32 (*continued*).

Station No.	Station Name	Lat. ° ′	Long. ° ′	Declination (D) Obs. ° ′	Calc. ° ′	Diff. ′	Northerly Comp. (X) Obs.	Calc.	Diff. (γ)	Westerly Comp. (Y) Obs.	Calc.	Diff. (γ)	Station No.
136	Hlabisa	28 18·5 S.	32 6·0 E.	22 9 W.	22 20 W.	11	·16914	·17179	265	·06886	·07058	172	136
137	Hluti	27 11·6 ..	31 35·0 ..	21 45 ..	22 17 ..	32	·17424	·17392	− 32	·06953	·07163	210	137
138	Hoetjes Bay	33 1·0 ..	17 57·0 ..	28 27 ..	28 32 ..	5	·16185	·16200	15	·08768	·08769	1	138
139	Holfontein	29 14·9 ..	27 22·5 ..	24 38 ..	24 26 ..	−12	·16819	·16852	33	·07710	·07636	− 74	139
140	Honey Nest Kloof	29 12·2 ..	24 33·0 ..	25 36 ..	25 14 ..	−22	·16793	·16821	28	·08048	·07864	−184	140
141	Honing Spruit	27 27·0 ..	27 25·0 ..	23 15 ..	23 27 ..	12	·17472	·17384	− 88	·07502	·07530	28	141
142	Hopefield	33 14·4 ..	18 21·0 ..	28 22 ..	28 30 ..	8	·16146	·16144	− 2	·08715	·08723	8	142
143	Howhoek	34 12·7 ..	19 10·0 ..	28 38 ..	28 37 ..	− 1	·15910	·15986	76	·08688	·08753	65	143
144	Huguenot	33 45·3 ..	19 0·0 ..	28 30 ..	28 33 ..	3	·16070	·16028	− 42	·08724	·08719	− 5	144
145	Humansdorp	34 2·0 ..	24 38·5 ..	26 58 ..	27 33 ..	35	·15777	·15700	− 77	·08029	·08189	160	145
146	Hutchinson	31 29·6 ..	23 15·0 ..	26 47 ..	26 59 ..	12	·16344	·16178	−166	·08250	·08244	− 6	146
147	Ibisi Bridge	30 24·4 ..	29 54·5 ..	24 27 ..	24 22 ..	− 5	·16541	·16328	−213	·07519	·07395	−124	147
148	Idutywa	32 0·8 ..	28 20·4 ..	25 56 ..	25 44 ..	−12	·15790	·16003	213	·07677	·07683	6	148
149	Igusi	19 40·8 ..	28 6·0 ..	17 19 ..	18 27 ..	68	·20652	·20570	− 82	·06441	·06858	417	149
150	Illovo River	30 6·1 ..	30 51·0 ..	23 49 ..	23 49 ..	0	·16468	·16408	− 60	·07270	·07239	− 31	150
151	Imvani	32 2·0 ..	27 5·0 ..	26 4 ..	26 4 ..	0	·15973	·16086	113	·07812	·07856	44	151
152	Indowane	29 57·5 ..	29 26·7 ..	24 14 ..	24 19 ..	5	·16482	·16568	86	·07417	·07485	68	152
153	Indwe	31 27·8 ..	27 21·0 ..	25 32 ..	25 25 ..	− 7	·16409	·16290	−119	·07841	·07759	− 82	153
154	Inoculation	20 49·7 ..	27 38·0 ..	19 12 ..	19 28 ..	16	·19947	·20059	112	·06944	·07105	161	154
155	Inyantué	18 32·5 ..	26 41·8 ..	18 27 ..	18 8 ..	−19	·21191	·20796	−395	·07073	·06803	−270	155
157	Kaalkop Farm	27 47·3 ..	28 58·3 ..	22 55 ..	22 48 ..	− 7	·17290	·17343	53	·07307	·07290	− 17	157
158	Kaapmuiden	25 31·7 ..	31 19·0 ..	20 58 ..	20 38 ..	−20	·18032	·18137	105	·06909	·06772	−137	158
159	Kalkbank	23 31·5 .	29 20·0 ..	20 9 ..	20 25 ..	16	·19042	·18765	−277	·06986	·06984	− 2	159
160	Kaloombies	22 39·3 ..	29 14·0 ..	20 2 ..	20 3 ..	1	·18930	·19016	86	·06905	·06952	47	160
161	Karree	28 52·5 ..	26 21·0 ..	23 27 ..	24 30 ..	63	·17015	·16902	−113	·07383	·07674	291	161
162	Kathoek	34 23·3 ..	20 20·0 ..	28 37 ..	28 33 ..	− 4	·15812	·15866	54	·08627	·08629	2	162
163	Kenhardt	29 18·0 ..	21 9·0 ..	26 22 ..	26 5 ..	−17	·16956	·17087	131	·08405	·08363	− 42	163
165	King William's Town	32 52·5 ..	27 25·0 ..	26 19 ..	26 39 ..	20	·15889	·15774	−115	·07859	·07861	2	165
166	Klaarstroom	33 20·0 ..	22 32·5 ..	27 56 ..	27 56 ..	0	·15802	·15767	− 35	·08388	·08416	28	166
167	Klerksdorp	26 52·3 ..	26 38·0 ..	23 42 ..	23 35 ..	− 7	·17510	·17399	−111	·07686	·07596	− 90	167
168	Klipfontein, C. C.	30 42·1 ..	22 23·5 ..	23 6 ..	26 42 ..	216	·16870	·16418	−452	·07181	·08286	1105	168
169	Klipfontein (Spelonken)	23 5·7 ..	30 10·0 ..	19 31 ..	19 56 ..	25	·19175	·18877	−298	·06798	·06847	49	169
170	Klipplaat	33 2·0 ..	24 26·0 ..	27 27 ..	27 20 ..	− 7	·15747	·15827	80	·08180	·08189	9	170
171	Knysna	34 1·7 ..	23 3·0 ..	27 53 ..	27 54 ..	1	·15770	·15782	12	·08342	·08327	− 15	171
172	Kokstad	30 32·8 ..	29 28·0 ..	24 40 ..	24 27 ..	−13	·16372	·16358	− 14	·07520	·07451	− 69	172
174	Komgha	32 35·6 ..	27 54·5 ..	26 12 ..	26 19 ..	7	·15955	·15831	−124	·07851	·07778	− 73	174
175	Kraal	26 25·1 ..	28 26·0 ..	22 6 ..	22 38 ..	32	·17667	·17682	15	·07172	·07366	194	175
176	Krantz Kloof	29 48·0 ..	30 54·0 ..	23 32 ..	23 38 ..	6	·16504	·16549	45	·07187	·07301	114	176
177	Krantz Kop	30 48·8 ..	20 45·4 ..	26 55 ..	26 53 ..	− 2	·16654	·16649	− 5	·08454	·08431	− 23	177
178	Kromm River	27 19·0 ..	28 18·8 ..	24 6 ..	22 57 ..	−69	·17290	·17488	198	·07733	·07400	−333	178

179	Krugers	29 57·1 ..	25 50·0 ..	25 0 ..	25 19 ..	19	·16757	·16607	−150	·07815	·07807	− 8	179
180	Kruispad	32 56·8 ..	20 33·3 ..	28 10 ..	28 10 ..	0	·15996	·15976	− 20	·08567	·08557	− 10	180
181	Kwambonambi	28 36·2 ..	32 5·0 ..	22 58 ..	22 29 ..	−29	·16999	·17044	45	·07202	·07056	−146	181
182	Laat Rivier	29 38·2 ..	21 19·3 ..	26 26 ..	26 15 ..	−11	·16829	·16959	130	·08364	·08358	− 6	182
183	Ladismith, C. C.	33 29·0 ..	21 17·0 ..	28 15 ..	28 15 .	0	·15834	·15854	20	·08506	·08520	14	183
184	L'Agulhas	34 50·0 ..	20 0·0 ..	28 41 ..	28 39 ..	− 2	·15764	·15857	93	·08627	·08683	56	184
185	Laingsburg	33 12·0 ..	20 52·0 ..	28 17 ..	28 12 ..	− 5	·15871	·15913	42	·08539	·08539	0	185
186	Lake Banagher	26 22·0 ..	30 19·0 ..	21 39 ..	21 40 ..	1	·17721	·17764	43	·07029	·07059	30	186
187	Langlaagte	26 11·8 ..	28 1·0 ..	23 5 ..	22 43 ..	−22	·17394	·17705	311	·07394	·07415	21	187
189	Libode	31 32·1 ..	29 1·5 ..	25 13 ..	25 12 ..	− 1	·16290	·16114	−176	·07671	·07561	−110	189
190	Lobatsi	25 13·8 ..	25 40·0 ..	22 28 ..			·18065			·07453			190
191	Lochard	19 55·3 ..	29 3·0 ..	18 17 ..	18 15 ..	− 2	·20420	·20623	203	·06748	·06801	53	191
192	Lydenburg	25 5·8 ..	30 26·0 ..	20 54 ..	20 47 ..	− 7	·18202	·18334	132	·06948	·06948	0	192
193	Machadodorp	25 39·9 ..	30 15·0 ..	20 50 ..	21 18 ..	28	·17827	·18094	267	·06784	·07049	265	193
194	Macheke	18 8·3 ..	31 51·0 ..	16 15 ..	15 46 ..	−29	·21528	·21422	−106	·06274	·06132	−142	194
195	Mafeking	25 52·0 ..	25 39·0 ..	23 5 ..			·17927			·07640			195
196	Magalapye	23 6·8 ..	26 50·0 ..	20 24 ..			·19212			·07142			196
197	Magnet Heights	24 44·8 ..	29 58·0 ..	19 45 ..	20 43 ..	58	·18139	·18415	276	·06512	·06962	450	197
198	Makwiro Siding	17 57·3 ..	30 25·0 ..	16 34 ..	16 32 ..	− 2	·21515	·21461	− 54	·06403	·06379	− 24	198
199	Malagas	34 18·5 ..	20 36·0 ..	28 41 ..	28 29 ..	−12	·15849	·15857	8	·08671	·08601	− 70	199
200	Malenje Siding	18 55·2 ..	32 15·0 ..	16 7 ..	16 1 ..	− 6	·21174	·21104	− 70	·06118	·06072	− 46	200
201	Malinde	18 45·0 ..	27 1·3 ..	18 2 ..	18 10 ..	8	·20786	·20766	− 20	·06764	·06806	42	201
202	Malmesbury	33 28·0 ..	18 43·0 ..	28 29 ..	28 32 ..	3	·16194	·16078	−116	·08786	·08722	− 64	202
203	Mandegos	19 7·0 ..	33 28·0 ..	15 58 ..	15 24 ..	−34	·21084	·21052	− 32	·06033	·05862	−171	203
204	Mapani Loep	22 17·5 ..	29 3·0 ..	19 4 ..	19 58 ..	54	·19405	·19116	−289	·06708	·06961	253	204
206	Marandellas	18 11·3 ..	31 32·9 ..	15 55 ..	16 0 ..	5	·21359	·21392	33	·06092	·06184	92	206
207	Maribogo	26 25·1 ..	25 15·0 ..	23 35 ..			·17748			·07750			207
208	Matetsi	18 12·5 ..	26 1·5 ..	17 58 ..	18 9 ..	11	·20638	·20824	186	·06692	·06813	121	208
209	Matjesfontein	33 14·2 ..	20 36·0 ..	28 23 ..	28 15 ..	− 8	·15855	·15930	75	·08569	·08562	− 7	209
210	Meyerton	26 33·2 ..	28 1·0 ..	22 42 ..	22 49 ..	7	·17664	·17645	− 19	·07391	·07416	25	210
211	Middelberg (Tzitzikama)	34 0·0 ..	24 9·0 ..	27 44 ..	27 38 ..	− 6	·15737	·15753	16	·08273	·08225	− 48	211
212	Middlepost	31 54·2 ..	20 14·0 ..	27 29 ..	27 28 ..	− 1	·16372	·16366	− 6	·08517	·08498	− 19	212
213	Middleton	32 57·8 ..	25 51·0 ..	26 56 ..	26 56 ..	0	·15817	·15818	1	·08037	·08051	14	213
215	Miller Siding	33 5·4 ..	24 8·0 ..	27 31 ..	27 25 ..	− 6	·15780	·15835	55	·08222	·08234	12	215
216	Miller's Point	34 14·2 ..	18 26·0 ..	28 41 ..	28 43 ..	2	·16018	·16083	65	·08764	·08852	88	216
217	Misgund	33 45·5 ..	23 32·0 ..	27 53 ..	27 42 ..	−11	·15755	·15798	43	·08338	·08308	− 30	217
218	Mission Station	23 12·7 ..	30 27·0 ..	19 32 ..	19 53 ..	21	·18765	·18841	76	·06655	·06814	159	218
219	Modder Spruit	28 28·9 ..	29 53·0 ..	23 10 ..	23 2 ..	− 8	·17064	·17103	39	·07303	·07306	3	219
220	Molteno	31 24·0 ..	26 21·0 ..	25 58 ..	25 58 ..	0	·16251			·07907			220
221	Mossel Bay	34 10·8 ..	22 9·5 ..	28 12 ..	28 6 ..	− 6	·15802	·15806	4	·08473	·08456	− 17	221
222	Mount Ayliff (near)	30 48·2 ..	29 31·5 ..	24 33 ..	24 28 ..	− 5	·16272	·16333	61	·07434	·07444	10	222
223	Mount Frere	30 53·5 ..	28 59·0 ..	24 34 ..	24 39 ..	5	·16432	·16346	− 86	·07510	·07517	7	223
224	Mount Moreland	29 38·4 ..	31 11·0 ..	23 12 ..	23 28 ..	16	·16603	·16617	14	·07117	·07210	93	224
224A	Movene	25 34·0 ..	32 7·0 ..	17 43 ..	20 14 ..	151	·18646	·18107	−539	·05954	·06570	616	224A
225	M'Phatele's Location	24 19·8 ..	29 41·0 ..	20 59 ..	20 38 ..	−21	·18228	·18535	307	·06994	·06979	− 15	225

TABLE 32 (*continued*).

Station No.	Station Name	Lat. ° ′	Long. ° ′	Declination (D) Obs. ° ′	Declination (D) Calc. ° ′	Declination (D) Diff. ′	Northerly Comp. (X) Obs.	Northerly Comp. (X) Calc.	Northerly Comp. (X) Diff. (γ)	Westerly Comp. (Y) Obs.	Westerly Comp. (Y) Calc.	Westerly Comp. (Y) Diff. (γ)	Station No.
226	Naauwpoort	31 14·0 S.	24 55·0 E.	26 2 W.	26 9 W.	7	·16359	·16311	− 48	·07988	·08023	35	226
227	Naboomspruit	24 31·3 ..	28 43·0 ..	21 3 ..	21 4 ..	1	·18561	·18495	− 66	·07146	·07122	− 24	227
228	Nelspoort	32 7·7 ..	23 1·0 ..	27 35 ..	27 17 ..	−18	·15772	·16099	327	·08241	·08310	69	228
229	Nelspruit	25 28·1 ..	30 58·5 ..	21 0 ..	20 47 ..	−13	·18010	·18175	165	·06910	·06855	− 55	229
230	Newcastle	27 45·3 ..	29 58·0 ..	22 33 ..	22 36 ..	3	·17404	·17302	−102	·07227	·07218	− 9	230
231	Newcastle (Trans.)	26 32·1 ..	30 27·0 ..	21 58 ..	21 41 ..	−17	·17603	·17683	80	·07101	·07032	− 69	231
232	Nooitgedacht	25 38·1 ..	30 31·0 ..	19 1 ..	21 8 ..	127	·18157	·18105	− 52	·06256	·06980	724	232
235	Nylstroom	24 42·4 ..	28 26·0 ..	21 12 ..	21 11 ..	− 1	·18501	·18445	− 56	·07175	·07147	− 28	235
235A	Orange River	29 38·0 ..	24 16·0 ..	25 24 ..	25 35 ..	11	·16769	·16698	− 71	·07962	·07925	− 37	235A
236	Orjida	33 26·0 ..	23 19·0 ..	27 49 ..	27 54 ..	5	·15784	·15757	− 27	·08306	·08330	24	236
237	Oudemuur	31 5·8 ..	20 19·1 ..	27 19 ..	27 4 ..	−15	·16479	·16613	134	·08514	·08482	− 32	237
238	Oudtshoorn	33 35·2 ..	22 12·5 ..	28 11 ..	28 2 ..	− 9	·15755	·15782	27	·08444	·08427	− 17	238
239	Paardevlei	30 36·1 ..	21 54·0 ..	27 1 ..	26 42 ..	−19	·16560	·16573	13	·08442	·08333	−109	239
241	Palapye	22 33·4 ..	27 7·0 ..	20 33 ..			·19327			·07248			241
242	Pampoenpoort	31 3·5 ..	22 39·1 ..	26 48 ..	26 47 ..	− 1	·16356	·16371	15	·08261	·08280	19	242
243	Payne's Farm	30 36·9 ..	29 47·5 ..	24 37 ..	24 24 ..	−13	·16207	·16304	97	·07424	·07404	− 20	243
244	Picene	25 40·8 ..	32 18·5 ..	20 48 ..	20 11 ..	−37	·18049	·18048	− 1	·06857	·06523	−334	244
245	Pienaar's River	25 12·7 ..	28 19·0 ..	23 22 ..	22 16 ..	−66	·18112	·17937	−175	·07822	·07367	−455	245
246	Pietersburg	23 50·3 ..	29 27·0 ..	20 12 ..	20 30 ..	18	·18634	·18676	42	·06856	·06984	128	246
247	Piet Potgietersrust	24 11·2 ..	29 1·0 ..	20 40 ..	20 48 ..	8	·18555	·18586	31	·07000	·07063	63	247
248	Piet Retief	27 0·5 ..	30 48·5 ..	22 15 ..	21 58 ..	−17	·17430	·17461	31	·07128	·07046	− 82	248
249	Pilgrim's Rest	24 56·8 ..	30 45·0 ..	20 59 ..	20 32 ..	−27	·18312	·18349	37	·07024	·06864	−160	249
250	Piquetberg	32 55·0 ..	18 43·0 ..	28 11 ..	28 25 ..	14	·16166	·16102	− 64	·08661	·08656	− 5	250
251	Pivaan's Poort	27 33·8 ..	30 28·0 ..	22 39 ..	22 30 ..	− 9	·17346	·17287	− 59	·07239	·07200	− 39	251
252	Platrand	27 6·4 ..	29 29·0 ..	22 36 ..	22 23 ..	−13	·17454	·17468	14	·07268	·07218	− 50	252
253	Plettenberg Bay	34 2·2 ..	23 21·0 ..	27 48 ..	27 51 ..	3	·15756	·15750	− 6	·08308	·08295	− 13	253
255	Pokwani (Transvaal)	24 52·2 ..	29 46·0 ..	21 18 ..	20 51 ..	−27	·18404	·18384	− 20	·07175	·06995	−180	255
256	Port Alfred	33 35·8 ..	26 54·0 ..	26 55 ..	27 2 ..	7	·15689	·15597	− 92	·07964	·07957	− 7	256
257	Port Beaufort	34 23·8 ..	20 49·0 ..	28 27 ..	28 26 ..	− 1	·15817	·15846	29	·08572	·08581	9	257
258	Port Elizabeth	33 58·0 ..	25 37·0 ..	27 39 ..	27 23 ..	−16	·15626	·15636	10	·08189	·08094	− 95	258
259	Port Shepstone	30 43·7 ..	30 27·0 ..	24 16 ..	24 17 ..	1	·16228	·16202	− 26	·07317	·07311	− 6	259
260	Port St Johns	31 37·8 ..	29 33·0 ..	25 18 ..	25 12 ..	− 6	·15864	·16031	167	·07500	·07506	6	260
261	Potchefstroom	26 42·8 ..	27 5·0 ..	23 0 ..	23 17 ..	17	·17561	·17485	− 76	·07457	·07537	80	261
262	Potfontein	30 12·2 ..	24 7·0 ..	25 39 ..	25 47 ..	8	·16618	·16335	−283	·07977	·07793	−184	262
263	Pretoria	25 45·3 ..	28 12·0 ..	22 16 ..	22 29 ..	13	·17931	·17817	−114	·07344	·07387	43	263
264	Prince Albert	33 13·2 ..	22 3·0 ..	28 4 ..	28 3 ..	− 1	·15831			·08440			264
265	Prince Albert Road	32 58·7 ..	21 42·0 ..	27 51 ..	28 1 ..	10	·15920	·15861	− 59	·08441	·08496	55	265
266	Queenstown	31 54·0 ..	26 52·0 ..	26 16 ..	26 0 ..	−16	·16185	·16141	− 44	·07988	·07864	−124	266
267	Randfontein	26 10·7 ..	27 42·0 ..	22 46 ..	22 52 ..	6	·17659	·17668	9	·07412	·07454	42	267

268	Rateldraai	28 45·7 ..	21 17·9 ..	24 54 ..	25 47 ..	53	·17411	·17232	−179	·08079	·08328	249	268
269	Rateldrift	31 31·6 ..	20 17·6 ..	27 32 ..	27 16 ..	−16	·16350	·16462	112	·08525	·08474	− 51	269
271	Richmond Road	31 13·0 ..	23 38·0 ..	27 12 ..	26 37 ..	−35	·16216	·16086	−130	·08331	·08156	−175	271
272	Rietkuil Farm	30 14·4 ..	29 22·0 ..	24 12 ..	24 30 ..	18	·16398	·16445	47	·07371	·07494	123	272
273	Rietpoort	31 4·4 ..	20 55·1 ..	26 54 ..	27 3 ..	9	·16520	·16569	49	·08378	·08450	72	273
274	Rietvlei	24 35·0 ..	30 40·0 ..	20 24 ..	20 24 ..	0	·18552	·18452	−100	·06900	·06856	− 44	274
275	Rietvlei, C. C.	33 32·0 ..	22 29·0 ..	27 59 ..	27 58 ..	− 1							275
276	Riversdale	34 5·0 ..	21 16·0 ..	28 17 ..	28 19 ..	2	·15848	·15837	− 11	·08528	·08532	4	276
277	Rivierplaats	32 8·5 ..	20 24·0 ..	27 17 ..	27 35 ..	18	·16290	·16285	− 5	·08402	·08499	97	277
278	Roadside	30 44·3 ..	20 25·5 ..	26 52 ..	26 52 ..	0	·16713	·16708	− 5	·08464	·08456	− 8	278
279	Robertson	33 48·8 ..	19 53·0 ..	28 24 ..	28 31 ..	7	·15961	·15901	− 60	·08631	·08637	6	279
280	Rodekrantz	24 38·0 ..	30 35·0 ..	20 53 ..	20 27 ..	−26	·18437	·18440	3	·07034	·06871	−163	280
281	Roodepoort	30 13·0 ..	23 22·0 ..	25 14 ..	26 21 ..	67	·16765	·16482	−283	·07902	·08170	268	281
282	Rooidam	29 50·7 ..	23 11·8 ..	25 0 ..	26 10 ..	70	·16910	·16620	−290	·07883	·08174	291	282
283	Revué	18 59·0 ..	33 3·0 ..	16 6 ..	15 34 ..	−32	·21133	·21097	− 36	·06098	·05933	−165	283
284	Rooipüts	29 17·4 ..	21 38·6 ..	27 0 ..	26 2 ..	−58	·16523	·17020	497	·08420	·08311	−109	284
285	Rooival	32 12·0 ..	21 58·3 ..	27 19 ..	27 26 ..	7	·16236	·16158	− 78	·08386	·08389	3	285
286	Rosmead Junction	31 39·6 ..	25 5·0 ..	26 14 ..	26 20 ..	6	·16292	·16215	− 77	·08031	·08025	− 6	286
287	Rouxville	30 31·6 ..	26 47·3 ..	25 7 ..	25 24 ..	17	·16557	·16453	−104	·07761	·07795	34	287
288	Rusapi	18 32·0 ..	32 8·0 ..	16 3 ..	15 51 ..	−12	·21314	·21262	− 52	·06135	·06087	− 48	288
289	Rustplaats	24 50·6 ..	30 38·0 ..	20 53 ..	20 32 ..	−21	·18350	·18378	28	·06990	·06875	−115	289
290	Ruyterbosch	33 55·7 ..	22 2·0 ..	28 0 ..	28 7 ..	7	·15896	·15875	− 21	·08454	·08454	0	290
291	Sabie River	25 6·1 ..	30 45·0 ..	21 4 ..	20 39 ..	−25	·18368	·18334	− 34	·07076	·06886	−190	291
292	Salisbury	17 50·3 ..	31 3·0 ..	16 8 ..	16 5 ..	− 3	·21261	·21527	266	·06147	·06267	120	292
293	Saxony	28 44·1 ..	27 44·4 ..	24 12 ..	24 1 ..	−11	·16931	·17049	118	·07609	·07574	− 35	293
294	Schietfontein	32 41·7 ..	20 46·6 ..	27 53 ..	27 50 ..	− 3	·16189	·16105	− 84	·08563	·08500	− 63	294
295	Schikhoek	27 24·6 ..	30 34·0 ..	23 16 ..	22 25 ..	−51	·17139	·17371	232	·07369	·07182	−187	295
296	Schoemanshoek	25 27·9 ..	30 21·0 ..	21 9 ..	21 7 ..	− 2	·17993	·18187	194	·06961	·07016	55	296
297	Schuilplaats	26 54·2 ..	29 47·0 ..	22 16 ..	22 11 ..	− 5	·17511	·17533	22	·07170	·07165	− 5	297
298	Secocoeni's Stad	24 28·3 ..	29 52·0 ..	19 18 ..	20 38 ..	80	·18942	·18496	−446	·06633	·06961	328	298
299	Seruli	21 55·7 ..	27 19·0 ..	20 26 ..	19 56 ..	−30	·19358	·19589	231	·07213	·07329	116	299
300	Shangani	19 45·8 ..	29 24·0 ..	18 38 ..	18 0 ..	−38	·21265	·20733	−532	·07166	·06740	−426	300
301	Shashi	21 23·2 ..	27 27·0 ..	19 29 ..	19 58 ..	29	·19982	19820	−162	·07067	·07220	153	301
302	Shela River	26 51·0 ..	30 43·0 ..	21 44 ..	21 55 ..	11	·17553	·17525	− 28	·06996	·07046	50	302
303	Shoshong Road	23 34·8 ..	26 34·0 ..	21 54 ..			·18324			·07367			303
304	Signal Hill	33 55·0 ..	18 24·3 ..	28 50 ..	28 39 ..	−11	·15959	·16091	132	·08787	·08818	31	304
305	Simonstown Rifle Range	34 12·0 ..	18 26·0 ..	28 42 ..	28 43 ..	1	·15972	·16084	112	·08744	·08700	− 44	305
305A	Simonstown (Glencairn)	34 10·8 ..	18 26·0 ..	28 37 ..	28 43 ..	6	·16022	·16085	63	·08743	·08846	103	305A
306	Sir Lowry's Pass	34 7·3 ..	18 55·0 ..	28 58 ..	28 38 ..	−20	·15862	·16023	161	·08780	·08774	− 6	306
307	Smaldeel	28 24·3 ..	26 44·0 ..	24 21 ..	24 10 ..	−11	·17018	·17075	57	·07704	·07637	− 67	307
308	Spitzkopje	25 18·2 ..	30 49·0 ..	20 36 ..	20 47 ..	11	·17994	·18256	262	·06766	·06891	125	308
309	Springfontein	30 16·7 ..	25 44·0 ..	25 57 ..	25 33 ..	−24	·16276	·16512	236	·07923	·07843	− 80	309
310	Springs	26 13·0 ..	28 27·0 ..	22 20 ..	22 31 ..	11	·17683	·17756	73	·07262	·07360	98	310
311	Stanford	34 26·7 ..	19 28·0 ..	28 43 ..	28 42 ..	− 1	·15898	·15918	20	·08710	·08728	18	311
312	Stanger	29 21·1 ..	31 15·0 ..	22 59 ..	23 15 ..	16	·16581	·16741	160	·07034	·07184	150	312
313	Steenkampspoort	32 6·3 ..	21 44·1 ..	27 23 ..	27 24 ..	1	·16282	·16280	− 2	·08435	·08404	− 31	313

TABLE 32 (*continued*).

Station No.	Station Name	Lat. ° ′	Long. ° ′	Declination (D) Obs. ° ′	Declination (D) Calc. ° ′	Declination (D) Diff. ′	Northerly Comp. (X) Obs.	Northerly Comp. (X) Calc.	Northerly Comp. (X) Diff. (γ)	Westerly Comp. (Y) Obs.	Westerly Comp. (Y) Calc.	Westerly Comp. (Y) Diff. (γ)	Station No.
314	Stellenbosch	33 56·0 S.	18 50·0 E.	28 45 W.	28 37 W.	− 8	·15924	·16042	118	·08739	·08763	24	314
315	Sterkstroom	31 34·5 ..	26 33·0 ..	25 58 ..	26 1 ..	3	·16268	·16196	− 72	·07924	·07889	− 35	315
316	Steynsburg	31 18·5 ..	25 48·0 ..	26 6 ..	26 1 ..	− 5	·16396	·16288	−108	·08032	·07951	− 81	315
317	Still Bay	34 22·0 ..	21 25·0 ..	28 22 ..	28 18 ..	− 4	·15810	·15823	13	·08535	·08521	− 14	317
318	Stormberg Junction	31 17·5 ..	26 16·0 ..	25 59 ..	25 55 ..	− 4	·16306	·16279	− 27	·07944	·07919	− 25	318
319	Storms River	33 58·0 ..	23 49·5 ..	27 42 ..	27 42 ..	0	·15752	·15771	19	·08267	·08256	− 11	319
320	Strandfontein	34 5·3 ..	18 34·0 ..	28 53 ..									320
321	Sutherland	32 25·0 ..	20 39·3 ..	27 31 ..	27 41 ..	10	·16403	·16193	−210	·08545	·08495	− 50	321
322	Swellendam	34 2·0 ..	20 27·0 ..	28 25 ..	28 28 ..	3	·15926	·15871	− 55	·08615	·08612	− 3	322
323	Taungs	27 34·8 ..	24 45·0 ..	24 5 ..			·17248			·07696			323
324	Thaba 'Nchu	29 10·7 ..	26 49·0 ..	24 27 ..	24 35 ..	8	·16884	·16859	− 25	·07679	·07685	6	324
325	Tinfontein	30 24·0 ..	26 54·8 ..	24 45 ..	25 16 ..	31	·16416	·16479	63	·07568	·07744	176	325
326	Toise River	32 27·3 ..	27 28·7 ..	26 28 ..	26 16 ..	−12	·15718	·15924	206	·07827	·07819	− 8	326
327	Touws River	33 21·0 ..	20 3·0 ..	28 50 ..	28 30 ..	−20	·15902	·15907	5	·08752	·08640	−112	327
328	Tsolo	31 18·2 ..	28 45·6 ..	24 50 ..	25 2 ..	12	·16338	·16222	−116	·07559	·07575	16	328
330	Tulbagh Road	33 19·3 ..	19 10·0 ..	28 29 ..	28 26 ..	− 3	·16096	·16025	− 71	·08734	·08688	− 46	330
331	Tweepoort	26 36·7 ..	30 43·0 ..	22 3 ..	21 36 ..	−27	·17850	·17639	−211	·07233	·06967	−266	331
332	Twee Rivieren	33 50·3 ..	23 56·5 ..	27 40 ..	28 3 ..	23	·15749	·15760	11	·08259	·08242	− 17	332
333	Twelfelhoek	27 27·4 ..	29 20·4 ..	22 1 ..	22 26 ..	25	·17425	·17410	− 15	·07046	·07236	190	333
334	Tygerfontein	34 10·0 ..	21 35·5 ..	28 13 ..	28 14 ..	1	·15832	·15817	− 15	·08495	·08477	− 18	334
335	Tygerkloof Drift	28 10·8 ..	28 35·2 ..	22 57 ..	23 21 ..	24	·17346	·17232	−114	·07344	·07432	88	335
336	Thirtyfirst	25 40·6 ..	29 37·6 ..	22 1 ..	21 38 ..	−23	·17920	·18098	178	·07196	·07210	14	336
337	Uitenhage	33 47·0 ..	25 24·0 ..	27 23 ..	27 21 ..	− 2	·15713	·15683	− 30	·08137	·08108	− 29	337
338	Uitkyk	25 49·5 ..	29 25·0 ..	21 24 ..	21 54 ..	30	·18115	·17982	−133	·07099	·07251	152	338
339	Uitspan Farm	31 41·2 ..	21 27·2 ..	27 39 ..	27 13 ..	−26	·16086	·16338	252	·08424	·08399	− 25	339
340	Umhlatuzi	28 51·7 ..	31 54·0 ..	22 29 ..	22 43 ..	14	·17188	·16928	−260	·07113	·07083	− 30	340
341	Umhlengana Pass	31 36·0 ..	29 19·6 ..	25 11 ..	25 12 ..	1	·16097	·16063	− 34	·07566	·07530	− 36	341
342	Umtali	18 59·2 ..	32 39·0 ..	15 50 ..	15 49 ..	− 1	·21170	·21086	− 84	·06005	·06003	− 2	342
343	Umtata	31 35·9 ..	28 47·1 ..	25 12 ..	25 18 ..	6	·16177	·16097	− 80	·07614	·07596	− 18	343
345	Umzinto	30 19·4 ..	30 39·0 ..	23 40 ..	24 1 ..	21	·16579	·16318	−261	·07267	·07269	2	345
346	Underberg Hotel	29 47·9 ..	29 30·5 ..	24 39 ..	24 12 ..	−27	·16421	·16643	222	·07537	·07477	− 60	346
347	Upington	28 27·7 ..	21 14·9 ..	27 4 ..	25 38 ..	−86	·17083	·17333	250	·08726	·08323	−403	347
348	Utrecht (West of)	27 39·9 ..	30 16·0 ..	22 41 ..	22 34 ..	− 7	·17328	·17315	− 13	·07244	·07210	− 34	348
349	Van Reenen	28 22·2 ..	29 24·5 ..	23 5 ..	22 58 ..	− 7	·17107	·17154	47	·07292	·07289	− 3	349
350	Van Wyk's Farm	33 49·4 ..	21 12·0 ..	28 11 ..	28 18 ..	7	·15887	·15847	− 40	·08515	·08534	19	350
351	Van Wyk's Vlei	30 22·3 ..	21 50·0 ..	27 5 ..	26 35 ..	−30	·16437	·16656	219	·08404	·08332	− 72	351
352	Victoria Falls	17 55·6 ..	25 51·0 ..	17 53 ..	18 0 ..	7	·21005	·20905	−100	·06774	·06782	8	352
353	Villiersdorp	33 59·5 ..	19 19·0 ..	28 38 ..	28 33 ..	− 5	·15902	·15975	73	·08682	·08707	25	353
354	Virginia	28 7·5 ..	26 55·0 ..	24 6 ..	24 2 ..	− 4	·17114	·17180	66	·07658	·07635	− 23	354
355	Vlaklaagte	26 50·6 ..	29 5·0 ..	22 23 ..	22 23 ..	0	·17622	·17617	− 5	·07260	·07268	8	355

356	Vogelvlei	29 8·3 ..	27 31·1 ..	24 5 ..	24 20 ..	15	·16819	·16886	67	·07517	·07607	90	356
357	Vondeling	33 19·8 ..	23 4·0 ..	27 46 ..	27 51 ..	5	·15802	·15759	– 43	·08321	·08372	51	357
358	Vredefort	27 1·2 ..	27 22·9 ..	22 36 ..	23 15 ..	39	·17797	·17481	–316	·07406	·07505	99	358
360	Vryburg	26 57·1 ..	24 43·0 ..	22 58 ..			·17660			·07481			360
361	Wakkerstroom	27 21·5 ..	30 9·0 ..	22 27 ..	22 23 ..	– 4	·17445	·17401	– 44	·07208	·07174	– 34	361
362	Wankie	18 22·3 ..	26 28·5 ..	16 6 ..	18 8 ..	122	·21322	·20833	–489	·06154	·06793	639	362
363	Warmbad (Waterberg)	24 53·0 ..	28 20·0 ..	21 21 ..	21 21 ..	0	·18339	·18396	57	·07169	·07194	25	363
364	Warmbad (Zoutpans.)	22 24·9 ..	29 12·0 ..	21 17 ..	19 58 ..	–79	·17913	·19082	1169	·06976	·06944	– 32	364
365	Warrenton	28 6·9 ..	24 52·0 ..	24 52 ..	24 28 ..	– 24	·17056	·17131	75	·07903	·07747	–156	365
366	Waschbank	28 18·8 ..	30 8·0 ..	23 16 ..	22 56 ..	–20	·17095	·17141	46	·07350	·07288	– 62	366
367	Waterworks	29 4·5 ..	26 28·0 ..	24 30 ..	24 35 ..	5	·16835	·16855	20	·07670	·07679	9	367
368	Welverdiend	26 22·7 ..	27 17·0 ..	23 0 ..	23 7 ..	7	·17477	·17576	99	·07416	·07509	93	368
369	Wepener	29 43·6 ..	27 3·7 ..	25 23 ..	24 48 ..	–35	·16536	·16669	133	·07847	·07676	–171	369
370	Williston	31 20·4 ..	20 55·2 ..	26 40 ..	27 8 ..	28	·16489	·16479	– 10	·08279	·08439	160	370
371	Willowmore	33 9·4 ..	23 30·0 ..	27 51 ..	27 45 ..	– 6	·15727	·15743	16	·08320	·08325	5	371
372	Winburg	28 31·2 ..	27 3·0 ..	24 13 ..	24 2 ..	–11	·17006	·17048	42	·07647	·07611	– 36	372
373	Winkeldrift	27 10·6 ..	27 7·6 ..	24 15 ..	23 26 ..	–49	·17312	·17422	110	·07800	·07542	–258	373
374	Witklip	23 16·5 ..	29 17·0 ..	21 58 ..	20 19 ..	–99	·18509	·18688	179	·07466	·06978	–488	374
375	Witmoss	32 33·0 ..	25 45·0 ..	26 27 ..	26 46 ..	19	·15994	·15913	– 81	·07951	·08010	59	375
377	Wolvehoek	26 54·9 ..	27 50·0 ..	23 1 ..	23 1 ..	0	·17524	·17542	18	·07447	·07445	– 2	377
378	North of Limpopo	22 7·2 ..	29 10·0 ..	20 0 ..	19 51 ..	– 9	·19068	·19168	100	·06939	·06933	– 6	378
379	Wonderfontein	25 48·3 ..	29 53·0 ..	23 35 ..	21 34 ..	–121	·18188	·18039	–149	·07943	·07148	–795	379
380	Woodville	33 56·3 ..	22 41·0 ..	27 57 ..	27 58 ..	1	·15796	·15786	– 10	·08380	·08370	– 10	380
381	Worcester	33 39·0 ..	19 26·0 ..	28 34 ..	28 28 ..	– 6	·15999	·15975	– 24	·08711	·08650	– 61	381
382	Zak Rivier	30 30·9 ..	20 31·0 ..	27 1 ..	26 48 ..	–13	·16666	·16799	133	·08497	·08471	– 26	382
383	Zand River	23 3·8 ..	29 34·0 ..	21 13 ..	20 7 ..	–66	·18863	·18894	31	·07323	·06928	–395	383
384	Zeekoegat	33 3·0 ..	22 31·0 ..	27 57 ..	27 55 ..	– 2	·15581	·15757	176	·08278	·08433	155	384
385	Zuurbraak	34 0·3 ..	20 39·0 ..	28 20 ..	28 26 ..	6	·15924	·15864	– 60	·08587	·08591	4	385
386	Zuurfontein	32 51·0 ..	18 35·0 ..	28 17 ..	28 25 ..	8	·16227	·16122	–105	·08729	·08666	– 63	386
387	Zuurpoort	32 2·9 ..	24 8·0 ..	26 53 ..	26 45 ..	– 8	·16114	·16143	29	·08169	·08124	– 45	387
388		23 42·7 ..	29 44·0 ..	20 48 ..	20 21 ..	–27	·18804	·18709	– 95	·07141	·06939	–202	388
389		24 8·0 ..	29 28·0 ..	20 24 ..	20 37 ..	13	·18660	·18594	– 66	·06939	·06994	55	389
390		25 9·8 ..	29 44·8 ..	21 34 ..			·18188			·07187			390
391		25 47·5 ..	29 36·0 ..	21 10 ..			·17921			·06939			391
393		28 54·6 ..	27 44·1 ..	24 6 ..	24 7 ..	1	·16867	·16950	83	·07542	·07570	28	393
394		28 31·6 ..	27 42·3 ..	23 58 ..			·17058			·07595			394
395		28 6·7 ..	29 3·1 ..	22 58 ..	23 0 ..	2	·17157	·17252	95	·07271	·07314	43	395
396		27 22·7 ..	29 0·0 ..	22 32 ..	22 34 ..	2	·17449	·17452	3	·07241	·07289	48	396
398	East of Komgha	32 32·6 ..	27 58·3 ..	26 7 ..	26 16 ..	9	·15701	·15844	143	·07589	·07768	179	398
399	Outspan	32 13·4 ..	28 10·4 ..	25 57 ..	25 56 ..	– 1	·16008	·15946	– 62	·07790	·07718	– 72	399
400	Bashee	31 42·0 ..	28 30·0 ..	25 58 ..	25 26 ..	–32	·15990	·16102	112	·07789	·07638	–151	400
402		31 0·6 ..	29 30·5 ..	24 18 ..	24 40 ..	22	·16381	·16256	–125	·07396	·07463	67	402
403		30 49·7 ..	29 15·5 ..	24 14 ..	24 23 ..	9	·16370	·16345	– 25	·07371	·07478	107	403
405	Ugie (on road to)	31 8·3 ..	28 26·2 ..	25 7 ..	24 57 ..	–10	·16271	·16312	41	·07629	·07602	– 27	405

Table 33.

Summary of observed and calculated values and anomalies of dip, horizontal intensity, vertical intensity, and total intensity.

Station No.	Station Name	Dip (θ) Obs. ° ′	Dip Calc. ° ′	Dip Diff. ′	Horizontal Intens. Obs.	Horizontal Intens. Calc.	(H) Diff. (γ)	Vertical Intens. Obs.	Vertical Intens. Calc.	(Z) Diff. (γ)	Total Intens. Obs.	Total Intens. Calc.	(T) Diff. (γ)	Station No.
1	Abelsdam	59 3 S.	58 53 S.	−10	·18790	·18826	36	·31333	·31161	−172	·36536	·36403	−133	1
2	Aberdeen, C. C.	60 25 ..	60 41 ..	16	·17991	·17902	− 89	·31692	·31898	206	·36443	·36595	152	2
3	Aberdeen (Trans.)	58 43 ..	58 25 ..	−18	·19217	·19343	126	·31627	·31488	−139	·37008	·36907	−101	3
4	Aberdeen Road	60 21 ..	60 54 ..	33	·18192	·17831	−361	·31929	·32051	122	·36745	·36681	− 64	4
5	Aberfeldy	58 4 ..	58 11 ..	7	·19425	·19353	− 72	·31169	·31212	43	·36727	·36732	5	5
6	Adelaide	61 22 ..	61 24 ..	2	·17743	·17757	14	·32497	·32554	57	·37026	·37090	64	6
7	Albert Falls	60 55 ..	60 53 ..	− 2	·18338	·18255	− 83	·32979	·32774	−205	·37727	·37516	−211	7
8	Alicedale	61 37 ..	61 43 ..	6	·17546	·17651	105	·32467	·32518	51	·36905	·37007	102	8
9	Aliwal North	60 22 ..	60 19 ..	− 3	·18205	·18224	19	·32001	·31995	− 6	·36819	·36819	0	9
10	Alma	59 22 ..	59 36 ..	14	·18857	·18838	− 19	·31821	·32069	248	·36989	·37205	216	10
11	Amabele Junction	61 22 ..	61 35 ..	13	·17815	·17684	−131	·32630	·32677	47	·37178	·37155	− 23	11
12	Amaranja	61 0 ..	61 34 ..	34	·18073	·17822	−251	·32608	·32849	241	·37282	·37379	97	12
13	Amatongas	53 38 ..	53 28 ..	−10	·21839	·22132	293	·29658	·29583	− 75	·36831	·36769	− 62	13
14	Ashton	59 41 ..	59 42 ..	1	·18135	·18105	− 30	·31015	·31009	− 6	·35927	·35922	− 5	14
15	Assegai Bosch				·17746	·17784	38							15
16	Avontuur	60 42 ..	60 45 ..	3	·17842	·17867	25	·31794	·31908	114	·36458	·36576	118	16
17	Ayrshire Mine	50 55 ..	51 0 ..	5	·22805	·22631	−174	·28078	·28013	− 65	·36173	·36018	−155	17
17A	Balmoral	58 10 ..	58 17 ..	7	·19362	·19361	− 1	·31186	·31324	138	·36709	·36831	122	17A
18	Bamboo Creek	53 46 ..	53 36 ..	−10	·21930	·22132	202	·29928	·29674	−254	·37102	36812	−290	18
19	Bankpan	59 43 ..	58 32 ..	−71	·18485	·19263	778	·31654	·31534	−120	·36656	·36912	256	19
20	Barberton	58 36 ..	58 44 ..	8	·19287	·19263	− 24	·31598	·31648	50	·37018	·37102	84	20
21	Barrington	60 8 ..	60 41 ..	33	·18308	·17887	−421	·31881	·31850	− 31	·36764	·36540	−224	21
22	Battlefields	52 31 ..	52 40 ..	9	·21872	·22032	160	·28521	·29005	484	·35942	·36359	417	22
23	Bavaria	59 16 ..	59 15 ..	− 1	·18843	·18861	18	·31694	·31754	60	·36872	·36927	55	23
24	Baviaanskrantz	58 42 ..	58 48 ..	6	·18903	·18887	− 16	·31090	·31174	84	·36386	·36447	61	24
25	Beaufort West	59 43 ..	59 51 ..	8	·18205	·18181	− 24	·31173	·31262	89	·36105	·36201	96	25
26	Beira	54 34 ..			·21673			·30458			·37383			26
27	Belleville	59 11 ..	59 14 ..	3	·18262	·18232	− 30	·30614	·30584	− 30	·35648	·35604	− 44	27
28	Berg River Mouth	58 29 ..	58 37 ..	8	·18435	·18434	− 1	·30063	·30132	69	·35265	·35309	44	28
29	Bethal	58 13 ..	58 35 ..	22	·19416	·19221	−195	·31335	·31546	211	·36861	·36897	36	29
30	Bethany	59 42 ..	59 41 ..	− 1	·18443	·18447	4	·31561	·31541	− 20	·36555	·36535	− 20	30
31	Bethesda Road	60 29 ..	60 29 ..	0	·18062	·18029	− 33	·31903	·31827	− 76	·36661	·36591	− 70	31
32	Bethlehem	59 17 ..	59 27 ..	10	·18931	·18740	−191	·31863	·31749	−114	·37062	·36864	−198	32
33	Bethulie	59 51 ..	60 3 ..	12	·18400	·18293	−107	·31678	·31740	62	·36634	·36622	− 12	33
34	Biesjespoort	59 38 ..	59 44 ..	6	·18445	·18252	−193	·31481	·31211	−270	·36486	·36211	−275	34
35	Birthday	59 7 ..	57 0 ..	−127	18566	·19908	1342	·31043	·30705	−338	36171	·36566	395	35
36	Blaauwbosch	58 52 ..	58 57 ..	5	·18496	·18481	− 15	·30621	·30709	88	·35777	·35846	69	36

37	Blaauwkrantz	60 58 ..	60 52 ..	− 6	·17813	·17851	38	·32091	·32012	− 79	·36704	·36652	− 52	37
38	The Bluff	61 22 ..	61 27 ..	5	·17924	·17966	42	·32830	·33010	180	·37404	·37607	203	38
39	Boschkopjes	56 49 ..	56 42 ..	− 7	·19957	·19908	− 49	·30517	·30537	20	·36463	·36514	51	39
40	Boschrand	59 3 ..	59 4 ..	1	·18842	·18815	− 27	·31421	·31374	− 47	·36637	·36584	− 53	40
41	Boston	60 54 ..	60 52 ..	− 2	·18242	·18222	− 20	·32775	·32705	− 70	·37509	·37420	− 89	41
42	Botha's Berg	57 57 ..	58 9 ..	12	·20347	·19526	− 821	·32498	·31413	−1085	·38343	·36941	−1402	42
43	Brak River	55 55 ..	56 20 ..	25	·20452	·20207	− 245	·30245	·30327	82	·36511	·36438	− 73	43
44	Brandboontjes	56 40 ..	56 54 ..	14	·20065	·19941	− 124	·30507	·30670	163	·36514	·36569	55	44
45	Bredasdorp	59 56 ..	59 57 ..	1	·18015	·18074	59	·31119	·31205	86	·35957	·36074	117	45
46	Breekkerrie	58 21 ..	58 29 ..	8	·18531	·18663	132	·30064	·30426	362	·35314	·35694	380	46
47	Britstown	59 6 ..	59 24 ..	18	·18097	·18342	245	·30238	·30947	709	·35244	·35984	740	47
48	Buffelsberg	57 11 ..	56 58 ..	− 13	·19484	·19939	455	·30214	·30681	467	·35951	·36583	632	48
49	Buffelshoek	57 5 ..	56 24 ..	− 41	·19809	·20186	377	·30601	·30366	− 235	·36452	·36469	17	49
50	Buffelsklip	60 38 ..	60 39 ..	1	·17867	·17882	15	·31752	·31812	60	·36433	·36503	70	50
51	Bulawayo	53 28 ..	53 58 ..	30	·21461	·21591	130	·28968	·29700	732	·36051	·36722	671	51
52	Bult and Baatjes	59 12 ..	58 40 ..	− 32	·19223	·19235	12	·32247	·31602	− 645	·37542	·37004	− 538	52
52A	Bulwer				·18080	·18226	146							52A
53	Burghersdorp	60 18 ..	60 24 ..	6	·18212	·18174	− 38	·31931	·31979	48	·36758	·36776	18	53
54	Bushmanskop	59 41 ..	59 46 ..	5	·18231	·18198	− 33	·31177	·31204	27	·36112	·36149	37	54
55	Butterworth	61 20 ..	61 37 ..	17	·17841	·17687	− 154	·32634	·32744	110	·37191	·37219	28	55
56	Caledon River	60 23 ..	60 12 ..	− 11	·18089	·18252	163	·31820	·31880	60	·36603	·36757	154	56
57	Calitzdorp	60 19 ..	60 20 ..	1	·17918	·17959	41	·31436	·31501	65	·36181	·36270	89	57
58	Camperdown	61 29 ..	61 9 ..	− 20	·17833	·18104	271	·32822	·32867	45	·37354	·37528	174	58
59	Cango	60 44 ..	60 29 ..	− 15	·17780	·17922	142	·31727	·31634	− 93	·36368	·36356	− 12	59
60	Cape Town (Royal Obs.)	59 6 ..	59 14 ..	8	·18271	·18236	− 35	·30533	·30584	51	·35584	·35604	20	60
61	Cathcart	61 46 ..	61 17 ..	− 29	·17477	·17841	364	·32549	·32633	84	·36944	·37185	241	61
62	Ceres Road	59 11 ..	59 12 ..	1	·18284	·18264	− 20	·30671	·30673	2	·35708	·35703	− 5	62
63	Charlestown	59 24 ..	59 25 ..	1	·18861	·18822	− 39	·31892	·31872	− 20	·37052	·37015	− 37	63
64	Clarkson	61 9 ..	61 4 ..	− 5	·17800	·17794	− 6	·32310	·32176	− 134	·36890	·36751	− 139	64
65	Coerney	61 30 ..	61 28 ..	− 2	·17626	·17668	42	·32464	·32469	5	·36939	·36968	29	65
66	Colenso	60 9 ..	60 7 ..	− 2	·18548	·18532	− 16	·32322	·32315	− 7	·37265	·37254	− 11	66
67	Colesberg	60 1 ..	59 56 ..	− 5	·18183	·18282	99	·31515	·31568	53	·36384	·36460	76	67
68	Connan's Farm	56 16 ..	57 45 ..	89	·19602	·18955	− 647	·29354	·29948	594	·35301	·35432	131	68
69	Cookhouse	61 14 ..	61 16 ..	2	·17732	·17779	47	·32299	·32427	128	·36846	·36991	145	69
70	Cotswold Hotel	61 28 ..	61 30 ..	2	·17769	·17866	97	·32681	·32870	189	·37199	·37414	215	70
71	Cradock	60 49 ..	60 50 ..	1	·17981	·17967	− 14	·32201	·32220	19	·36881	·36881	0	71
72	Cream of Tartarfontein	55 45 ..	56 7 ..	22	·20406	·20295	− 111	·29969	·30209	240	·36258	·36305	47	72
73	Crocodile Pools	56 14 ..			·19995			·29907			·35975			73
74	Dalton	61 24 ..	60 52 ..	− 32	·17974	·18291	317	·32966	·32795	− 171	·37549	·37548	− 1	74
75	Dambiesfontein	59 16 ..	59 5 ..	− 11	·18355	·18422	67	·30873	·30805	− 68	·35918	·35889	− 29	75
76	Dannhauser	59 44 ..	59 48 ..	4	·18597	·18682	85	·31868	·32120	252	·36896	·37168	272	76
77	Dargle Road	60 39 ..	60 48 ..	9	·18468	·18282	− 186	·32843	·32708	− 135	·37679	·37450	− 229	77
78	Darling	58 56 ..	58 56 ..	0	·18351	·18331	− 20	·30461	·30372	− 89	·35560	·35468	− 92	78
79	De Aar	59 52 ..	59 35 ..	− 17	·18447	·18296	− 151	·31781	·31086	− 695	·36746	·36079	− 667	79

TABLE 33 (*continued*).

No.	Station Name	Dip (θ) Obs. ° ′	Dip Calc. ° ′	Dip Diff. ′	Horizontal Intens. (H) Obs.	H Calc.	H Diff. (γ)	Vertical Intens. (Z) Obs.	Z Calc.	Z Diff. (γ)	Total Intens. (T) Obs.	T Calc.	T Diff. (γ)	Station No.
80	De Doorns	59 20 S.	59 16 S.	− 4	·18205	·18215	10	·30700	·30711	11	·35693	·35722	29	80
81	Deelfontein Farm	60 6 ..	60 4 ..	− 2	·18243	·18306	63	·31726	·31794	68	·36597	·36701	104	81
82	Deelfontein	59 25 ..	59 25 ..	0	·18706	·18708	2	·31650	·31658	8	·36766	·36701	− 65	82
83	De Jager's Farm	59 34 ..	59 33 ..	− 1	·18799	·18742	− 57	·32000	·32029	29	·37112	·37016	− 96	83
84	Dewetsdorp	59 37 ..	59 51 ..	14	·18528	·18456	− 72	·31603	·31603	0	·36632	·36666	34	84
85	Draghoender	58 16 ..	58 14 ..	− 2	·18723	·18767	44	·30276	·30235	− 41	·35599	·35587	− 12	85
86	Drew	59 51 ..	59 45 ..	− 6	·18130	·18104	− 26	·31215	·31110	−105	·36096	·36007	− 89	86
87	Driefontein	59 6 ..	58 34 ..	−32	·18788	·19243	455	·31393	·31525	132	·36585	·36905	320	87
88	Driehoek	59 25 ..	59 31 ..	6	·18898	·18839	− 59	·31973	·31990	17	·37143	·37133	− 10	88
89	East London	61 48 ..	61 59 ..	11	·17600	·17501	− 99	·32824	·32767	− 57	·37245	·37155	− 90	89
90	Elandshoek	59 52 ..	58 29 ..	−83	·18867	·19394	527	·32504	·31552	−952	·37583	·37058	−525	90
91	Elandskloof Farm	58 57 ..	59 3 ..	6	·18830	·18740	− 90	·31277	·31229	− 48	·36507	·36527	20	91
92	Elim	59 55 ..	59 51 ..	− 4	·18065	·18076	11	·31184	·31164	− 20	·36039	·36029	− 10	92
93	Ellerton	56 54 ..	56 55 ..	1	·19819	·19945	126	·30403	·30660	257	·36292	·36554	262	93
94	Elliot	61 5 ..	61 0 ..	− 5	·17998	·18046	48	·32569	·32662	93	·37216	·37308	92	94
95	Elsburg	58 26 ..	58 23 ..	− 3	·19140	·19222	82	·31152	·31268	116	·36562	·36677	115	95
96	Emmasheim	59 29 ..	59 25 ..	− 4	·18697	·18723	26	·31718	·31724	6	·36820	·36835	15	96
97	Estcourt	60 12 ..	60 19 ..	7	·18376	·18498	122	·32086	·32486	400	·36976	·37361	385	97
98	Ferreira	59 31 ..	59 30 ..	− 1	·18480	·18525	45	·31393	·31469	76	·36429	·36517	88	98
99	Fish River	60 44 ..	60 39 ..	− 5	·18018	·18023	5	·32151	·32072	− 79	·36856	·36792	− 64	99
100	Forty-one mile Siding	51 47 ..	51 36 ..	−11	·22465	·22431	− 34	·28518	·28392	−126	·36300	·36201	− 99	100
101	Fountain Hall	60 20 ..	60 33 ..	13	·18500	·18408	− 92	·32481	·32615	134	·37377	·37419	42	101
102	Francistown	56 30 ..	54 43 ..	−107	·20380	·21178	798	·30790	·29582	−1208	·36924	·36386	−538	102
103	Fraserburg	59 16 ..	59 23 ..	7	·18290	·18323	33	·30752	·30948	196	·35781	·35968	187	103
104	Fraserburg Road	59 59 ..	60 7 ..	8	·18251	·18029	−222	·31591	·31403	−188	·36484	·36235	−249	104
105	Gamtoos River Bridge	61 14 ..	61 21 ..	7	·17736	·17694	− 42	·32306	·32384	78	·36855	·36892	37	105
106	George Town	60 29 ..	60 35 ..	6	·17970	·17907	− 63	·31741	·31768	27	·36474	·36486	12	106
107	Gemsbokfontein	59 27 ..	59 32 ..	5	·18326	·18284	− 42	·31049	·31080	31	·36058	·36086	28	107
108	Ginginhlovu	61 3 ..	61 5 ..	2	·18251	·18249	− 2	·32995	·32972	− 23	·37705	·37710	5	108
109	Glenallen	41 53 ..	58 32 ..	999	·23457	·18660	−4797	·21182	·30400	9218	·31606	·35677	4071	109
110	Glenconnor	61 19 ..	61 17 ..	− 2	·17637	·17723	86	·32237	·32332	95	·36746	·36872	126	110
111	Globe and Phœnix	52 44 ..	53 11 ..	27	·21945	·21932	− 13	·28841	·29486	645	·36241	·36834	593	111
112	Goedgedacht	58 44 ..	58 43 ..	− 1	·19138	·19142	4	·31519	·31600	81	·36873	·36918	45	112
113	Gordon's Bay	59 24 ..	59 25 ..	1	·18100	·18181	81	·30605	·30768	163	·35559	·35740	181	113
114	Graaff Reinet	60 35 ..	60 41 ..	6	·17931	·17949	18	·31801	·31964	163	·36507	·36664	157	114
115	Grahamstown	61 44 ..	61 42 ..	− 2	·17589	·17606	17	·32711	·32631	− 80	·37140	·37086	− 54	115
116	Grange	61 6 ..	60 59 ..	− 7	·18097	·18185	88	·32783	·32788	5	·37446	·37487	41	116
117	Graskop	59 18 ..	59 20 ..	2	·18865	·18859	− 6	·31773	·31810	37	·36951	·36980	29	117
118	Greylingstad	58 57 ..	58 43 ..	−14	·18969	·19153	184	·31507	·31492	− 15	·36777	·36859	82	118

119	Greytown	60 29 ..	60 37 ..	8	·18690	·18402	−288	·33012	·32687	−325	·37935	·37512	−423	119
120	Grobler's Bridge	58 26 ..	58 32 ..	6	·19310	·19324	14	·31428	·31553	125	·36887	·36996	109	120
121	Groenkloof	59 57 ..	60 4 ..	7	·18451	·18324	−127	·31894	·31879	− 15	·36846	·36805	− 41	121
122	Groenplaats	59 1 ..	58 59 ..	− 2	·19076	·18987	− 89	·31769	·31577	−192	·37056	·36848	−208	122
123	Grootfontein	60 2 ..	59 55 ..	− 7	·18044	·18075	31	·31284	·31196	− 88	·36035	·36067	32	123
124	Gwaai	52 27 ..	52 53 ..	26	·21749	·21759	10	·28292	·28744	452	·35686	·36059	373	124
125	Gwelo	54 23 ..	53 38 ..	−45	·22034	·21807	−227	·30757	·29813	−944	·37836	·36985	−851	125
126	Hamaan's Kraal	57 30 ..	57 58 ..	28	·19710	·19413	−297	·30939	·31014	75	·36683	·36630	− 53	126
127	Hankey	61 13 ..	61 17 ..	4	·17773	·17713	− 60	·32349	·32258	− 91	·36912	·36861	− 51	127
128	Hartley	51 56 ..	52 2 ..	6	·22273	·22229	− 44	·28441	·28675	234	·36124	·36339	215	128
129	Hector Spruit	58 11 ..	58 47 ..	36	·19673	·19296	−377	·31709	·31685	− 24	·37316	·37193	−123	129
130	Heidelberg (C. C.)	60 7 ..	60 12 ..	5	·18041	·18011	− 30	·31396	·31378	− 18	·36209	·36177	− 32	130
131	Heilbron	58 51 ..	58 55 ..	4	·18998	·18954	− 44	·31432	·31442	10	·36727	·36707	− 20	131
132	Helvetia	59 59 ..	59 57 ..	− 2	·18336	·18265	− 71	·31737	·31737	0	·36654	·36677	23	132
133	Hermanus	59 38 ..	59 39 ..	1	·18138	·18114	− 24	·30957	·30965	8	·35879	·35880	1	133
134	Hermon	59 5 ..	59 7 ..	2	·18309	·18282	− 27	·30573	·30575	2	·35635	·35624	− 11	134
135	Highlands	59 56 ..	59 45 ..	−11	·18742	·18797	55	·32374	·32185	−189	·37408	·37289	−119	135
136	Hlabisa	60 54 ..	60 48 ..	− 6	·18262	·18475	213	·32810	·32960	150	·37550	·37811	261	136
137	Hluti	59 44 ..	59 46 ..	2	·18760	·18804	44	·32148	·32211	63	·37220	·37317	97	137
138	Hoetjes Bay	58 35 ..	58 40 ..	5	·18407	·18411	4	·30136	·30141	5	·35312	·35303	− 9	138
139	Holfontein	59 50 ..	59 50 ..	0	·18502	·18469	− 33	·31832	·31780	− 52	·36818	·36779	− 39	139
140	Honey Nest Kloof	59 2 ..	59 2 ..	0	·18623	·18644	21	·31035	·31057	22	·36193	·36165	− 28	140
141	Honing Spruit	58 42 ..	58 57 ..	15	·19005	·18886	−119	·31257	·31218	− 39	·36582	·36599	17	141
142	Hopefield	58 47 ..	58 48 ..	1	·18348	·18379	31	·30276	·30275	− 1	·35402	·35407	5	142
143	Howhoek	59 32 ..	59 31 ..	− 1	·18128	·18151	23	·30816	·30874	58	·35752	·35818	66	143
144	Huguenot	59 14 ..	59 16 ..	2	·18286	·18234	− 52	·30715	·30679	− 36	·35747	·35690	− 57	144
145	Humansdorp	61 16 ..	61 15 ..	− 1	·17702	·17721	19	·32289	·32310	21	·36823	·36837	14	145
146	Hutchinson	59 34 ..	59 41 ..	7	·18316	·18237	− 79	·31209	·31164	− 45	·36188	·36140	− 48	146
147	Ibisi Bridge	61 2 ..	61 19 ..	17	·18170	·17929	−241	·32824	·32822	− 2	·37518	·37394	−124	147
148	Idutywa	61 44 ..	61 34 ..	−10	·17559	·17753	194	·32653	·32756	103	·37075	·37261	186	148
149	Igusi	53 1 ..	53 21 ..	20	·21633	·21682	49	·28724	·29171	447	·35960	·36355	395	149
150	Illovo River	61 19 ..	61 31 ..	12	·18001	·17907	− 94	·32902	·32998	96	·37504	·37566	62	150
151	Imvani	61 28 ..	61 5 ..	−23	·17781	·17860	79	·32699	·32609	− 90	·37225	·37213	− 12	151
152	Indowane	61 1 ..	60 49 ..	−12	·18074	·18192	118	·32628	·32619	− 9	·37300	·37315	15	152
153	Indwe	60 56 ..	60 53 ..	− 3	·18186	·18076	−110	·32719	·32606	−113	·37433	·37269	−164	153
154	Inoculation	54 0 ..	54 12 ..	12	·21121	·21264	143	·29071	·29515	444	·35933	·36383	450	154
155	Inyantué	51 9 ..	51 47 ..	38	·22341	·21884	−457	·27736	·27784	48	·35616	·35357	−259	155
156	Kaalfontein	58 13 ..	58 17 ..	4	·19459	·19279	−180	·31405	·31200	−205	·36944	·36681	−263	156
157	Kaalkop Farm	59 27 ..	59 19 ..	− 8	·18770	·18810	40	·31802	·31792	− 10	·36928	·36930	2	157
158	Kaapmuiden	58 29 ..	58 42 ..	13	·19310	·19307	− 3	·31490	·31650	160	·36939	·37143	204	158
159	Kalkbank	56 22 ..	56 45 ..	23	·20283	·20044	−239	·30490	·30553	63	·36620	·36546	− 74	159
160	Kaloombies	56 35 ..	56 12 ..	−23	·20150	·20253	103	·30540	·30262	−278	·36589	·36405	−184	160
161	Karree	59 20 ..	59 24 ..	4	·18548	·18582	34	·31279	·31437	158	·36366	·36515	149	161
162	Kathoek	60 3 ..	60 1 ..	− 2	·18012	·18056	44	·31259	·31255	− 4	·36079	·36103	24	162

TABLE 33 (*continued*).

No.	Station Name	Dip (θ) Obs. ° ′	Dip (θ) Calc. ° ′	Diff. ′	Horizontal Intens. (H) Obs.	Calc.	Diff. (γ)	Vertical Intens. (Z) Obs.	Calc.	Diff. (γ)	Total Intens. (T) Obs.	Calc.	Diff. (γ)	Station No.
163	Kenhardt	58 1 S.	57 55 S.	− 6	·18925	·18886	− 39	·30307	·30066	−241	·35736	·35498	−238	163
164	Kimberley	58 49 ..	58 51 ..	2	·18727	·18732	5	·30932	·30984	52	·36159	·36153	− 6	164
165	King William's Town	61 39 ..	61 43 ..	4	·17726	·17616	−110	·32851	·32699	−152	·37329	·37144	−185	165
166	Klaarstroom	60 28 ..	60 36 ..	8	·17906	·17897	− 9	·31606	·31716	110	·36326	·36436	110	166
167	Klerksdorp	58 12 ..	58 29 ..	17	·19123	·18987	−136	·30841	·31008	167	·36289	·36357	68	167
168	Klipfontein (C. C.)	59 16 ..	59 2 ..	−14	·18340	·18468	128	·30848	·30767	− 81	·35889	·35899	10	168
169	Klipfontein (Spel.)	56 21 ..	56 43 ..	22	·20344	·20035	−309	·30563	·30543	− 20	·36714	·36508	−206	169
170	Klipplaat	61 11 ..	61 0 ..	−11	·17745	·17800	55	·32254	·32130	−124	·36815	·36735	− 80	170
171	Knysna	60 51 ..	60 43 ..	− 8	·17840	·17879	39	·31987	·31903	− 84	·36624	·36580	− 44	171
172	Kokstad	61 12 ..	61 12 ..	0	·18016	·17994	− 22	·32771	·32771	0	·37397	·37462	65	172
173	Komati Poort	59 45 ..	58 51 ..	−54	·18869	·19270	401	·32354	·31717	−637	·37456	·37222	−234	173
174	Komgha	61 35 ..	61 44 ..	9	·17782	·17631	−151	·32865	·32741	−124	·37367	·37190	−177	174
175	Kraal	58 39 ..	58 35 ..	− 4	·19068	·19167	99	·31300	·31372	72	·36651	·36765	114	175
176	Krantz Kloof	61 21 ..	61 19 ..	− 2	·18001	·18029	28	·32947	·32951	4	·37544	·37580	36	176
177	Krantz Kop	58 27 ..	58 38 ..	11	·18677	·18583	− 94	·30419	·30527	108	·35693	·35721	28	177
178	Kromm River	58 56 ..	58 56 ..	0	·18940	·18966	26	·31438	·31530	92	·36703	·36789	86	178
179	Krugers	59 53 ..	59 47 ..	− 6	·18490	·18389	−101	·31875	·31579	−296	·36850	·36532	−318	179
180	Kruispad	59 46 ..	59 36 ..	−10	·18146	·18149	3	·31135	·30983	−152	·36033	·35898	−135	180
181	Kwambonambi	60 38 ..	60 59 ..	21	·18454	·18355	− 99	·32795	·33012	217	·37631	·37805	174	181
182	Laat Rivier	58 13 ..	58 9 ..	− 4	·18793	·18796	3	·30330	·30212	−118	·35685	·35577	−108	182
183	Ladismith (C. C.)	60 6 ..	60 7 ..	1	·17974	·18013	39	·31260	·31310	50	·36058	·36115	57	183
184	L'Agulhas	60 5 ..	60 1 ..	− 4	·17970	·18035	65	·31230	·31294	64	·36033	·36131	98	184
185	Laingsburg	59 56 ..	59 49 ..	− 7	·18022	·18096	74	·31131	·31136	5	·35971	·36006	35	185
186	Lake Banagher	58 55 ..	58 48 ..	− 7	·19056	·19112	56	·31610	·31650	40	·36910	·37011	101	186
187	Langlaagte	58 56 ..	58 20 ..	−36	·18907	·19223	316	·31383	·31180	−203	·36639	·36657	18	187
188	Letjesbosch	59 39 ..	59 53 ..	14	·18346	·18156	−190	·31332	·31273	− 59	·36309	·36190	−119	188
189	Libode	61 31 ..	61 33 ..	2	·18006	·17802	−204	·33187	·32811	−376	·37758	·37334	−424	189
190	Lobatsi	57 0 ..			·19542			·30093			·35881			190
191	Lochard	53 34 ..	53 56 ..	22	·21506	·21717	211	·29134	·29830	696	·36212	·36920	708	191
192	Lydenburg	58 5 ..	58 4 ..	− 1	·19483	·19557	74	·31279	·31329	50	·36852	·36927	75	192
193	Machadodorp	58 44 ..	58 26 ..	−18	·19073	·19392	319	·31412	·31519	107	·36750	·37000	250	193
194	Macheke	52 9 ..	52 9 ..	0	·22424	·22353	− 71	·28857	·28747	−110	·36545	·36369	−176	194
195	Mafeking	57 12 ..			·19487			·30237			·35974			195
196	Magalapye	55 34 ..			·20496			·29895			·36247			196
197	Magnet Heights	57 59 ..	57 37 ..	−22	·19272	·19694	422	·30821	·31027	206	·36351	·36758	407	197
198	Makwiro Siding	51 36 ..	51 51 ..	15	·22447	·22327	−120	·28321	·28575	254	·36138	·36280	142	198
199	Malagas	60 5 ..	60 7 ..	2	·18066	·18035	− 31	·31396	·31324	− 72	·36225	·36138	− 87	199
200	Malenje Siding	53 3 ..	53 3 ..	0	·22040	·22109	69	·29302	·29328	26	·36665	·36649	− 16	200
201	Malinde	51 50 ..	52 5 ..	15	·21859	·21856	− 3	·27811	·28067	256	·35373	·35577	204	201

202	Malmesbury	59 6 ..	59 4 ..	− 2	·18220	·18295	75	·30443	·30507	64	·35479	·35479	0	202
203	Mandegos	53 26 ..	53 22 ..	− 4	·21930	·22133	203	·29563	·29523	− 40	·36810	·36741	− 69	203
204	Mapani Loep.	55 48 ..	55 57 ..	9	·20532	·20354	−178	·30213	·30126	− 87	·36529	·36344	−185	204
205	Mara	57 16 ..	56 31 ..	−45	·20181	·20131	− 50	·31396	·30431	−965	·37322	·36485	−837	205
206	Marandellas	52 55 ..	52 11 ..	−44	·22211	·22336	125	·29385	·28771	−614	·36836	·36382	−454	206
207	Maribogo	57 18 ..			·19366			·30176			·35857			207
208	Matetsi	51 41 ..	51 14 ..	−27	·21696	·21913	217	·27456	·27231	−225	·34993	·34945	− 48	208
209	Matjesfontein	59 51 ..	59 45 ..	− 6	·18028	·18112	84	·30926	·31076	150	·35881	·35946	65	209
210	Meyerton	58 30 ..	58 31 ..	1	·19148	·19147	− 1	·31247	·31264	17	·36647	·36662	15	210
211	Middelberg (Tzitzikama)	61 2 ..	61 3 ..	1	·17779	·17795	16	·32118	·32167	49	·36711	·36748	37	211
212	Middlepost	58 50 ..	59 2 ..	12	·18455	·18400	− 55	·30503	·30705	202	·35651	·35755	104	212
213	Middleton	61 22 ..	61 22 ..	0	·17742	·17742	0	·32497	·32500	3	·37024	·37035	11	213
214	Mill River	60 38 ..	60 40 ..	2	·17839	·17881	42	·31702	·31826	124	·36377	·36517	140	214
215	Miller Siding	61 0 ..	60 54 ..	− 6	·17794	·17828	34	·32100	·32043	− 57	·36703	·36664	− 39	215
216	Miller's Point	59 17 ..	59 21 ..	4	·18259	·18192	− 67	·30732	·30675	− 57	·35747	·35650	− 97	216
217	Misgund	60 49 ..	60 50 ..	1	·17825	·17847	22	·31916	·31998	82	·36556	·36631	75	217
218	Mission Station	57 6 ..	56 51 ..	−15	·19910	·19974	64	·30776	·30621	−155	·36655	·36535	−120	218
219	Modder Spruit	60 0 ..	60 0 ..	0	·18560	·18572	12	·32148	·32232	84	·37120	·37216	96	219
220	Molteno	60 36 ..	60 33 ..	− 3										220
221	Mossel Bay	60 33 ..	60 38 ..	5	·17930	·17912	− 18	·31756	·31763	7	·36467	·36470	3	221
222	Mount Ayliff, near	61 16 ..	61 19 ..	3	·17890	·17960	70	·32641	·32826	185	·37222	·37421	199	222
223	Mount Frere	61 2 ..	61 11 ..	9	·18067	·18011	− 56	·32629	·32766	137	·37298	·37389	91	223
224	Mount Moreland	61 20 ..	61 17 ..	− 3	·18064	·18074	10	·33041	·32960	− 81	·37656	·37608	− 48	224
224A	Movene	59 7 ..	59 0 ..	− 7	·19573	·19200	−373	·32726	·31771	− 955	·38133	·37251	−882	224A
225	M'Phatele's Location	59 12 ..	57 18 ..	−114	·19535	·19822	287	·32748	·30854	−1894	·38133	·36683	−1450	225
226	Naauwpoort	60 2 ..	60 8 ..	6	·18166	·18176	10	·31506	·31658	152	·36369	·36509	140	226
227	Naboomspruit	57 5 ..	57 9 ..	4	·19889	·19903	14	·30723	·30768	45	·36600	·36678	78	227
228	Nelspoort	60 18 ..	59 52 ..	−26	·17795	·18191	396	·31199	·31287	88	·35916	·36241	325	228
229	Nelspruit	58 28 ..	58 34 ..	6	·19290	·19378	88	·31436	·31591	155	·36884	·37098	214	229
230	Newcastle	59 39 ..	59 37 ..	− 2	·18845	·18745	−100	·32186	·32005	−181	·37296	·37093	−203	230
231	Newcastle (Trans.)	58 57 ..	58 56 ..	− 1	·18982	·19081	99	·31529	·31698	169	·36802	·37030	228	231
232	Nooitgedacht	57 22 ..	58 30 ..	68	·19204	·19379	175	·29990	·31552	1562	·35612	·37036	1424	232
233	Norval's Pont	59 50 ..	59 59 ..	9	·18320	·18285	− 35	·31524	·31634	110	·36462	·36522	60	233
234	'Nqutu Road	60 13 ..	59 56 ..	−17	·18584	·18653	69	·32472	·32235	−237	·37414	·37252	−162	234
235	Nylstroom	57 29 ..	57 13 ..	−16	·19843	·19878	35	·31129	·30807	−322	·36915	·36702	−213	235
235A	Orange River	59 7 ..	59 10 ..	3	·18559	·18570	11	·31058	·31100	42	·36182	·36155	− 27	235A
236	Orjida	60 45 ..	60 45 ..	0	·17836	·17859	23	·31848	·31897	49	·36503	·36498	− 5	236
237	Oudemuur	58 42 ..	58 40 ..	− 2	·18549	·18564	15	·30507	·30545	38	·35707	·35719	12	237
238	Oudtshoorn	60 25 ..	60 29 ..	4	·17944	·17916	− 28	·31609	·31669	60	·36347	·36410	63	238
239	Paardevlei	58 44 ..	58 50 ..	6	·18588	·18525	− 63	·30614	·30647	33	·35821	·35814	− 7	239
240	Paarl	59 14 ..	59 16 ..	2										240
241	Palapye	55 8 ..			·20639			·29621			·36101			241
242	Pampoenpoort	59 19 ..	59 17 ..	− 2	·18324	·18373	49	·30883	·30926	43	·35909	·35993	84	242
243	Payne's Farm	61 34 ..	61 22 ..	−12	·17826	·17916	90	·32925	·32841	− 84	·37438	·37410	− 28	243

TABLE 33 (*continued*).

No.	Station Name	Dip (θ) Obs.	Dip Calc.	Dip Diff.	Horizontal Intens. (H) Obs.	H Calc.	H Diff.	Vertical Intens. (Z) Obs.	Z Calc.	Z Diff.	Total Intens. (T) Obs.	T Calc.	T Diff.	Station No.
		° ′	° ′	′			(γ)			(γ)			(γ)	
244	Picene	59 0 S.	59 7 S.	7	·19308	·19138	−170	·32134	·31820	−314	·37488	·37279	−209	244
245	Pienaar's River	57 49 ..	57 53 ..	4	·19729	·19452	−277	·31349	·31016	−333	·37041	·36643	−398	245
246	Pietersburg	56 57 ..	56 57 ..	0	·19855	·19963	108	·30516	·30663	147	·36407	·36597	190	246
247	Piet Potgietersrust	56 59 ..	57 2 ..	3	·19831	·19893	62	·30519	·30709	190	·36395	·36637	242	247
248	Piet Retief	59 33 ..	59 23 ..	−10	·18831	·18877	46	·32034	·31897	−137	·37103	·37157	54	248
249	Pilgrim's Rest	57 50 ..	57 56 ..	6	·19612	·19551	−61	·31184	·31207	23	·36838	·36817	−21	249
250	Piquetberg	58 56 ..	58 49 ..	−7	·18340	·18381	41	·30443	·30335	−108	·35540	·35463	−77	250
251	Pivaan's Poort	59 43 ..	59 40 ..	−3	·18796	·18766	−30	·32187	·32064	−123	·37273	·37159	−114	251
252	Platrand	59 12 ..	59 3 ..	−9	·18907	·18959	52	·31715	·31659	−56	·36925	·36885	−40	252
253	Plettenberg Bay	60 55 ..	60 48 ..	−7	·17812	·17864	52	·32024	·31969	−55	·36644	·36625	−19	253
254	Plumtree	53 47 ..	53 59 ..	12	·21219	·21392	173	·28975	·29452	477	·35913	·36408	495	254
255	Pokwani (Transvaal)	57 41 ..	57 38 ..	−3	·19753	·19694	−59	·31226	·31034	−192	·36949	·36769	−180	255
256	Port Alfred	61 42 ..	61 53 ..	11	·17595	·17553	−42	·32678	·32756	78	·37113	·37167	54	256
257	Port Beaufort	60 12 ..	60 14 ..	2	·17990	·18011	21	·31412	·31420	8	·36199	·36214	15	257
258	Port Elizabeth	61 35 ..	61 33 ..	−2	·17640	·17639	−1	·32525	·32521	−4	·37001	·36988	−13	258
259	Port Shepstone	61 53 ..	61 42 ..	−11	·17801	·17775	−26	·33315	·32974	−341	·37772	·37476	−296	259
260	Port St Johns	62 2 ..	61 49 ..	−13	·17547	·17692	145	·33036	·32878	−158	·37406	·37347	−59	260
261	Potchefstroom	58 22 ..	58 28 ..	6	·19079	·19161	82	·30973	·31080	107	·36377	·36452	75	261
262	Potfontein	59 21 ..	59 21 ..	0	·18433	·18421	−12	·31106	·31027	−79	·36158	·36060	−98	262
263	Pretoria	58 3 ..	58 8 ..	5	·19377	·19332	−45	·31069	·31116	47	·36617	·36639	22	263
264	Prince Albert	60 21 ..	60 23 ..	2										264
265	Prince Albert Road	60 14 ..	60 6 ..	−8	·18010	·18018	8	·31473	·31367	−106	·36263	·36176	−87	265
266	Queenstown	61 2 ..	60 57 ..	−5	·18049	·18014	−35	·32605	·32533	−72	·37268	·37172	−96	266
267	Randfontein	58 19 ..	58 17 ..	−2	·19151	·19205	54	·31029	·31060	31	·36463	·36555	92	267
268	Rateldraai	57 46 ..	57 40 ..	−6	·19194	·18986	−208	·30440	·29897	−543	·35979	·35404	−575	268
269	Rateldrift	58 51 ..	58 53 ..	2	·18439	·18462	23	·30496	·30616	120	·35638	·35711	73	269
270	Richmond (Natal)	61 5 ..	61 9 ..	4	·18091	·18071	−20	·32750	·32828	78	·37414	·37477	63	270
271	Richmond Road	59 54 ..	59 46 ..	−8	·18231	·18199	−32	·31620	·31248	−372	·36498	·36130	−368	271
272	Rietkuil Farm	61 5 ..	60 58 ..	−7	·17978	·18164	186	·32546	·32650	104	·37180	·37297	117	272
273	Rietpoort	58 55 ..	58 49 ..	−6	·18523	·18520	−3	·30717	·30635	−82	·35869	·35785	−84	273
274	Rietvlei	57 30 ..	57 41 ..	11	·19793	·19643	−150	·31068	·31080	12	·36838	·36757	−81	274
275	Rietvlei (C. C.)	60 29 ..	60 33 ..	4										275
276	Riversdale	60 11 ..	60 20 ..	9	·17997	·17986	−11	·31402	·31477	75	·36194	·36234	40	276
277	Rivierplaats	58 49 ..	59 12 ..	23	·18456	·18345	−111	·30495	·30805	310	·35646	·35820	174	277
278	Roadside	58 18 ..	58 30 ..	12	·18734	·18626	−108	·30322	·30448	126	·35643	·35669	26	278
279	Robertson	59 41 ..	59 34 ..	−7	·18145	·18154	9	·31032	·30952	−80	·35946	·35894	−52	279
280	Rodekrantz	57 43 ..	57 42 ..	−1	·19734	·19642	−92	·31237	·31084	−153	·36947	·36762	−185	280
281	Roodepoort	57 58 ..	59 3 ..	65	·18534	·18448	−86	·29623	·30725	1102	·34947	·35853	906	281
282	Rooidam	58 41 ..	58 49 ..	8	·18657	·18546	−111	·30655	·30572	−83	·35886	·35767	−119	282

283	Revué				·21995	·22150	155							283
284	Rooiputs	58 16 ..	58 3 ..	−13	·18545	·18836	291	·29994	·30136	142	·35265	·35534	269	284
285	Rooival	59 27 ..	59 38 ..	11	·18274	·18242	− 32	·30961	·31110	149	·35952	·36081	129	285
286	Rosmead Junction	60 21 ..	60 27 ..	6	·18170	·18081	− 89	·31909	·31889	− 20	·36717	·36662	− 55	286
287	Rouxville	60 13 ..	60 16 ..	3	·18286	·18258	− 28	·31950	·31965	15	·36813	·36810	− 3	287
288	Rusapi	52 35 ..	52 37 ..	2	·22179	·22250	71	·28992	·29047	55	·36502	·36514	12	388
289	Rustplaats	57 50 ..	57 51 ..	1	·19639	·19588	− 51	·31225	·31159	− 66	·36889	·36798	− 91	289
290	Ruyterbosch	60 19 ..	60 32 ..	13	·18004	·17926	− 78	·31586	·31676	90	·36356	·36404	48	290
291	Sabie River	58 17 ..	58 9 ..	− 8	·19694	·19518	−176	·31866	·31374	−492	·37461	·36953	−508	291
292	Salisbury	51 43 ..	51 46 ..	3	·22132	·22246	114	·28041	·28498	457	·35722	·36251	529	292
293	Saxony	59 47 ..	59 35 ..	−12	·18562	·18622	60	·31871	·31737	−134	·36883	·36798	− 85	293
294	Schietfontein	59 32 ..	59 33 ..	1	·18314	·18223	− 91	·31132	·31028	−104	·36119	·35982	−137	294
295	Schikhoek	59 49 ..	59 36 ..	−13	·18656	·18796	140	·32076	·32038	− 38	·37106	·37151	45	295
296	Schoemanshoek	58 30 ..	58 22 ..	− 8	·19293	·19446	153	·31484	·31499	15	·36924	·37013	89	296
297	Schuilplaats	59 7 ..	59 0 ..	− 7	·18922	·18990	68	·31659	·31666	7	·36882	·36917	35	297
298	Secocoeni's Stad	56 27 ..	57 25 ..	58	·20070	·19769	− 301	·30266	·30922	656	·36315	·36710	395	298
299	Seruli	54 56 ..	55 4 ..	8	·20658	·20878	220	·29429	·29918	489	·35956	·36477	521	299
300	Shangani	54 41 ..	53 56 ..	−45	·22431	·21803	−628	·31660	·29949	−1711	·38801	·37045	−1756	300
301	Shashi	57 0 ..	54 38 ..	−142	·21195	·21068	−127	·32639	·29702	−2937	·38916	·36417	−2499	301
302	Shela River	59 10 ..	59 20 ..	10	·18896	·18916	20	·31657	·31876	219	·36867	·37073	206	302
303	Shoshong Road	56 40 ..			·19750			·30028			·35941			303
304	Signal Hill	59 7 ..	59 12 ..	5	·18271	·18244	− 27	·30550	·30554	4	·35596	·35581	− 15	304
305	Simonstown (R. R.)	59 16 ..	59 20 ..	4	·18209	·18200	− 9	·30625	·30647	22	·35623	·35640	17	305
305A	Simonstown (Glenc.)	59 12 ..	59 20 ..	8	·18252	·18200	− 52	·30618	·30647	29	·35646	·35640	− 6	305A
306	Sir Lowry's Pass	59 25 ..	59 25 ..	0	·18108	·18181	73	·30639	·30768	129	·35590	·35740	150	306
307	Smaldeel	59 7 ..	59 17 ..	10	·18680	·18667	− 13	·31233	·31403	170	·36393	·36544	151	307
308	Spitzkopje	58 48 ..	58 26 ..	−22	·19224	·19438	214	·31742	·31538	−204	·37109	·37076	− 33	308
309	Springfontein	60 15 ..	59 56 ..	−19	·18102	·18327	225	·31671	·31641	− 30	·36630	·36550	− 80	309
310	Springs	58 22 ..	58 24 ..	2	·19116	·19247	131	·31033	·31288	255	·36448	·36739	291	310
311	Stanford	59 41 ..	59 41 ..	0	·18127	·18117	− 10	·31001	·31011	10	·35911	·35929	18	311
312	Stanger	61 16 ..	61 10 ..	− 6	·18011	·18166	155	·32852	·32950	98	·37466	·37646	180	312
313	Steenkampspoort	59 24 ..	59 31 ..	7	·18335	·18274	− 61	·31003	·31039	36	·36012	·36028	16	313
314	Stellenbosch	59 20 ..	59 19 ..	− 1	·18192	·18215	23	·30682	·30687	5	·35657	·35674	17	314
315	Sterkstroom	60 49 ..	60 43 ..	− 6	·18095	·18086	− 9	·32399	·32315	− 84	·37109	·36988	−121	315
316	Steynsburg	60 15 ..	60 24 ..	9	·18258	·18153	−105	·31944	·31900	− 44	·36794	·36745	− 49	316
317	Still Bay	60 18 ..	60 30 ..	12	·17967	·17960	− 7	·31500	·31608	108	·36263	·36340	77	317
318	Stormberg Junction				·18138	·18014	−124							318
319	Storms River	61 1 ..	60 58 ..	− 3	·17790	·17819	29	·32116	·32092	− 24	·36714	·36694	− 20	319
320	Strandfontein	59 14 ..	59 19 ..	5										320
321	Sutherland	59 16 ..	59 23 ..	7	·18495	·18282	−213	·31099	·30925	−174	·36188	·35901	−287	321
322	Swellendam	59 53 ..	59 57 ..	4	·18107	·18062	− 45	·31214	·31193	− 21	·36087	·36052	− 35	322
323	Taungs	58 22 ..			·18887			·30662			·36011			323
324	Thaba 'Nchu	59 37 ..	59 41 ..	4	·18548	·18498	− 50	·31636	·31622	− 14	·36672	·36649	− 23	324
325	Tinfontein	60 33 ..	60 20 ..	−13	·18077	·18214	137	·32016	·31960	− 56	·36767	·36820	53	325

TABLE 33 (*continued*).

No.	Station Name	Dip (θ) Obs. ° ′	Dip (θ) Calc. ° ′	Diff. ′	Horizontal Intens. (H) Obs.	Calc.	Diff. (γ)	Vertical Intens. (Z) Obs.	Calc.	Diff. (γ)	Total Intens. (T) Obs.	Calc.	Diff. (γ)	Station No.
326	Toise River	61 47 S.	61 30 S.	−17	·17559	·17744	185	·32716	·32681	− 35	·37132	·37186	54	326
327	Touws River	59 32 ..	59 32 ..	0	·18151	·18108	− 43	·30844	·30864	20	·35789	·35811	22	327
328	Tsolo	60 57 ..	61 19 ..	22	·18002	·17916	− 86	·32417	·32765	348	·37079	·37343	264	328
329	Tugela	61 9 ..	61 8 ..	− 1	·18139	·18204	65	·32926	·32967	41	·37592	·37678	86	329
330	Tulbagh Road	59 2 ..	59 7 ..	5	·18313	·18291	− 22	·30519	·30595	76	·35591	·35646	55	330
331	Tweepoort	58 51 ..	59 4 ..	13	·19260	·19022	−238	·31865	·31751	−114	·37233	·37066	−167	331
332	Twee Rivieren	60 59 ..	60 57 ..	− 2	·17783	·17832	49	·32059	·32081	22	·36661	·36695	34	332
333	Twelfelhoek	59 29 ..	59 19 ..	−10	·18796	·18837	41	·31889	·31745	−144	·37015	·36906	−109	333
334	Tygerfontein	60 18 ..	60 20 ..	2	·17967	·17953	− 14	·31500	·31603	103	·36263	·36355	92	334
335	Tygerkloof Drift	59 22 ..	59 27 ..	5	·18836	·18753	− 83	·31809	·31819	10	·36967	·36926	− 41	335
336	Thirtyfirst	58 25 ..	58 14 ..	−11	·19197	·19460	263	·31224	·31434	210	·36654	·36917	263	336
337	Uitenhage	61 25 ..	61 26 ..	1	·17673	·17673	0	·32437	·32442	5	·36939	·36939	0	337
338	Uitkyk	58 10 ..	58 17 ..	7	·19456	·19415	− 41	·31338	·31428	90	·36887	·36913	26	338
339	Uitspan Farm	59 34 ..	59 16 ..	−18	·18157	·18353	196	·30907	·30923	16	·35851	·35957	106	339
340	Umhlatuzi	60 38 ..	61 6 ..	28	·18602	·18270	−332	·33057	·33019	− 38	·37932	·37769	−163	340
341	Umhlengana Pass	61 50 ..	61 43 ..	− 7	·17787	·17736	− 51	·33222	·32850	−372	·37685	·37341	−344	341
342	Umtali	53 0 ..	53 10 ..	10	·22005	·22118	113	·29200	·29393	193	·36564	·36680	116	342
343	Umtata	61 35 ..	61 30 ..	− 5	·17879	·17818	− 61	·33052	·32785	−267	·37578	·37317	−261	343
344	Umtwalumi	61 45 ..	61 44 ..	− 1	·17735	·17725	− 10	·33007	·33013	6	·37469	·37490	21	344
345	Umzinto	61 22 ..	61 35 ..	13	·18102	·17846	−256	·33156	·32991	−165	·37776	·37527	−249	345
346	Underberg Hotel	61 2 ..	60 44 ..	−18	·18068	·18253	185	·32639	·32602	− 37	·37307	·37329	22	346
347	Upington	57 42 ..	57 30 ..	−12	·19183	·19055	−128	·30111	·29782	−329	·35710	·35341	−369	347
348	Utrecht (West of)	59 34 ..	59 39 ..	5	·18780	·18752	− 28	·31967	·32049	82	·37075	·37138	63	348
349	Van Reenen				·18596	·18630	34							349
350	Van Wyk's Farm	60 3 ..	60 12 ..	9	·18025	·18005	− 20	·31282	·31378	96	·36105	·36170	65	350
351	Van Wyk's Vlei	58 47 ..	58 41 ..	− 6	·18461	·18582	121	·30463	·30553	90	·35613	·35762	149	351
352	Victoria Falls	51 24 ..	50 54 ..	−30	·22070	·21983	− 87	·27646	·26978	−668	·35376	·34784	−592	352
353	Villiersdorp	59 31 ..	59 27 ..	− 4	·18118	·18176	58	·30778	·30850	72	·35715	·35812	97	353
354	Virginia	59 2 ..	59 12 ..	10	·18749	·18734	− 15	·31244	·31393	149	·36439	·36580	141	354
355	Vlaklaagte	58 54 ..	58 49 ..	− 5	·19059	·19074	15	·31593	·31561	− 32	·36898	·36857	− 41	355
356	Vogelvlei	60 0 ..	59 48 ..	−12	·18422	·18489	67	·31909	·31783	−126	·36844	·36794	− 50	356
357	Vondeling	60 34 ..	60 40 ..	6	·17858	·17871	13	·31650	·31830	180	·36340	·36515	175	357
358	Vredefort	57 37 ..	58 42 ..	65	·19276	·18996	−280	·30394	·31232	838	·35991	·36548	557	358
359	Vredefort Road	59 4 ..	58 48 ..	−16	·17948	·18958	1010	·29950	·31150	1200	·34916	·36483	1567	359
360	Vryburg	58 2 ..			·19179			·30737			·36230			360
361	Wakkerstroom	59 24 ..	59 28 ..	4	·18874	·18823	− 51	·31914	·31917	3	·37078	·37056	− 22	361
362	Wankie	51 6 ..	51 33 ..	27	·22192	·21917	−275	·27503	·27564	61	·35340	·35213	−127	362
363	Warmbad (Waterb.)	57 11 ..	57 16 ..	5	·19691	·19870	179	·30536	·30824	288	·36334	·36721	387	363
364	Warmbad (Zoutp.)	57 17 ..	56 4 ..	−73	·19223	·20309	1086	·29924	·30184	260	·35566	·36367	801	364

365	Warrenton	58 49 ..	58 34 ..	−15	·18798	·18855	57	·31061	·30851	−210	·36305	·36103	−202	365
366	Waschbank	59 59 ..	59 58 ..	− 1	·18608	·18613	5	·32209	·32243	34	·37197	·37235	38	366
367	Waterworks	59 37 ..	59 32 ..	− 5	·18500	·18531	31	·31554	·31509	− 45	·36577	·36565	− 12	367
368	Welverdiend	58 44 ..	58 20 ..	−24	·18985	·19134	149	·31266	·31050	−216	·36579	·36472	−107	368
369	Wepener	60 10 ..	60 2 ..	− 8	·18303	·18347	44	·31915	·31831	− 84	·36792	·36772	− 20	369
370	Williston	59 3 ..	58 57 ..	− 6	·18451	·18467	16	·30769	·30720	− 49	·35875	·35831	− 44	370
371	Willowmore	60 49 ..	60 46 ..	− 3	·17794	·17847	53	·31860	·31908	48	·36493	·36562	69	371
372	Winburg	59 17 ..	59 24 ..	7	·18646	·18634	− 12	·31384	·31512	128	·36504	·36576	72	372
373	Winkeldrift	58 44 ..	58 45 ..	1	·18988	·18944	− 44	·31271	·31211	− 60	·36585	·36507	− 78	373
374	Witklip	56 31 ..	56 35 ..	4	·19958	·20108	150	·30173	·30468	295	·36177	·36505	328	374
375	Witmoss	61 15 ..	61 11 ..	− 4	·17852	·17809	− 43	·32541	·32364	−177	·37115	·36953	−162	375
376	Wolvefontein	61 19 ..	61 11 ..	− 8	·17600	·17753	153	·32170	·32215	45	·36669	·36824	155	376
377	Wolvehoek	58 36 ..	58 42 ..	6	·19040	·19030	− 10	·31193	·31317	124	·36544	·36642	98	377
378	North of Limpopo	55 52 ..	55 53 ..	1	·20291	·20379	88	·29930	·30086	156	·36162	·36319	157	378
379	Wonderfontein	58 0 ..	58 22 ..	22	·19847	·19391	−456	·31761	·31497	−264	·37453	·36951	−502	379
380	Woodville	60 37 ..	60 38 ..	1	·17881	·17896	15	·31755	·31811	56	·36444	·36514	70	380
381	Worcester	59 24 ..	59 20 ..	− 4	·18199	·18223	24	·30762	·30775	13	·35742	·35772	30	381
382	Zak Rivier	58 17 ..	58 23 ..	6	·18707	·18692	− 15	·30270	·30410	140	·35587	·35689	102	382
383	Zand River	56 21 ..	56 32 ..	11	·20234	·20118	−116	·30397	·30442	45	·36515	·36478	− 37	383
384	Zeekoegat	60 38 ..	60 37 ..	− 1	·17639	·17896	257	·31346	·31678	332	·35969	·36404	435	384
385	Zuurbraak	59 59 ..	60 1 ..	2	·18092	·18046	− 46	·31315	·31249	− 66	·36166	·36089	− 77	385
386	Zuurfontein	58 44 ..	58 44 ..	0	·18426	·18399	− 27	·30346	·30275	− 71	·35501	·35419	− 82	386
387	Zuurpoort	60 23 ..	60 27 ..	4	·18072	·18003	− 69	·31791	·31727	− 64	·36569	·36498	− 71	387
388		56 35 ..	57 57 ..	82	·20114	·19953	−161	·30485	·30670	185	·36523	·36589	66	388
389		56 55 ..	57 8 ..	13	·19908	·19894	− 14	·30558	·30764	206	·36471	·36645	174	389
390		58 14 ..	58 0 ..	−14	·19557	·19616	59	·31583	·31360	−223	·37148	·36930	−218	390
391		58 30 ..	58 17 ..	−13	·19228	·19426	198	·31363	·31449	86	·36781	·36922	141	391
392		58 9 ..	57 49 ..	−20	·19547	·19598	51	·31464	·31143	−321	·37042	·36788	−254	392
393		59 54 ..	59 45 ..	− 9	·18476	·18533	57	·31873	·31803	− 70	·36841	·36835	− 6	393
394		59 32 ..	59 29 ..	− 3	·18697	·18667	− 30	·31873	·31677	−196	·36875	·36769	−106	394
395		59 39 ..	59 31 ..	− 8	·18634	·18737	103	·31825	·31910	85	·36879	·36998	119	395
396		59 8 ..	59 8 ..	0	·18892	·18891	− 1	·31608	·31682	74	·36824	·36879	55	396
397		58 58 ..	58 49 ..	− 9	·18777	·18885	108	·31210	·31186	− 24	·36422	·36455	33	397
398	East of Komgha	61 33 ..	61 44 ..	11	·17730	·17639	− 91	·32712	·32744	32	·37209	·37197	− 12	398
399	Outspan	61 19 ..	61 37 ..	18	·17803	·17715	− 88	·32529	·32748	219	·37078	·37236	158	399
400	Bashee	61 35 ..	61 30 ..	− 5	·17786	·17829	43	·32871	·32757	−114	·37376	·37296	− 80	400
401		61 2 ..	61 52 ..	50	·18449	·17756	−693	·33329	·32653	−676	·38096	·37361	−735	401
402		61 7 ..	61 27 ..	20	·17973	·17893	− 80	·32577	·32837	260	·37205	·37400	195	402
403		61 18 ..	61 15 ..	− 3	·17953	·17990	37	·32786	·32796	10	·37381	·37407	26	403
404		61 19 ..	61 15 ..	− 4	·17793	·17965	172	·32530	·32765	235	·37079	·37366	287	404
405	Ugie (on road to)	61 12 ..	61 6 ..	− 6	·17971	·18019	48	·32690	·32715	25	·37303	·37345	42	405

The results given in Table 21 are shown in various charts; these have been drawn by entering the names of the stations and putting alongside each the value—reduced to the epoch—of the element whose isomagnetic was required. The isomagnetic lines were then drawn in by determining points on them with the help of a graduated rod and a pair of compasses and connecting these points by a freehand curve. The variation between any two stations on opposite sides of such a curve was assumed for this purpose to be uniform between them. Isomagnetics drawn in this way are very complicated in appearance and may give an exaggerated importance to the results of isolated stations. To avoid this as much as possible sinuosities were taken out which depended on the results at a single station only.

The charts are not very satisfactory means of studying the magnetic disturbances when drawn as these have been for regular increments of the element concerned. In many cases interesting deviations of the lines are omitted; some of the more important omissions are shown in smaller maps given in the later discussion of the disturbances.

The following is a list of the charts prepared in the manner described above:

Chart II. Isogonics (D).
Chart III. Isoclinals (θ).
Chart IV. Lines of equal horizontal intensity (H).
Chart V. Lines of equal total intensity (T).
Chart VI. Lines of equal vertical intensity (Z).
Chart VII. Lines of equal northerly intensity (X).
Chart VIII. Lines of equal westerly intensity (Y).

The study of the magnetic disturbances in the different parts of South Africa has been carried out after the method employed by Thorpe and Rücker in their magnetic survey of the British Isles. Before going into the details of these disturbances it is necessary to consider the results in tables 32 and 33 to get a clearer idea of their meaning. The calculated values of Z, X, and Y given there are positive when Z is measured towards the zenith, X towards the north, Y towards the west. The differences tabulated are in each case found in such a manner that a positive value means that the actual value is in defect of the calculated one by that amount. These differences represent the residual field after taking away the normal field of the district as derived from the values of the elements at the mean stations given in table 22 and the rates of variation with latitude and longitude as given in tables 23 and 24. The direction of the disturbing horizontal force is given by the equation

$$\tan\theta = \frac{Y_{\text{cal.}} - Y_{\text{obs.}}}{X_{\text{cal.}} - X_{\text{obs.}}} = \frac{\Delta Y}{\Delta X},$$

where θ is the angle made with the axis of X; further, this direction is the one in which the south pole of a magnet will point. In general the differences given in the above tables are denoted for brevity by prefixing Δ to the element considered.

A centre of attraction is so called because it attracts a south pole; it forms a peak.

A centre of repulsion is so called because it repels a south pole; it forms a hollow.

A line or region of attraction is so named when it attracts a south pole; such a line or region is a ridge.

A valley is a line or district which repels a south pole.

The intersection of two ridges is a peak, of two valleys is a hollow, of a valley and a ridge is a col.

The lines marking the positions of the ridges and the valleys were usually obtained in two ways. The direction of the disturbing horizontal force at a given station was calculated—as has been explained on the previous page—and marked on a map. Lines were then drawn, provided the stations were not far apart, which had throughout their course the disturbing horizontal forces directed towards them; these lines marked the positions of the ridges. Other lines were drawn which had throughout their course the disturbing forces directed from them; these marked the positions of the valleys. In this method the elements made use of are in reality the declination and the horizontal intensity.

After the ridges and the valleys had been indicated in position in this way the anomalies of the elements at places in the neighbourhood of these lines were taken into consideration and their agreement or otherwise noted. There are three principal anomalies which give readily the information required. At all stations to the east of a ridge running approximately parallel to the magnetic meridian the south pole of the declination magnet will be attracted towards the ridge and since the declination is westerly such an attraction will decrease its value. These stations will have a value of that element less than that which they would have had, had there been no ridge. For stations to the west of such a ridge the attraction which it exercises will cause the declination to be in excess of the normal. A good example of the effect produced on the declination values by a ridge is seen in the case of the one—shown on Map 7—which starts to the west of Witmoss, goes from there in a north-westerly direction having Witmoss, Cradock, Fish River and Bethesda Road on the east with values of the declination at each place lower than the normal, and Graaff Reinet, Zuurpoort on the west with values above it.

In the case of a valley just the opposite holds; that is, stations on the east have a declination greater and on the west less than the normal. An example of this is seen in the valley—shown on Map 5—which passes between Modder Spruit on the north and Colenso on the south and then passes south to a point between Dargle Road on the east and Boston on the west. Modder Spruit, Greytown, Dargle Road, and Albert Falls on the east have greater values of the declination, whereas Boston, Fountain Hall, Estcourt, and Colenso have smaller values than the normal.

In the case of a ridge running approximately perpendicular to the magnetic meridian, the declination anomalies in its immediate neighbourhood give little help. In this case the anomalies of the horizontal intensity and of the dip are of importance. The effect can be most easily seen by considering the result obtained when a mass of matter of the same magnetism as that of the south pole of the earth is introduced into a uniform horizontal field and extended along a region whose length is perpendicular to the direction of the field. As a magnet is brought from the geographical south towards this ridge the north pole of the magnet is repelled, that is the dip is increased and at the same time the horizontal intensity is decreased. On the northern side of the ridge the opposite state of affairs holds, that is the dip is in defect and the horizontal intensity is in excess. An example of a ridge so situated is the one—shown on Map 14—starting between Kaalfontein on the north and Langlaagte, Randfontein, Elsburg and Springs on the south, then continuing in a north-easterly direction to a station where observations were taken on 31/8/03. At Pretoria, Kaalfontein, Aberfeldy, Balmoral and Uitkyk stations on the north the horizontal intensity values are greater than the normal and those of the dip less; at Randfontein, Langlaagte, Elsburg, Springs, and two stations near Uitkyk, all places to the south of the ridge, the values of the horizontal intensity are below the normal and those of the dip—except at Springs—above it.

In the case of a valley situated across the direction of the earth's field the effects on the dip and the horizontal intensity are the opposite to those described above, that is to the south of such a valley the dip is smaller and the horizontal intensity greater than the normal values. Such a valley, which in this case coincides roughly with a line of no horizontal intensity disturbance, and is shown on Map 7, starts from near Port Alfred and proceeds nearly due west as far as Wolvefontein where it turns to the south between Hankey and Humansdorp. Wolvefontein, Glenconnor, Coerney, Alicedale and Grahamstown stations to the north of the valley have values of the horizontal intensity less than the normal for this region and values of the dip which are greater than it. At Hankey, Gamtoos River Bridge and Port Alfred stations to the south of the valley the values of the dip are below the normal and those of the horizontal intensity above it. At Uitenhage which is also to the south of this line the dip is in defect and the horizontal intensity has a normal value.

The second method of determining the positions of the ridges and the valleys is to show for each station the deviation of the declination and the horizontal intensity from the normal and then to draw lines of no horizontal intensity anomaly and no declination anomaly. These lines give again the approximate positions of the ridges and the valleys; the exact nature is determined by the direction of the disturbing horizontal intensities at stations in the neighbourhood of these lines. This second method is extremely useful in elucidating the distribution of the valleys and the ridges in cases where it is complicated. It also enables the positions of peaks and hollows to be determined; these are at the intersections of lines of no declination anomaly and no horizontal intensity anomaly, and whether it is a peak or a hollow is indicated by the direction of the disturbing horizontal intensities at the surrounding stations. A peak of positive magnetic matter has these forces directed towards it, while they are directed away from a hollow, where there is negative magnetic matter.

Natal and Cape Colony disturbances.

Natal and Zululand ridge and connected disturbances. A ridge extends from the east coast from a point between Umhlatuzi and Ginginhlovu railway stations and passes north-west to approximately 31° 30′ E. and 28° 55′ S., then turns south-west and continues in that direction to a point a little south of Boston, its course there changes and it turns northwards passing between Boston and Bulwer, continuing in the same direction west of Fountain Hall, Estcourt, Colenso and Modder Spruit. Just a little to the north-west of the latter place the direction of the ridge passes to the east, turning to the north again near Waschbank and continuing to the west of Dannhauser. There is a peak between Umhlatuzi and Ginginhlovu at the intersection of a line of no declination anomaly with one of no horizontal intensity anomaly, and it occurs in a district of maximum vertical force. A second peak lies between Boston and Bulwer about 29° 40′ S. and 29° 20′ E., where another ridge breaks off from the main one and passes to the north of Bulwer and Underberg. The directions of the horizontal disturbances at Bulwer, Underberg, Boston and Fountain Hall point nearly towards this peak. A valley starts between Modder Spruit and Colenso and passes south to intersect the main ridge near Boston, thence continuing between Camperdown and Richmond and finally intersecting the Griqualand East and South Natal valley near Durban.

When a region or line of attraction such as the main ridge spoken of above extends roughly east and west the horizontal intensity to the south of it will be less than the normal value and to the north greater. In this case Umhlatuzi, Greytown, Albert Falls, Dargle Road and Boston all to the north of the ridge have horizontal intensities greater than the normal, whereas Grange, Camperdown, Dalton, Stanger and Tugela on the south have smaller values.

The dips to the north and the south of such a line of attraction should be less and greater respectively than the normal dip.

The above stations all have values of the dip such as would be expected from such an arrangement of magnetic matter with the exception of Albert Falls and Boston. The irregularity in this instance might have been explained by the ridge which passes to the north of Underberg and of Bulwer continuing to the east of the peak and north of Boston; this however does not agree with the direction of the disturbing horizontal force at Boston, which would require the ridge to be to the south of Boston; the latter is more likely to be correct because the dip anomaly is small, only 2′.

The second part of the ridge is not so certainly marked; its position throughout a considerable portion of its length depends on observations at a single line of stations only. Here the magnetic matter runs north and south, and since in this region and in South Africa generally the declination is west, stations to the east of such a line ought to have a declination less than the normal, those to the west a greater. A glance at Map 4 will show that at Boston, Fountain Hall, Estcourt,

Colenso and Dannhauser the values of the declination are smaller than the normal and these places should therefore lie to the east of a ridge. Modder Spruit and Waschbank have declinations greater than the normal, which fix their positions as on the west of the ridge.

The valley which passes between Modder Spruit and Waschbank begins at a col—that is at a point where on the ridge the vertical force has a minimum value—; the

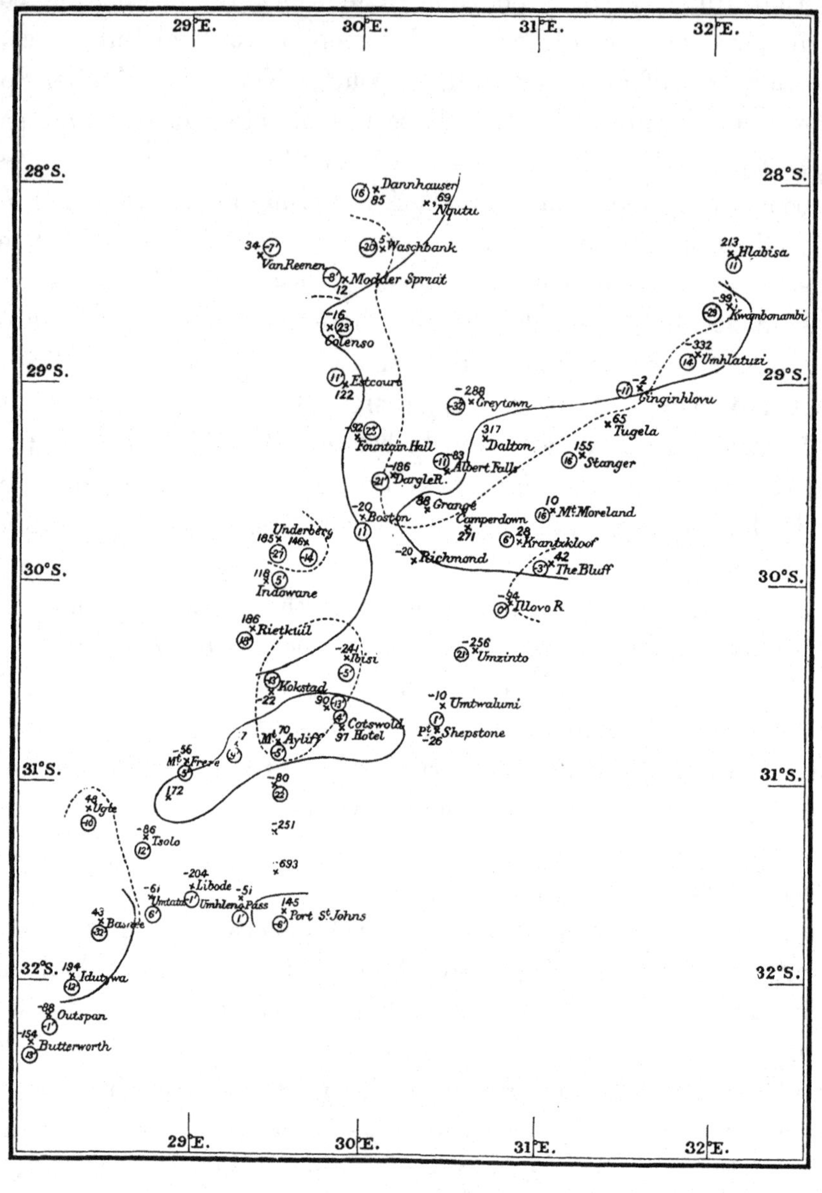

Map 4.

Showing horizontal intensity (ΔH) and declination (ΔD) anomalies for Natal and the Transkei. The ΔH are in terms of γ, the ΔD in minutes of arc. A negative sign denotes that the observed value is greater than the calculated. The continuous lines are lines of no ΔH. The dotted lines are lines of no ΔD.

existence of such a line of repulsion is supported by the values of the declination at Greytown and Dargle Road which are greater than the normal and which should therefore be to the east of a line or region of matter repelling a south pole.

The magnitudes of disturbing horizontal forces can be calculated from the results $X_c - X_o$, and $Y_c - Y_o$ given in columns 10 and 13 of table 32 respectively; and their directions are determined by the equation given on p. 82. In the case of a ridge defined as has been done above the directions of these forces must point to it.

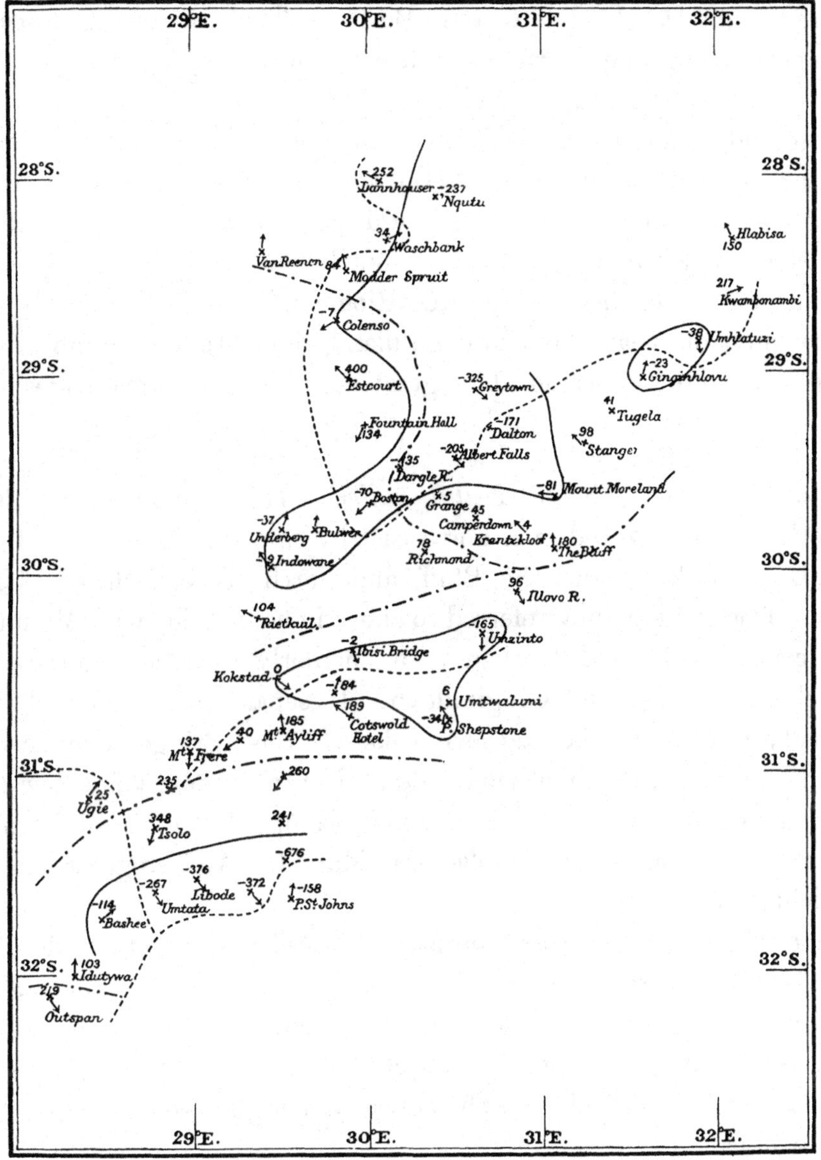

Map 5. Natal and Transkei.

Showing the vertical disturbances in terms of γ, the ridges - - - - - -, the valleys ·—·—·, the lines of no vertical disturbance ———, and the directions of the horizontal disturbances ↑. A negative sign denotes that the actual vertical intensity is greater than the calculated value.

A glance at Map 5—which gives the ridges and valleys, the vertical intensity disturbances and the directions of the disturbing horizontal forces for Natal and the Transkei—shows that the directions of these disturbing horizontal forces at Kwambonambi, Umhlatuzi, Ginginhlovu, Stanger, Albert Falls, Boston, Estcourt, Fountain Hall, Colenso, Modder Spruit and Waschbank are towards the main ridge which we have named the Zululand-Natal ridge. It will be seen from the same map that the valley intersecting the main ridge between Modder Spruit and Colenso has the disturbing horizontal forces directed away from it at Van Reenen, Modder Spruit, Colenso, Greytown, Dargle Road and Boston. The disturbing force at Mount Moreland appears to be due to the existence of another valley line—to be described immediately—passing to the north and east.

The second ridge thrown off from the main one in the neighbourhood of Boston has the disturbing horizontal forces at Bulwer and at Underberg directed towards it, it is probably continued to the west of Indowane and Rietkuil; there are not sufficient stations in this region to definitely settle the point. We shall see that there is also evidence for its continuance towards Richmond.

It is very probable that the Natal-Zululand ridge from Ginginhlovu to Boston owes its origin to magnetic matter at no great depth and not very widespread at that depth.

Griqualand East and South Natal valley. This region of repulsion passes between Rietkuil and Kokstad in a north-east by east direction—see Map 5—south of Richmond, Krantzkloof and the Bluff, and north of Ibisi Bridge, Umzinto and Illovo River. This is the valley referred to above in connection with Mount Moreland; it probably continues beyond Durban in a northerly direction and to the east of Mount Moreland, Stanger and Ginginhlovu. It separates a region of higher horizontal intensity containing the stations Kokstad, Ibisi Bridge, Umzinto and Illovo River from a region on the north where the horizontal intensity is below the normal; in this latter district are the stations Rietkuil, Camperdown, Krantzkloof, the Bluff, Mount Moreland, Stanger and Tugela—see Map 4. An irregular station in this region is Richmond.

The directions of the disturbing horizontal forces at the stations determining the position of this valley are all away from it and a glance at Map 5 enables us to see that it passes through a region where the vertical intensity has a value below the normal.

The dip results are such as to be expected from the position of this valley line, viz. lower than the normal value to the south and higher to the north. Richmond is in its dip also irregular.

The irregularity at Richmond can be accounted for by the valley which intersects the Natal-Zululand ridge near Boston, continuing between Richmond on one side and Grange and Camperdown on the other, and then passing south-east between Richmond and Grange, following a line of no horizontal intensity disturbance, the result at Richmond being more influenced by the ridge than by the valley.

A consideration of Map 6 gives support to the distribution of ridges and of valleys as described above. In this map the various lines of equal horizontal intensity are shown for differences of 100γ. It will be seen that the lines are very close together in the region of the Natal-Zululand ridge, that they are widely separated along the valley starting between Modder Spruit and Colenso and passing across the main ridge near Boston. The existence of the subsidiary ridge Underberg-Boston is also supported by the lines of equal horizontal intensity, the one marked 400 and corresponding to an actual horizontal intensity of ·18100 stretching in towards Richmond to approach the 300 line which bends over in the neighbourhood of

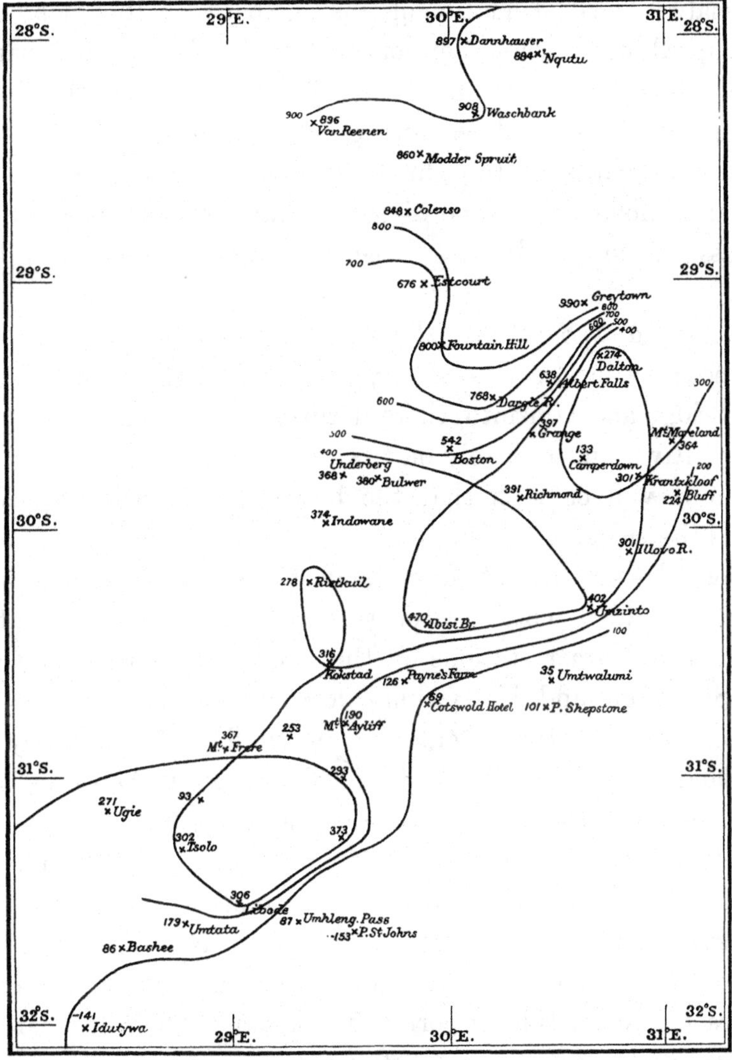

Map 6. NATAL AND TRANSKEI.

Lines of equal horizontal intensity differing by 100γ. The actual intensity in γ is obtained by adding the figure attached to a station to 17700.

Camperdown towards it. The isodynamic lines are also widely separated from each other in the region of the Griqualand East and South Natal valley.

The Natal-Zululand ridge coincides throughout its course with a line of no horizontal intensity disturbance except in the neighbourhood of Estcourt and Modder Spruit; its position near these two stations has been fixed by the declination anomalies and by the directions of the disturbing horizontal intensities. The peak between Ginginhlovu and Umhlatuzi coincides with the intersection of the lines of no declination and no horizontal intensity disturbances; the peak near Boston is in a neighbourhood where similar lines approach each other.

The valley which intersects the main ridge near Boston coincides with a line of no declination anomaly. The continuation of this line between Richmond on one side and Grange and Camperdown on the other coincides with a line of no horizontal intensity anomaly which intersects one of no declination anomaly in the neighbourhood of Durban. The place so marked is further distinguished by the fact that the disturbing horizontal forces at the three stations in its neighbourhood—the Bluff, Krantz Kloof, and Illovo River—are directed away from it, indicating the existence there of a mass of negatively magnetised matter. Compare in this connection Maps 4 and 5.

Griqualand East and South Natal ridge. This region of attraction begins near Mount Frere and runs in a north-easterly by east direction, keeping to the south of Kokstad, Ibisi Bridge and Umzinto, there it turns, continuing north-east by north to the east of Illovo River and probably on towards Durban.

From Map 5 it will be seen that the horizontal disturbances are in all cases directed towards the ridge.

The values of ΔH—or $H_c - H_o$ from table 33, column 8—are in accordance with the position of the ridge, viz., greater than would be expected at Ibisi Bridge, Umzinto, Kokstad and other stations on the north, but smaller on the south. Two stations, Port Shepstone and Umtwalumi, are exceptions, being on the south and having a greater intensity than would be expected. The anomaly is very small. A probable explanation of this and other anomalies at Port Shepstone and Umtwalumi is given in the account of the Griqualand East valley, p. 91.

The dip is less than the normal at Mount Frere, Ibisi Bridge, Umzinto and Illovo River, all on the north of this region of attraction, and greater at Umtwalumi and Port Shepstone on its south border. Stations No. 403, Cotswold Hotel and Mount Ayliff, in this region have small anomalous dip disturbances; their position however with respect to the ridge is correct as judged by their declination disturbances and by the direction of the horizontal disturbances there.

The directions of the horizontal disturbances near this region are all towards the ridge.

The lines of equal horizontal intensity—see Map 6—show very markedly the position of this ridge.

The Griqualand East valley. This region of repulsion is connected with the Eastern Province valley. It passes to the south of Toise River between Elliot and Bashee and keeps south of Ugie, Mount Frere—where it is intersected by the ridge spoken of in the previous paragraph, see Map 5—Cotswold Hotel and Port Shepstone. This valley is intersected by a second ridge passing northwards between Ugie and Tsolo and afterwards turning first west then in a south-westerly direction to join the Central Karroo system of ridges near a peak west of Witmoss and Cookhouse. This same ridge passes south between Umtata and Bashee, intersecting later the Port St Johns-Transkei ridge.

The disturbances of the horizontal intensities ΔH—given on Map 4—show that this element has a lower value at stations lying on the northern side of this valley region and higher than the normal at stations on the south. Port Shepstone and Umtwalumi are exceptions, the irregularity however being small again. It is probable that the exceptional position of these two stations when considered in relation to the distribution of ridges and valleys so far considered can be accounted for by a subsidiary ridge branching off from the East Griqualand and Natal ridge at a point on it between Payne's Farm and Ibisi. Such a ridge would mean the existence of a peak there, which is to be expected, as that point is the place of intersection of a line of no declination anomaly with a line of no horizontal intensity anomaly—see Map 4; the position of the peak is roughly indicated by the direction of the horizontal disturbances at Ibisi Bridge and at Payne's Farm. This ridge would continue—through a district where the vertical force is above the normal—in a south-westerly direction, having Port Shepstone on the east. Such a region of attraction is supported by the declination anomaly at Port Shepstone. The indications, though all pointing the same way, are so slight in all these cases that it has not been thought advisable to show this line on Map 5.

The values of the dip in the neighbourhood of this valley are below the normal at Tsolo and two other stations on the south, and are above the normal at Elliot, Ugie and Port Shepstone on the north. There is a small irregularity at one station—Mount Ayliff—on the north, where the dip, instead of being higher than the normal, is 3′ lower. There is another irregular station in this neighbourhood—Cotswold Hotel—also on the north, where the dip is lower by 2′. The ΔH and the direction of the horizontal disturbances for these stations are however consistent with their position—as given in Map 5—with respect to the valley.

The position of the intersecting ridge is fixed in this part of its course by Indwe, Elliot, Ugie, Tsolo, Bashee and Umtata. As it winds about it will be seen from Map 5 that Indwe, Ugie, Bashee, are to the west, and Elliot, Tsolo and Umtata to the east of it in places where the direction of ridge is roughly north and south, and where therefore the declinations observed on the west should be greater and those on the east less than the normal value. The declinations at the above stations are in accordance with this.

Port St Johns-Transkei ridge. This begins a little to the north of Port St Johns, runs west for a short distance, then turns south, passing between Port St Johns and Umhlengana Pass. It then again turns west, continuing so to a point between Umtata and Bashee; from there it goes south-west, ending near Komgha, where it is intersected by the Eastern Province valley for the second time—the first point of intersection with this valley being between Idutywa and Outspan.

This ridge is very clearly marked in the neighbourhood of Port St Johns. The dip for example at the latter place is 13′ in excess of the normal whereas at station No. 401 about 20 miles to the north it is in defect by 50′; again the horizontal intensity is lower than the normal by 145γ at Port St Johns and higher by 693γ at the other station. The values of the various magnetic elements at Idutywa, Outspan and Butterworth are in agreement with their position with respect to that branch of the Eastern Province valley which intersects the St Johns ridge between Idutywa and Outspan. At Idutywa the horizontal intensity is less than the normal, and at Butterworth and Outspan it is greater; in the case of the dip at these three stations the reverse holds, it being greater than the normal at Idutywa and less at the other two stations. The value of the declination at Idutywa is governed by the position it holds with respect to the St Johns ridge; it lies to the west of this, and its declination is greater than the normal.

The directions of the horizontal disturbances are towards the ridge and away from the valley at the various stations referred to above. See Maps 5 and 6.

Eastern Province valley. This valley, as we have already said, is connected with the Griqualand East valley. It has one branch passing south of Toise River in a south-westerly direction, and south of Idutywa—see Map 7. Another branch—not shown in the figure—passes between Komgha and a station about six miles to the east, running in a north-westerly direction to a point east of Toise River, where it joins the branch referred to at the beginning of the paragraph. From this point the first branch runs south-west to a little north of Port Alfred, there it turns south, passing between Hankey on the east and Humansdorp on the west.

The directions of the horizontal disturbing forces are away from the valley throughout its course.

The horizontal intensity results also agree with the position of this valley. Amabele Junction, King William's Town, Komgha, East London and Port Alfred stations, which are to the south of the valley, have all intensities greater than the normal value; Toise River, Grahamstown, Alicedale, Coerney, Glenconnor, Wolvefontein, Cathcart stations to the north, have values less than it.

The values of the dip are greater than the normal at Glenconnor, Coerney, Wolvefontein, Grahamstown, Toise River and Cathcart stations, which are to the north, and are less at Port Alfred, East London, King William's Town, Amabele Junction and Komgha, points on the south.

The vertical force results show that the valleys pass through districts where it is below the normal.

The only other point in this connection worthy of notice is the possible existence of a ridge extending almost east and west to the south of King William's Town and East London, turning afterwards to the north to join the Port St Johns-Transkei ridge. See Map 7.

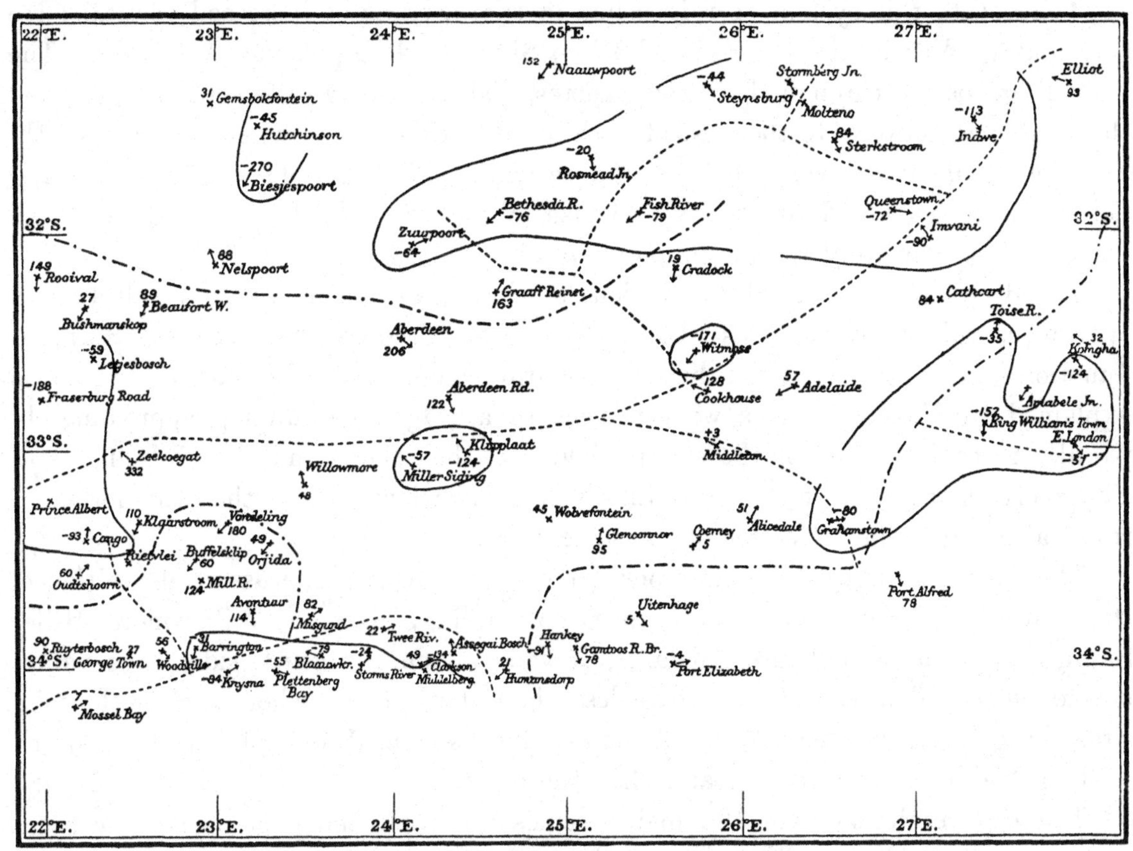

Map 7. CENTRAL, SOUTHERN, AND EASTERN CAPE COLONY.

Showing the vertical disturbances in terms of γ, the ridges, the valleys, the lines of no vertical disturbance, and the directions of the horizontal disturbances. Lines of no vertical disturbance ——— Ridges - - - - - - Valleys ·—·—·—· A negative sign denotes that the actual vertical intensity is greater than the calculated value.

Great Karroo and connected ridges. The Great Karroo ridge intersects the Eastern Province valley between Grahamstown and Port Alfred. It passes to the east of the former town, then bends round to the north-west, passing to the north of Middleton; it then continues almost due west, having Aberdeen Road, Prince Albert Road, and Schietfontein on the north, and Klipplaat, Miller Siding, Willowmore, Zeekoegat, Grootfontein, Laingsburg and Kruispad on the south. Between Klipplaat and Middleton another ridge passes to the north-east between Witmoss and Cookhouse to the north of Adelaide, continuing in the same direction so as to pass between Imvani and Queenstown and between Indwe and Elliot. It then turns east, passing to the south a little further on, and thereafter joining the ridge line passing between Ugie and Tsolo referred to on page 91.

To the west of Witmoss and Cookhouse and about midway between them another ridge runs northward and later westward, having Witmoss, Cradock, Fish River, Rosmead Junction, and Bethesda Road on one side and Graaff Reinet and Zuurpoort on the other. Map 7.

The key to this system of ridges lies in the districts of increased vertical force surrounding Witmoss in one case and Miller Siding and Klipplaat in the other. The main ridge passes through these two centres. There is very satisfactory evidence of the existence of a peak to the west of Cookhouse, and Witmoss, where we have firstly a region of increased vertical force and secondly the directions of the disturbing horizontal forces at Middleton, Cookhouse, Witmoss, Adelaide, and Cradock all pointing approximately to the supposed peak.

The declination results at the various stations are such as would result from a system of ridges arranged as above. Witmoss, Cradock, Fish River, Rosmead Junction and Bethesda Road have declinations less than the normal, which is explained by their situations with respect to a ridge line running approximately parallel to the direction of the earth's field; on the other hand Graaff Reinet and Zuurpoort on the west of the same line have declinations greater than the normal, a result also due to the attraction of the same ridge.

The horizontal force determinations agree very well with the assumed position of the main ridge. Kruispad, Laingsburg, Grootfontein, Zeekoegat, Willowmore, Miller Siding, Klipplaat, Cookhouse, Adelaide, Imvani, all on the south side, have values of the horizontal intensity which are less than the normal, whereas Schietfontein, Fraserburg Road, Aberdeen Road, Witmoss, Queenstown, Indwe, places to the north of the ridge, have intensities greater than the normal.

The dip results are—in the main—consistent with the existence of such an arrangement of ridges. The values ought to be greater than the normal at places on the south and smaller on the north. At Prince Albert Road the value of the dip is such as to suggest that the ridge passes to the north of it rather than to the south. The same position is suggested by the ΔH at this station. Other stations at which the dip anomalies are irregular are Cookhouse and Adelaide in the south, Witmoss, Queenstown and Indwe in the north.

The directions of the horizontal disturbing forces are towards the ridges at all the stations in this system.

Beaufort West-Cradock valley and connected disturbances. The Beaufort West-Cradock valley begins between Rivierplaats and Sutherland on the west, and runs in a westerly direction between Fraserburg and Steenkampspoort, Nelspoort and Beaufort West, Zuurpoort and Aberdeen, Cradock and Fish River, the first station of each pair being on the north side of the valley.

There is finally closely connected with this valley a ridge which begins between Graaff Reinet and Fish River, having Rosmead Junction, Steynsburg, Stormberg Junction, and Sterkstroom on the north, with Fish River and Queenstown on the south. This ridge intersects the Graaff Reinet-Zuurpoort ridge referred to on the previous page and the main Great Karroo ridge near Queenstown. The existence of peaks near Graaff Reinet and Queenstown is suggested by the intersection of the lines of no disturbance of horizontal intensity and of declination near both places.

The evidence for the existence of the Beaufort West-Cradock valley can be put under the following heads.

Firstly the assumed region of repulsion runs through a part of the country where the values of the vertical intensity are below the normal; and it may be noted in this connection that there is at Steenkampspoort and Bushmanskop evidence of the existence of a col.

Secondly the directions of the disturbing horizontal forces point away from the valley at stations on both sides of it throughout its course.

Thirdly the values of the horizontal intensity at Sutherland, Steenkampspoort, Rooival, Bushmanskop, Beaufort West, Aberdeen, and Cradock, all on the south side of the valley, are greater than the normal; the values of the same element at Fraserburg, Nelspoort, Graaff Reinet, Fish River on the north side are less. The values of ΔH at some of these stations are small numerically; at Fish River, for example, it is $+5\gamma$, and at Graaff Reinet $+18\gamma$; at Cradock it is -13γ; it is impossible to lay stress on differences of this size except in so far as they are in accordance with other indications in the same direction. A station which does not have a value of ΔH, such as might be expected from its situation with respect to this valley, is Rivierplaats, where the value is greater instead of less than the normal, and by an amount considerably more than the probable error.

Fourthly the values of the dip at Fish River and at Nelspoort, to the north of the valley, are greater, and at Steenkampspoort, Rooival, Bushmanskop, Beaufort West, Aberdeen and Cradock, on the south of the valley, are less than the normal.

The anomalies of the declination do not give much information in the case of a valley—such as the Beaufort West-Cradock—where the direction is approximately perpendicular to the magnetic meridian. In the present case additional information is given with respect to several of the stations, viz. Rooival, Bushmanskop and Beaufort West; all these have declinations less than the normal, indicating—when the other deviations are also considered—that they lie to the east of a ridge. This receives support from the fact that there is—as already mentioned—evidence of the existence of a col in the neighbourhood of Rooival.

When we consider the second valley running approximately parallel to the magnetic meridian and to the east of Naauwpoort, Rosmead Junction, Fish River, Cradock and Adelaide, we find the declinations at these stations less than the normal, and that at Steynsburg, a place on the other side of the valley, greater.

The effect of the Eastern Province, the Great Karroo, and the Beaufort West-

Cradock and connected disturbances on the isomagnetics for horizontal intensity is easily seen in Map 8.

The chief ridges and valleys mentioned above are of course related to the lines of no declination and no horizontal intensity anomalies. The Griqualand East and South Natal valley coincides in part of its course with a line of no declination anomaly, and in another part near Durban with one of normal values of the horizontal intensity.

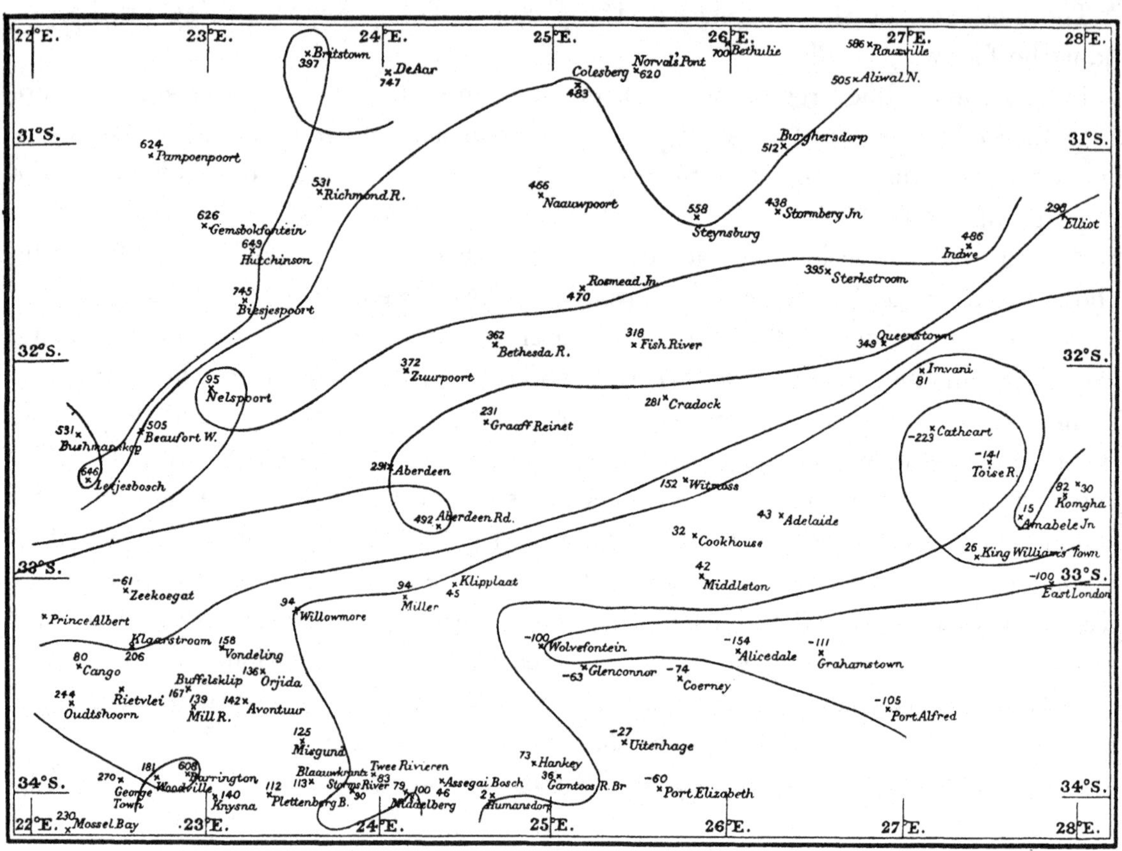

Map 8. CENTRAL, EASTERN, AND SOUTHERN CAPE COLONY.

Lines of equal horizontal intensity differing by 100γ. The actual intensity in terms of γ is obtained by adding to 17700 the figure attached to a station.

The Griqualand East and South Natal ridge has the same position as another line of normal values of the horizontal intensity; the subsidiary ridge referred to on page 91 has the evidence for its existence further strengthened by the fact that its position is identical with a line of no horizontal intensity anomaly.

The Griqualand East valley coincides with a line of no horizontal intensity anomaly. It may be noted that the continuation of this valley towards the Eastern Province valley does not correspond with a line of no anomaly; on the whole the existence of this part is more hypothetical than usual; lack of stations at which observations have been made gives a more than usual freedom to the imagination.

There is further in this neighbourhood an indication of another valley between Elliot and Ugie, the declination at the station on the road to Ugie being in excess, and that at Elliot in defect, the directions of the disturbing horizontal forces fit in with a valley in this position; it has not been indicated on Maps 5 or 7, because there was no means of determining by any of the usual methods its exact course.

The intersecting ridge referred to on page 91 coincides in part of its course with a line of no declination anomaly and in the other part with one of no horizontal intensity anomaly.

The Port St Johns and Transkei ridge coincides with a line of no horizontal intensity anomaly in the neighbourhood of Port St Johns; its further course towards the south does not however coincide with any normal line; there are no stations to the east of the line; its course in this part is fixed by the direction of the total disturbing horizontal intensity.

The various other ridges and valleys are coincident with lines either of no declination or of no horizontal intensity anomaly. In particular the Great Karroo ridge is marked by two lines, one of no horizontal intensity and the other of no declination anomaly; the two run almost side by side from near Fraserburg Road to Grahamstown.

The Little Karroo valley. This valley begins at a col whose position is given by the intersection of two lines, one joining Misgund and Plettenberg Bay, the other approximately joining Avontuur and Blaauwkrantz. In the first part of its course it runs almost due north, then turns west, passing to the north of Vondeling and Orjida; near Klaarstroom it turns south, having Rietvlei and Klaarstroom on the west; it turns west again to the south of Oudtshoorn, and continues in that direction to a point on the north of Robertson. See Maps 7 and 9. It coincides with a line of no declination anomaly throughout its whole course, and for a part of it with one of no horizontal intensity anomaly.

At Misgund, which lies to the east of this region of repulsion, the declination is greater than the normal. The declination is less than the normal at Orjida and Vondeling, places to the west of the valley. Klaarstroom and Rietvlei have declinations agreeing with the normal as calculated; the valley passes south in this part through these two stations. The values of the declination at Oudtshoorn and at Cango are determined not by their positions relative to the Little Karroo valley, but by the fact that they lie to the west of a cross ridge starting from the Great Karroo ridge between Prince Albert and Zeekoegat and running roughly parallel to the magnetic meridian to intersect the South Coast ridge—to be described later—between Barrington and Woodville. At Woodville ΔD has a small positive value —1′—while at Barrington its value is very large and positive.

When we consider the horizontal intensity it is found that at Calitzdorp and Ladismith, places to the north of the valley, the values of this element are less than the normal, whereas at Van Wyk's Farm and Ashton on the south the horizontal intensity is greater than the normal. The value of the horizontal intensity at

Robertson is not in accord with its position with respect to this valley; the ΔH is however very small, viz. 9γ. The value at Oudtshoorn is also irregular. The system of ridges and valleys is very complicated in this neighbourhood. There is a line of no declination disturbance passing about midway between Robertson and Worcester, turning afterwards east so as to pass between Touws River and Robertson; this same line passes on the south of Robertson towards Swellendam. Again on the east of Robertson and between it and Ashton there is a line of no disturbance of horizontal intensity. If we look upon Robertson as more influenced by the ridge Riversdale-Worcester, and particularly by that branch of it which passes between Robertson and Touws River, then the ΔH and the $\Delta \theta$ are both in accord with such a distribution. Further the declination at Robertson is less than the normal, a fact which agrees with the further course of the ridge.

The dip results on the whole give no additional evidence for the position of this valley. At Ashton and Van Wyk's Farm on the south its value is less than the normal as we should expect, and so it is at Ladismith and Calitzdorp. The anomalies are very small at all these places.

The disturbing horizontal forces point away from the valley throughout its course and towards the ridge. The latter passes through a region of increased and the former through one of decreased vertical intensity. See Maps 7 and 9.

Ridges and valleys in the south-west of Cape Colony. A main ridge begins between Still Bay and Tygerfontein; it continues almost due west to a point south of Robertson, following throughout a line of no declination disturbance. South of Drew a subsidiary ridge passes between Kathoek and Malagas and then turns between the latter place and Port Beaufort, joining the main ridge again between Heidelberg and Riversdale. The main ridge turns north and passes to the west of Robertson, joining the Little Karroo valley; from there it follows a line of no disturbance of horizontal intensity, and takes the form of the letter S passing north of Hex River, east of Ceres Road, south of Huguenot and east of Malmesbury. In the western part of the S a valley begins which passes in a northerly direction to the west of Hermon and Tulbagh Road, following a line of no declination disturbance. The ridge referred to above sends a branch westwards to the north of Stellenbosch, Cape Town, and Signal Hill; there it turns south and continues in that direction, having Simonstown and Miller's Point on the east. A little to the east of Cape Town a valley begins which has Stellenbosch, Sir Lowry's Pass and Howhoek on one side, and Hermanus and Stanford on the other. There are indications of other ridges and valleys along the south coast, but most of them depend on the indications of one station only, and have not been drawn on the map. See Map 9.

When we consider the evidence on which the above system of ridges and valleys rests, it is found that the declination at Swellendam, Drew and Robertson is less than the normal, due to the fact that they all lie to the east of the main ridge, whereas Villiersdorp and Worcester, which lie to the west, have declinations greater in value. Other stations whose declinations depend on their position with respect to a ridge are Ceres Road and Tulbagh Road on the west of it, where the declination is greater than the normal; again at Huguenot, which lies to the east, the declination is below normal, and the same is the case at Simonstown and Miller's Point. In the Cape Peninsula there are three stations which are peculiarly situated, viz. Cape Town, Signal Hill and Strandfontein. There is a line of no horizontal intensity which forms a valley; this intersects a line of no declination disturbance a little to

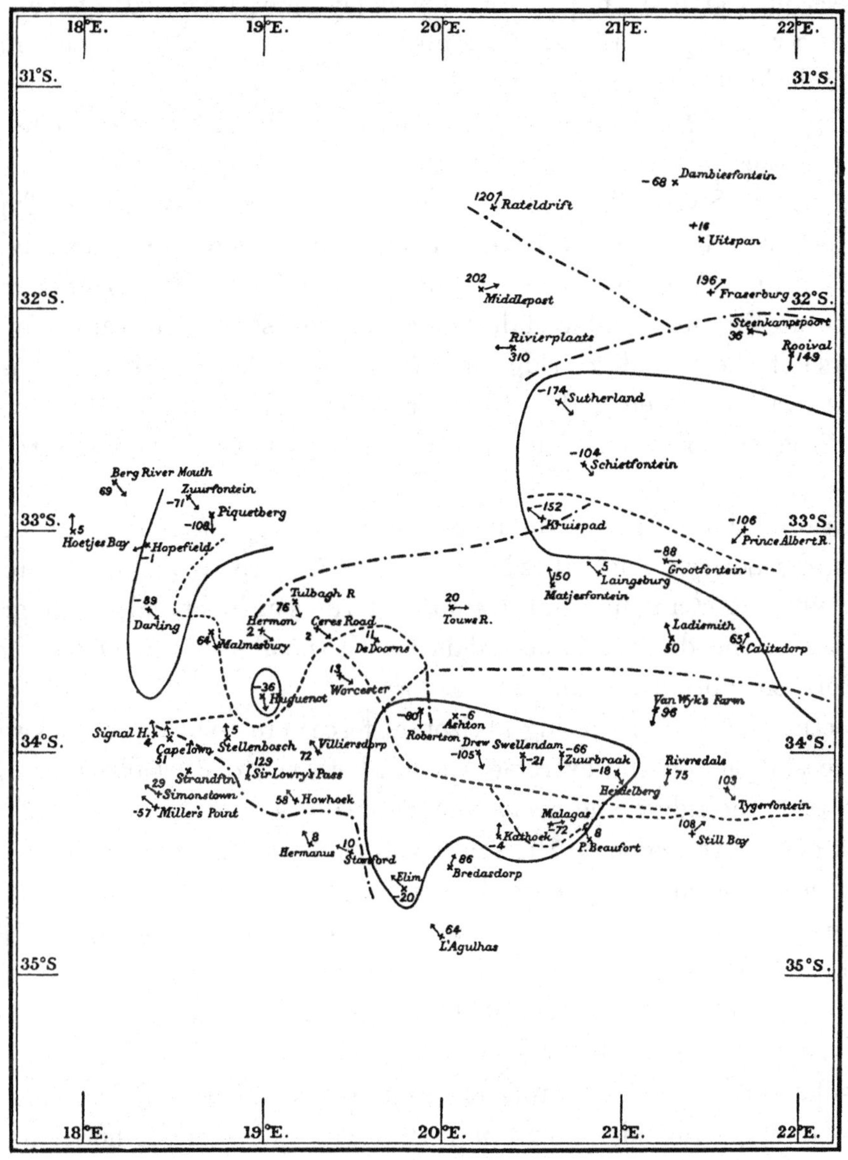

Map 9. South-Western Cape Colony.

Map showing the vertical disturbances in terms of γ, the ridges - - - -, the valleys ·—·—·, the lines of no vertical disturbance ———, and the directions of the horizontal disturbances ↑. A negative sign denotes that the actual vertical intensity is greater than the calculated value.

the north of the first two of these stations. The directions of the horizontal disturbing forces point towards the region of intersection. The declination results at all three stations are greater than the normal, which is what we should expect were they situated to the east of a valley. So far as this evidence goes it would seem that the western side of the valley had the greater effect. On the other hand the horizontal intensities and the dips are such as to be expected from stations to the south of a valley. The most likely distribution of disturbances seems to be that given in Map 9.

The effect of the eastern part of the valley line described above is seen at Stellenbosch, Sir Lowry's Pass and Howhoek, places to the east of this region, with values of the declination greater than the normal.

From the values of the horizontal intensity at the different stations we obtain additional evidence for the various disturbances described above. For example at Ashton, Drew, Swellendam, Zuurbraak, Heidelberg, Riversdale, Tygerfontein, Ruyterbosch, Ceres Road, Touws River, Hermon and Huguenot, places lying on the north of the main ridge, the value of the horizontal intensity is greater than the normal, while at De Doorns and Malmesbury, on the south, the values are less.

The course of the subsidiary ridge in the neighbourhood of Kathoek is also well marked. At Kathoek itself and at Port Beaufort—which are south of this ridge—the value of the horizontal intensity is less than the normal, whereas at Malagas on the north its value is greater.

In the case of those stations influenced by the valley the horizontal intensities at Stellenbosch, Sir Lowry's Pass, Howhoek and Bredasdorp to the north are less than the normal, while at Hermanus and Stanford on the south the values are greater.

The values of the dip are in accordance with the above view of the disturbances except at Port Beaufort, Drew and Bredasdorp.

The directions of the disturbing horizontal forces are—with the exception of the two or three stations on the extreme south where the disturbances are not clearly defined—towards the ridges and away from the valleys.

There is one other region on the south still to describe. It is a ridge, apparently a continuation of the main one described on page 98. It runs south of Woodville and Barrington, where a cross ridge goes to the north (see page 97); it goes north of Knysna and Plettenberg Bay, passes between Misgund on the north and Blaauwkrantz on the south, continues to the north of Twee Rivieren and Assegai Bosch, and turns south to the west of Humansdorp.

The results for the western and southern parts of Cape Colony are shown in another way in Maps 10 and 11. In Map 10(a) the values of the horizontal intensity along the parallels 34° S. and 34° 30′ S. are plotted vertically, and longitude is measured horizontally. A glance at this figure shows—after correcting the values of the horizontal intensity for those places not exactly on the particular parallel concerned to the value for the correct parallel—the decrease in the value of this element with increasing longitude. Points of interest obtained from the comparison of the two lines are the value at Hermon, where the horizontal intensity is in excess, and that at Sir Lowry's Pass, where it is in defect of the normal value, facts which are in

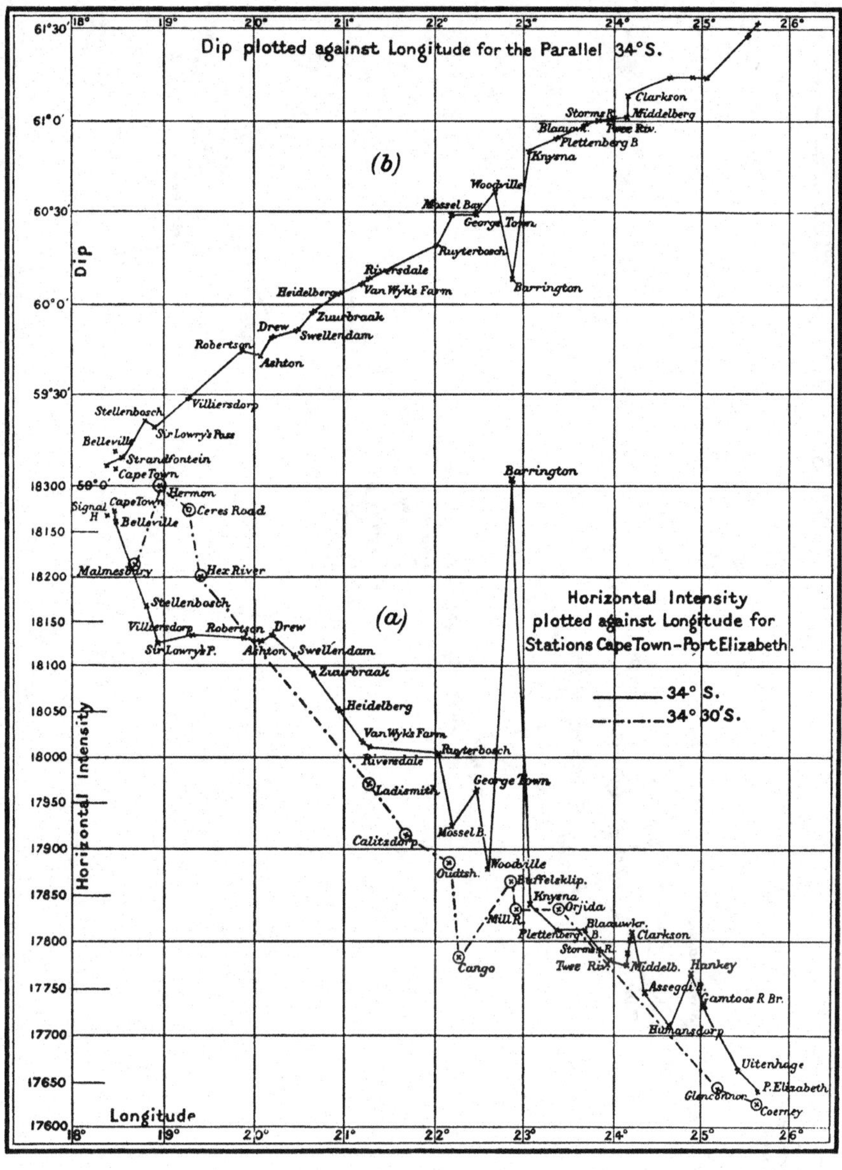

Map 10.

Horizontal intensity in γ plotted against longitude for stations Cape Town—Port Elizabeth.
——— 34° S. ·—·—· 34° 30′ S.

accordance with the conclusions obtained from the consideration of the distribution of ridges and valleys. The next point is seen in connection with the values at George Town and Cango, two stations with approximately the same longitude, the first having

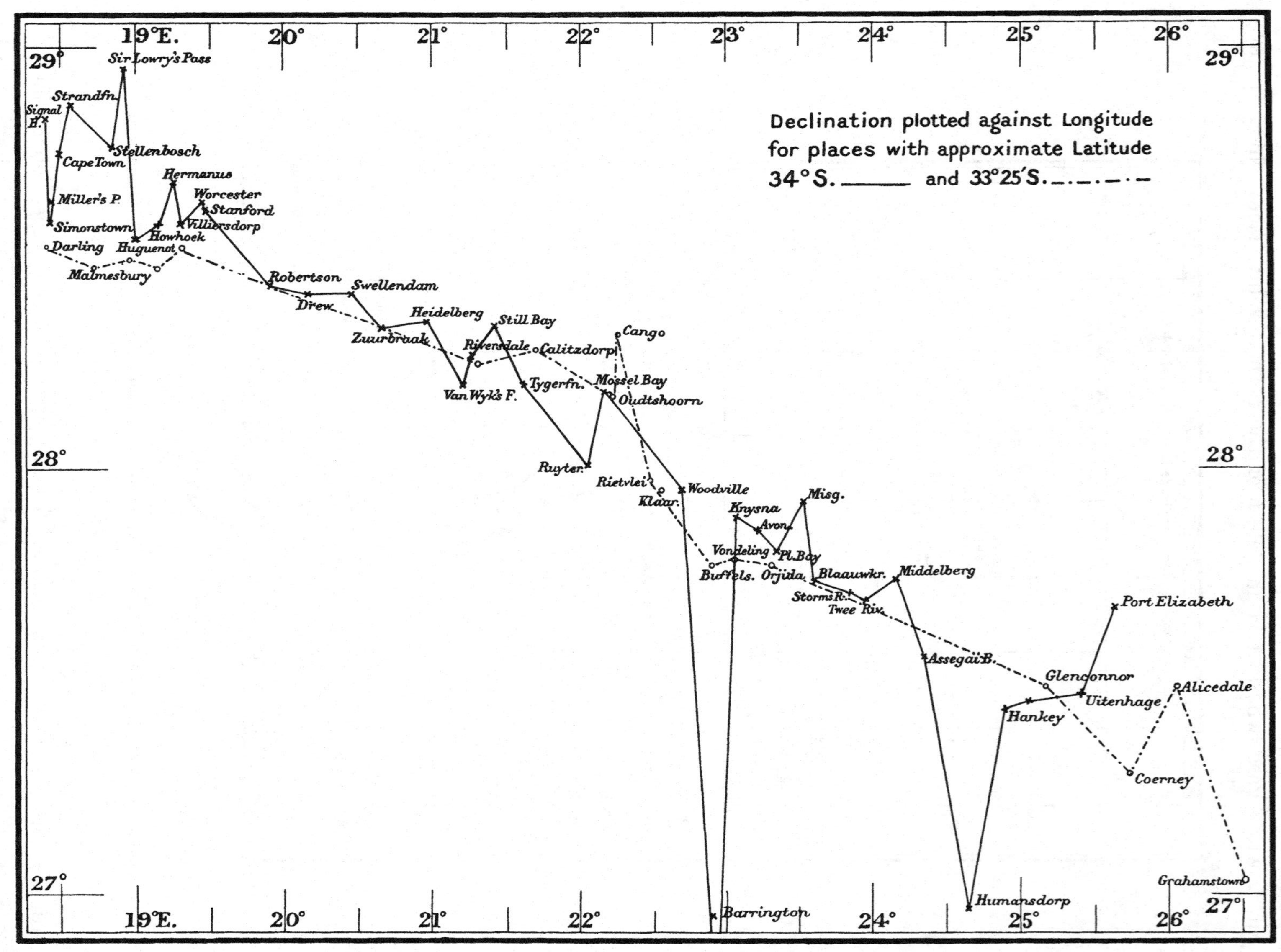

Map 11. Declination plotted against longitude for places with approximate latitude 34° S. — and 33° 25′ S. —·—

a value of the horizontal intensity greater, the second less than the normal for the parallel concerned. Another point of some interest is the value of the horizontal intensity at Barrington; it is greatly in excess of the normal. There is nothing in the geological formation of the country to indicate the source of an abnormal value of the magnetic elements here; there is no doubt, however, that the station is abnormal. The three observed elements have values differing greatly from the normal.

In Map 10(b) the dip—corrected for those stations not on the parallel—is given for 34° S., and its value is entered vertically with longitude measured horizontally. The dip at Barrington is seen to be considerably below the normal value.

In Map 11 the declination is plotted vertically and the longitude horizontally for stations with latitude 34° S. and those for latitude 33° 25′ S. Stations not on these parallels but appearing on the curve have had their declinations corrected by applying the variation with latitude as given on page 53. An examination of the value of the declination along 34° S. shows that there is a centre of attraction a little to the west of Huguenot; the value of the declination to the west of this is higher than the normal, viz. at Sir Lowry's Pass, and to the east at Huguenot it is less. From Map 9 it will be seen that this corresponds to the peak, which is indicated a little to the south and west of Huguenot by the intersection of two ridges there. There is another centre of attraction probably a little to the east of Tygerfontein; the value of the declination at Still Bay on the west is greater, and at Ruyterbosch on the east less than the normal. A glance at Map 9 shows that this centre probably coincides with the peak, due to the intersection of the two ridges in that neighbourhood. This peak appears in Map 9 to the west of the position given to it in Map 11.

The further consideration of the results along this parallel shows anomalous values of the declination at Barrington and at Humansdorp.

The values of this element along the parallel 33° 25′ S. show that from Darling on the west to Oudtshoorn on the east the declination has an approximately normal value throughout, and that its rate of decrease with longitude is slightly less than along the line 34° S. Between Oudtshoorn and Cango however the declination suddenly increases—instead of decreasing as was to be expected—and then falls below the normal at Buffelsklip. From there it remains constant with increasing longitude for some distance, and then decreases as one would expect it to do.

Ridges and valleys in the north-west of Cape Colony. The valley already referred to on page 94 under the name of the Beaufort West-Cradock valley continues unbroken to the west of a line joining Fraserburg and Steenkampspoort; at this point however a second valley breaks off towards the north, passes to the west of Fraserburg, Uitspan and Dambiesfontein, and later turns north-west between Williston and Rietpoort. The original Beaufort West-Cradock valley continues in a westerly direction after Steenkampspoort before it bifurcates; one part—referred to on page 94—passes south-west between Rivierplaats and Sutherland, the other goes north-west between Middlepost and Rateldrift. (See Map 12.)

The values of the horizontal intensity are such as were to be expected at Fraserburg and Steenkampspoort; the fact that at Rivierplaats the horizontal intensity disturbance has the particular value spoken of on page 95, is probably due to the existence of this second valley north of Middlepost; the values at Middlepost and at Rateldrift are in accordance with their position with regard to the valley. Again

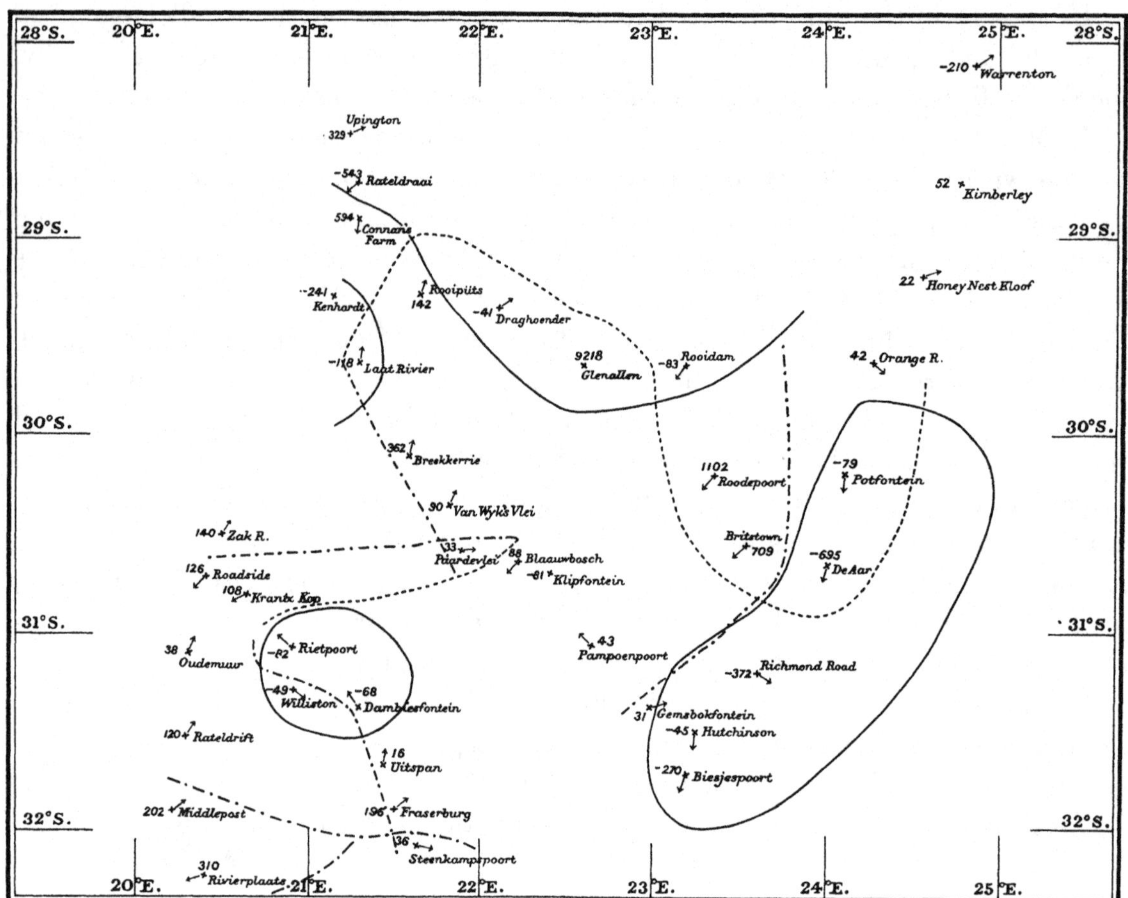

Map 12. North-Western Cape Colony.

Showing the vertical disturbances in terms of γ, the ridges, the valleys, the lines of no vertical disturbance and the directions of the total horizontal disturbances. Lines of no vertical disturbance ——— Ridges - - - - Valleys ·—·—· A negative sign denotes that the actual vertical intensity is greater than the calculated value.

the existence of a valley running to the west of Uitspan and Fraserburg should produce at these stations a value of the declination greater than the normal; a glance at Map 13 will show that this is the case. The further course assigned to this valley between Rietpoort and Williston is shown by the values of the horizontal intensities there.

Another valley in this part of Cape Colony begins between Pampoenpoort and Gemsbokfontein, runs in a north-westerly direction to a point between Britstown and De Aar; its course coincides approximately with a line of no declination disturbance and one of no disturbance of horizontal intensity. (See Map 13.) Throughout this part of its course this valley runs perpendicular to the magnetic meridian and the values of the horizontal intensity at Gemsbokfontein, Hutchinson, Biesjespoort, Richmond Road and De Aar are greater than the normal value, while at Pampoenpoort

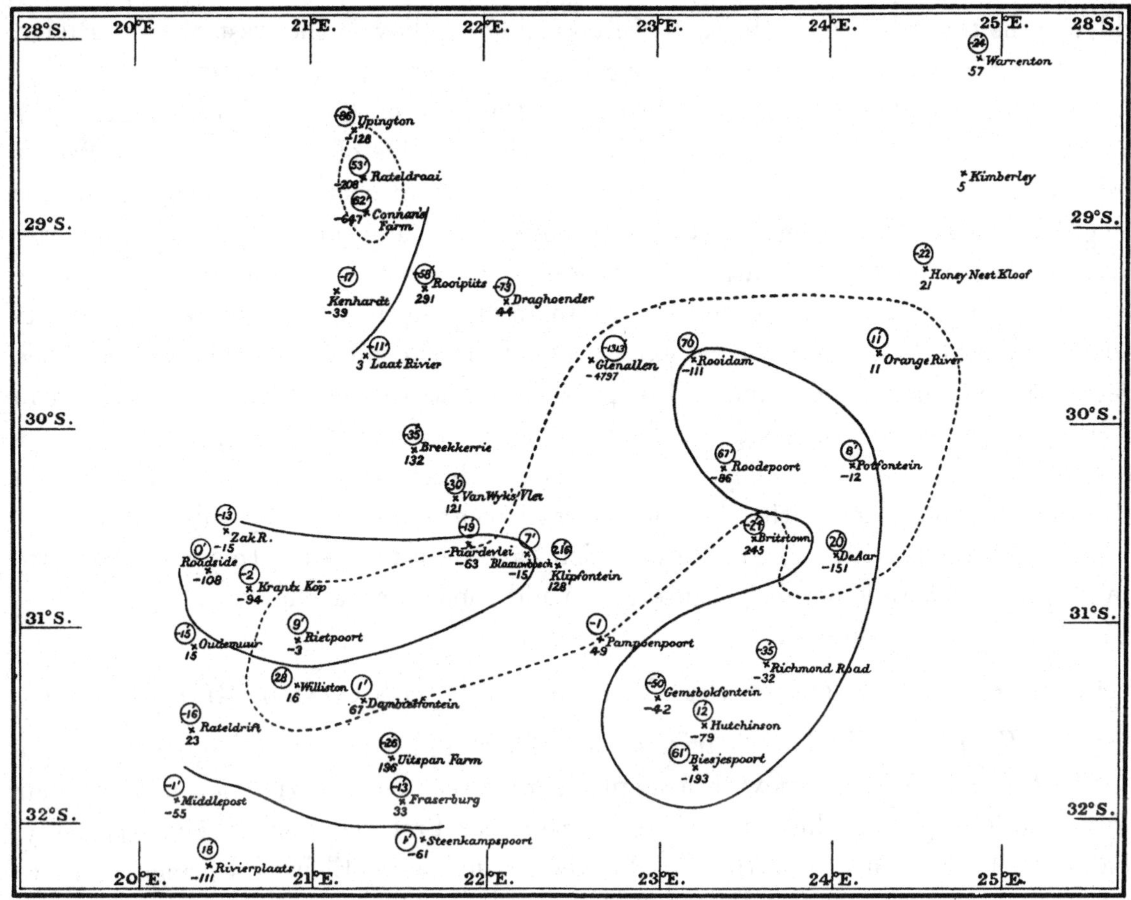

Map 13. North-Western Cape Colony.

Showing horizontal intensity (ΔH) and declination (ΔD) anomalies. The ΔH are in γ, the ΔD in minutes of arc. A negative sign denotes that the observed is greater than the calculated value. The continuous lines are lines of no ΔH. The dotted lines are lines of no ΔD.

and Britstown, places on the other side of the valley, the values are below it. This valley continues—from a point about midway between Britstown and De Aar—almost due north, with Potfontein and Orange River Road on the east and Britstown and Roodepoort on the west. The direction of the disturbing horizontal forces is away from the valley at all these places. This same valley in the first part of its course has on its southern side values of the vertical intensity greater and on its northern side values less than the normal.

A little to the south of Britstown a ridge intersects the valley described in the last paragraph. It passes in one direction to the south of De Aar, and a little to the east of that place turns to the north, continuing east of De Aar, Potfontein, Orange River Road, and Honey Nest Kloof; in the other direction this ridge runs north-west past Roodepoort and Rooidam, it then adopts a more westerly course, leaving Draghoender and Rooipüts on the south. Between the latter place and Rateldraai the ridge turns towards the south, finally intersecting a valley to the west of Laat Rivier.

Along the whole ridge the disturbing horizontal forces point towards it.

At Roodepoort and Rooidam, places to the east of the attracting region, the declination is less than the normal. Britstown has a value of the declination different from what is to be expected from its position with respect to this ridge; in this neighbourhood the magnetic disturbances are too complicated to be satisfactorily disentangled from the observations at stations so far apart as they are here; of course it is quite easy to draw ridges and valleys to account for the disturbances, the results however would be to too great an extent a matter of conjecture. The values of the declination at Draghoender and at Rooipüts are in accordance with their position in relation to the ridge; the value at Laat Rivier is such as is to be expected from its position to the east of a valley running west of Breekkerrie, Van Wyk's Vlei, and Paardevlei; the values at the three latter stations are influenced by this valley.

The few further ridges and valleys which have been marked in this part are shown on Map 12. There is nothing of interest to add concerning them.

Magnetic disturbances in the Transvaal and the Orange River Colony.

The Kaalfontein-Wonderfontein ridge. This region of attraction starts to the south of Pretoria, runs east with a slight inclination to the north to a point not far from Middelberg. In this part of its course it has Pretoria, Kaalfontein, Aberfeldy, Balmoral, Uitkyk on the north; Randfontein, Langlaagte, Elsburg, Springs, and two other stations on the south. Near Middelberg this ridge turns to the south, bending again to the north between Grobler's Bridge and Wonderfontein and continuing in this direction between Machadodorp and Botha's Berg. The direction of the ridge as described above coincides with a line of no horizontal disturbance. Near the middle point of a triangle whose corners are at Machadodorp, Wonderfontein, and Grobler's Bridge, a second ridge goes off to the north-east, having Schoemanshoek and Spitzkopje on the south side and Sabie River on the north. This ridge coincides with a line of no declination disturbance; the intersection of the lines of no horizontal disturbance and of no declination disturbance marks the position of a peak towards which the disturbing horizontal forces in the neighbourhood point.

The various elements which can be used to indicate the existence of the above

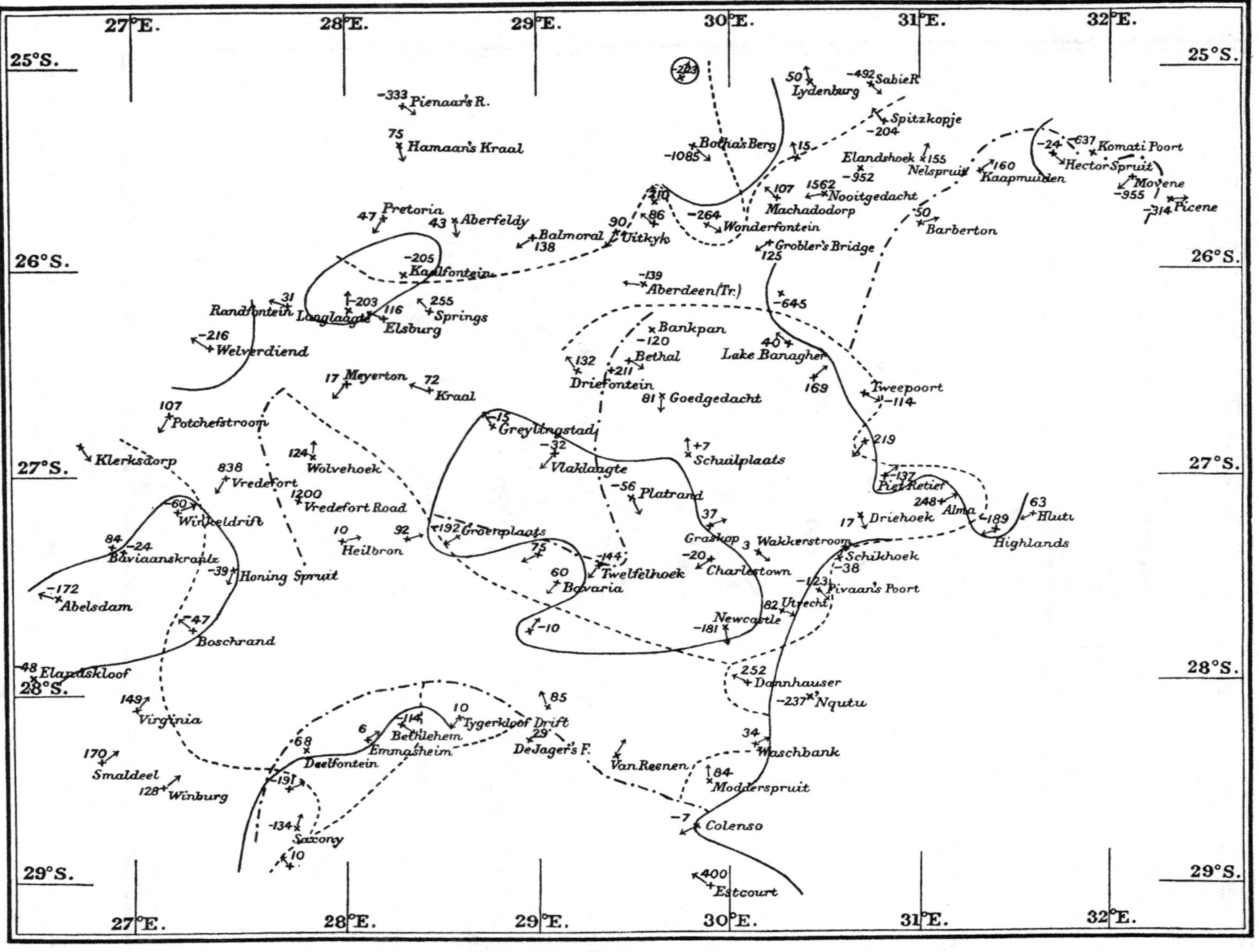

Map 14. TRANSVAAL.

Showing the vertical force anomalies in γ, the ridges - - - -, the valleys ·—·—·, the lines of no vertical disturbance and the directions of the horizontal disturbing forces ↑. A negative sign denotes that the actual vertical intensity is greater than the calculated value.

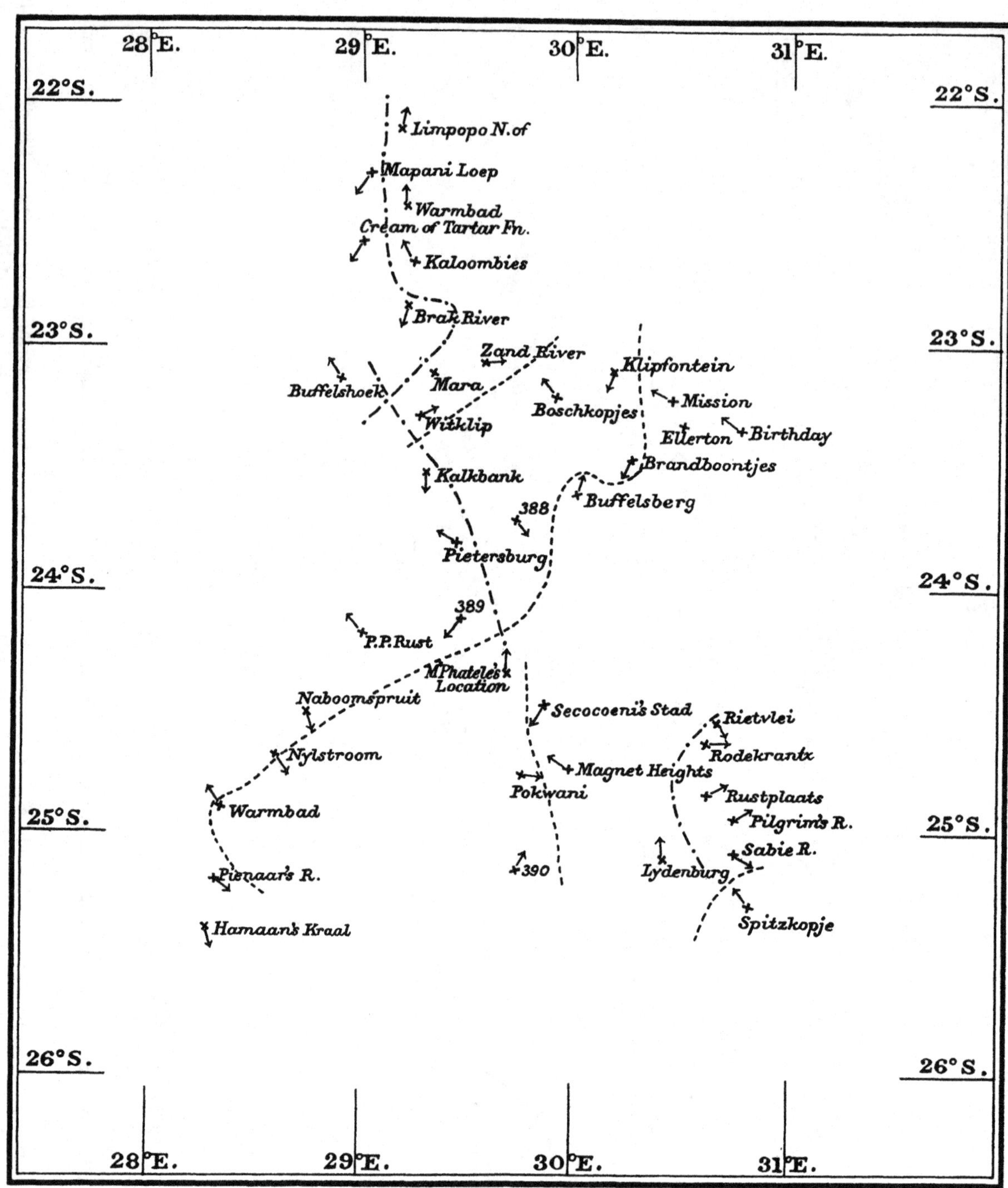

Map 15. North Transvaal.

Showing the ridges - - - -, the valleys ·—·—·, and the directions of the horizontal disturbing forces ↑.

ridge are first the horizontal intensity; its values at Pretoria, Kaalfontein, Aberfeldy, Uitkyk, Wonderfontein stations to the north are in excess of the normal value for this region, whereas at Randfontein, Langlaagte, Elsburg, Springs, Aberdeen (Transvaal), and Grobler's Bridge on the south the values are in defect.

Secondly the dip: the effect on the value of the dip should be just the opposite, viz. in defect to the north and in excess to the south; this is found to be the case at all the stations except Springs, where however the disturbance is small, and Grobler's Bridge.

Thirdly the declination results are in accordance with the position given to the ridge. Those stations to the west have a greater declination than the normal, viz. Botha's Berg, station No. 390, Schoemanshoek, Lydenburg, and Sabie River, whereas on the east at Machadodorp, Nooitgedacht and Spitzkopje the declination is below the normal for the district.

Further the directions of the horizontal disturbing forces are towards the ridge throughout its course. A glance at Map 17—in which the lines of equal horizontal intensity are shown—shows further how these lines are bunched together along this ridge.

That part of the ridge which passes north between Machadodorp and Botha's Berg continues in the same direction, passing between Pokwani (Transvaal) and Magnet Heights to the west of Secocoeni's Stad and to the east of M'Phatele's Location. A little to the north of this—see Map 15—another ridge intersects this one; that part going to the north-east passes north of Buffelsberg, south of Brandboontjes and east of Klipfontein; in the other direction this intersecting ridge passes south of Piet Potgietersrust, Naboomspruit, and Nylstroom, turning a little to the west of Warmbad and passing between that station and Pienaar's River.

The ridges referred to in the last paragraph coincide from a point south of Pokwani to M'Phatele's Location and from there to Pienaar's River with a line of no declination disturbance, and from M'Phatele's Location to Klipfontein with a line of no horizontal force disturbance. Near Pokwani (Transvaal) and Secocoeni's Stad the line of no declination disturbance is intersected twice by a line of no horizontal force disturbance —see Map 16. The directions of the horizontal disturbing forces at Magnet Heights, Pokwani (Transvaal), No. 390, and Secocoeni's Stad point towards the more southerly of the two intersections, marking clearly a peak; the disturbing horizontal forces at Secocoeni's Stad and at M'Phatele's Location point away from the other intersection, indicating that there we have a centre of repulsion. The values of the declination at No. 390, Pokwani (Transvaal) and M'Phatele's Location are greater than the normal, while at Magnet Heights and Secocoeni's Stad the values are smaller, results which are to be expected from the situations of these stations relative to the ridge.

There are in this part of the Transvaal two other well marked regions; one is a valley which begins between Lydenburg and Sabie River runs north and to the west of Pilgrim's Rest, Rustplaats, Rodekrantz, and Rietvlei. At all these stations—with

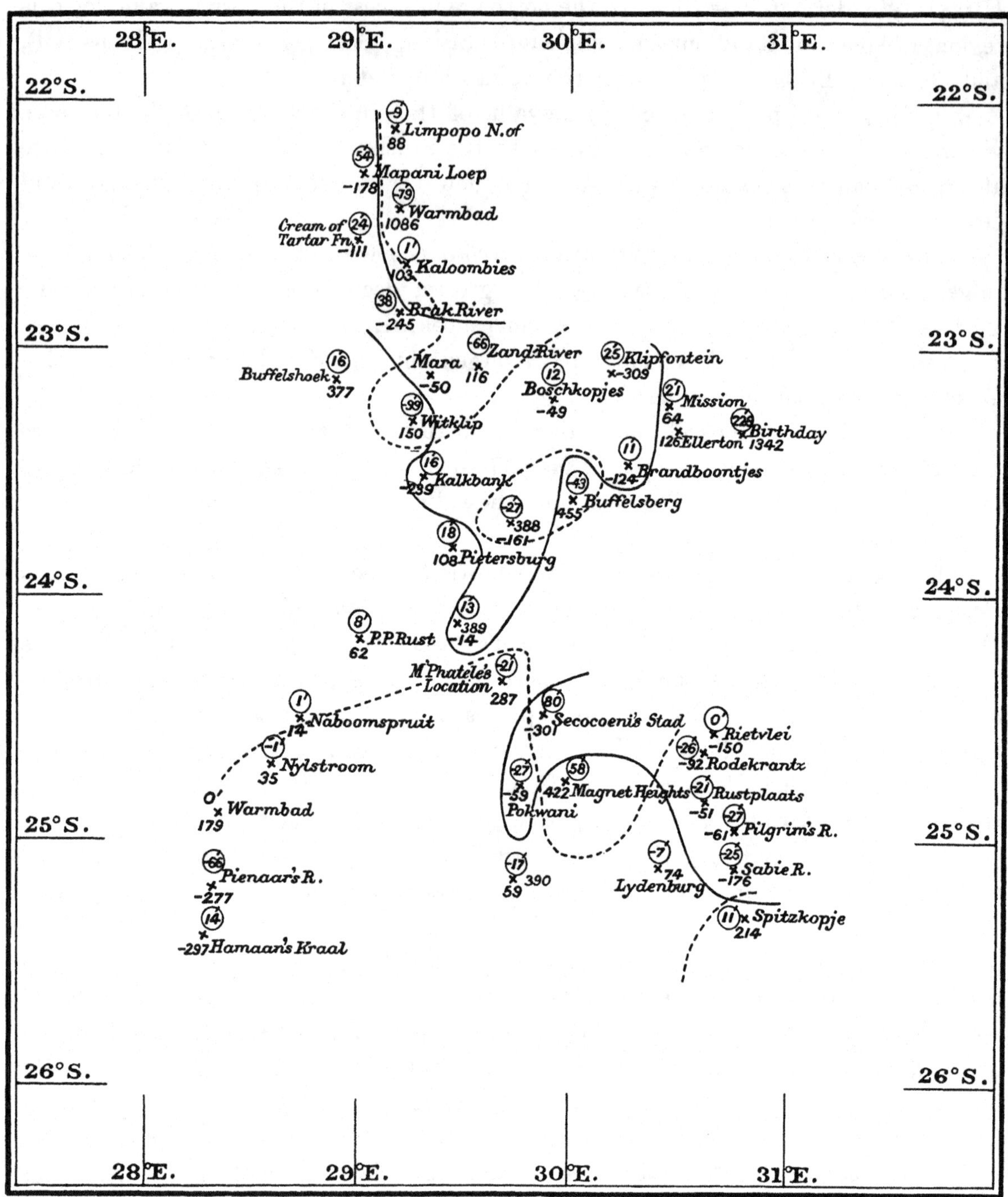

Map 16. NORTH TRANSVAAL.

Showing the anomalies in the horizontal intensity and the declination. The former are expressed in terms of γ, the latter are in minutes of arc. A negative sign denotes that the observed is greater than the calculated value. The continuous lines are lines of no horizontal intensity disturbance. The dotted lines are lines of no disturbance of declination.

the exception of Rietvlei where $\Delta D=0$—the values of the declination are greater than the normal and the directions of the disturbing horizontal forces are away from the valley (see Map 15).

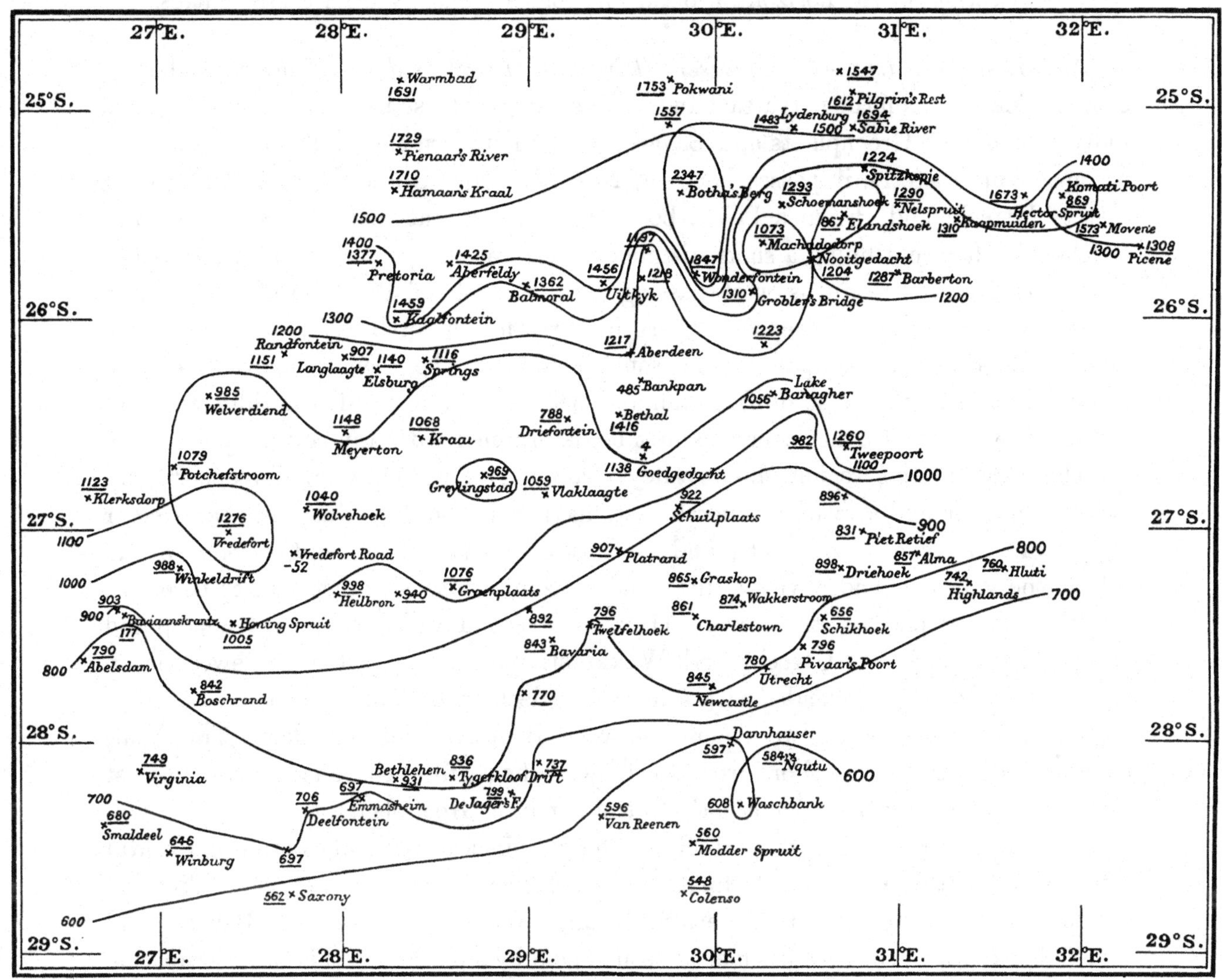

Map 17. TRANSVAAL.

Lines of equal horizontal intensity differing by 100γ. The actual intensity in terms of γ is obtained by adding to 18000 the figure attached to a station.

The second region is also a valley; it begins a little to the east of station No. 389 and continues northwards to the east of Pietersburg, Kalkbank, Brak River, Cream of Tartarfontein, and Mapani Loep; at all these stations the declination values are less than the normal, while at station No. 378, Warmbad, Zand River, Witklip, places to the east of this line of repulsion, the values of the declination are larger than the normal. This valley in its northernmost part coincides with lines of no declination disturbance and no disturbance of horizontal intensity (see Maps 17 and 15). There are two centres of repulsion, one between Witklip and Kalkbank, the other between Mara and Buffelshoek; these two centres coincide with the two intersections of a line of no horizontal disturbance with one of no declination disturbance.

The directions of the horizontal disturbing forces are away from these centres.

Meyerton—Bavaria—Dannhauser—Pivaan's Poort ridge. The ridge begins between Meyerton on the north-east and Wolvehoek on the south (see Map 14); it runs south-east having Groenplaats and Bavaria on the northern side, its course after that becomes more easterly, it passes between Newcastle and Dannhauser then continues east of Utrecht and Pivaan's Poort, bends again to the east between Schikhoek and Driehoek; after continuing a short distance to the east, it goes north between Highlands and Alma, then turns west passing to the north of Piet Retief; its further course takes it east of Tweepoort where its direction is north-west, then west, with Lake Banagher and Bankpan on the south side and Aberdeen (Transvaal) on the north; a little to the west of Bankpan and Aberdeen (Transvaal) the ridge turns to the south-west; its further course cannot be determined with any certainty.

There are in this region three valleys, one beginning between Tweepoort and Lake Banagher and continuing north-east, having Barberton, Kaapmuiden, Hector Spruit and Movene on one side, and Nelspruit, Komati Poort and Picene on the other; the valley coincides with a line of no horizontal intensity disturbance throughout the greater part of its course. There is a second valley between Graskop and Charlestown which has Utrecht and Wakkerstroom on the east and Newcastle on the west; this valley coincides with a line of no declination disturbance.

The third valley starts at a point between Bankpan and Aberdeen (Transvaal) running south to within a short distance of Twelfelhoek, from there it runs north-west by west, intersecting the ridge a little to the north of Groenplaats.

The main ridge in the first part of its course coincides with lines of no horizontal intensity disturbance and no declination disturbance. The values of the declination are less than the normal at Meyerton, Kraal, Groenplaats, No. 396, Bavaria and Twelfelhoek, due to the attraction of the south pole by this ridge; this same attraction produces an excess in the declination at Heilbron, Kromm River, Kaalkop Farm, stations on the other side of the ridge. That part of the ridge which starts from the neighbourhood of Newcastle and follows the course described in a previous paragraph, corresponds throughout nearly the whole of its length with a line of no declination disturbance. At Utrecht and Pivaan's Poort the declination is greater than the normal, a result which was to be expected from the position of these two places on the west of the ridge. Highlands, Hluti, Shela River, Aberdeen (Transvaal) stations all to the east of the ridge have declinations below the normal value, while at Alma, Piet Retief, and Tweepoort stations to the west the declinations are above the normal. Lake Banagher is irregular. The declinations at Wakkerstroom, Charlestown, Graskop have values such as are to be expected from their

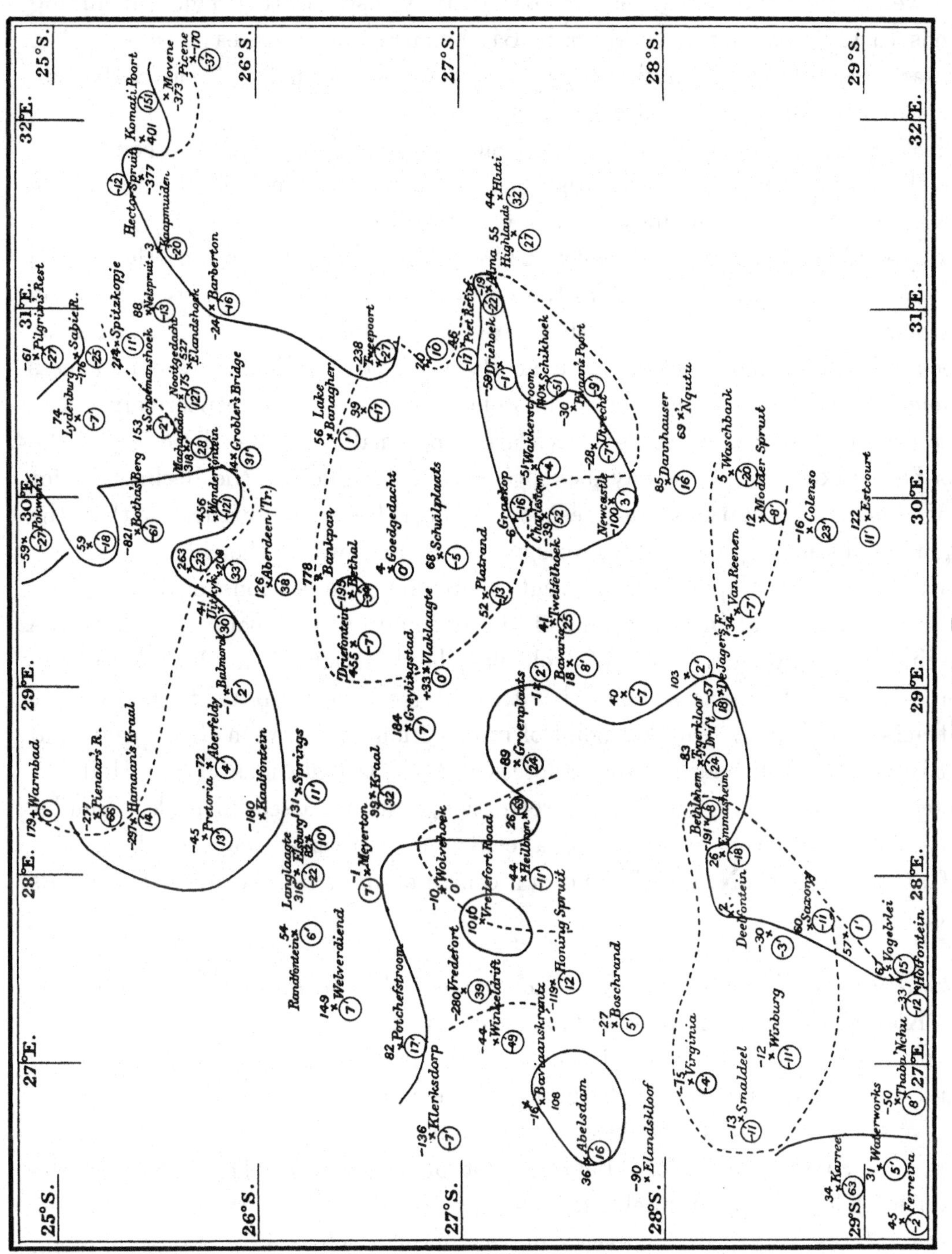

Map 18. TRANSVAAL.

Showing horizontal intensity and declination anomalies. The ΔH are in terms of γ. The ΔD are in minutes of arc. A negative sign denotes that the observed value is greater than the calculated. The continuous lines are lines of no ΔH. The dotted lines are lines of no ΔD.

positions with respect to the valley in their neighbourhood. The values of the declination at the stations near the third valley mentioned above are in accordance with their situations near the valley except at Driefontein.

When we consider the values of the horizontal intensity in this region, including the stations to the east of the valley from Barberton to the north-east, we find that the values at Nelspruit and Komati Poort are less than the normal, while at Barberton, Hector Spruit and Movene they are greater.

The declination results at Picene and Movene also give support to the course laid down for the valley; Picene to the east has a declination considerably above the normal, whereas at Movene on the west it is very much below the normal.

The values of the horizontal intensity at Groenplaats, No. 396, Newcastle, on the north side of the ridge, are greater than the normal, whereas at Dannhauser, Kaalkop Farm, Kromm River, on the south, they are below the normal. The values at Wolvehoek and Meyerton, however, and at Bavaria and Twelfelhoek are in excess and in defect of the normal instead of in defect and in excess respectively. The deviations from the normal at these places are very small, and the direction of the ridge has been taken in such a way as to give correct values of the declination for the situation of various stations with respect to the ridge; and further so that the direction of the disturbing horizontal forces should point towards the ridge.

A consideration of Map 18 shows that there are intersections of lines of no anomaly of the horizontal intensity and of the declination at a point a little to the east of a line joining Meyerton and Wolvehoek; the direction of the disturbing horizontal forces at these two places shows that this intersection is a peak, that is a point which attracts a south pole. Another point of intersection lies between Groenplaats and Kromm River; the disturbing horizontal forces at these two stations and at Heilbron are directed towards the region of this intersection, marking a second peak. Similar considerations indicate the existence of a centre of repulsion near Graskop, of a centre of attraction near Newcastle, and of a centre of attraction between Highlands and Alma.

Klerksdorp-Virginia-Saxony ridge and connected disturbances. This begins between Klerksdorp and Potchefstroom, goes south-east to a point between Winkeldrift and Vredefort; there it turns south, passing to the west of Honing Spruit and Boschrand, and east of Virginia; it then turns towards the east, passing north of Winburg and south of Deelfontein, where it again turns to the south, passing east of Saxony, and No. 393, but west of Vogelvlei. The values of the declination at Klerksdorp, Winkeldrift, Virginia, Smaldeel and Winburg, stations to the west of the ridge, are in excess of the normal, whereas at Potchefstroom, Vredefort, Honing Spruit, Boschrand, stations to the east, they are below.

The directions of the disturbing horizontal forces are towards the ridge throughout its course.

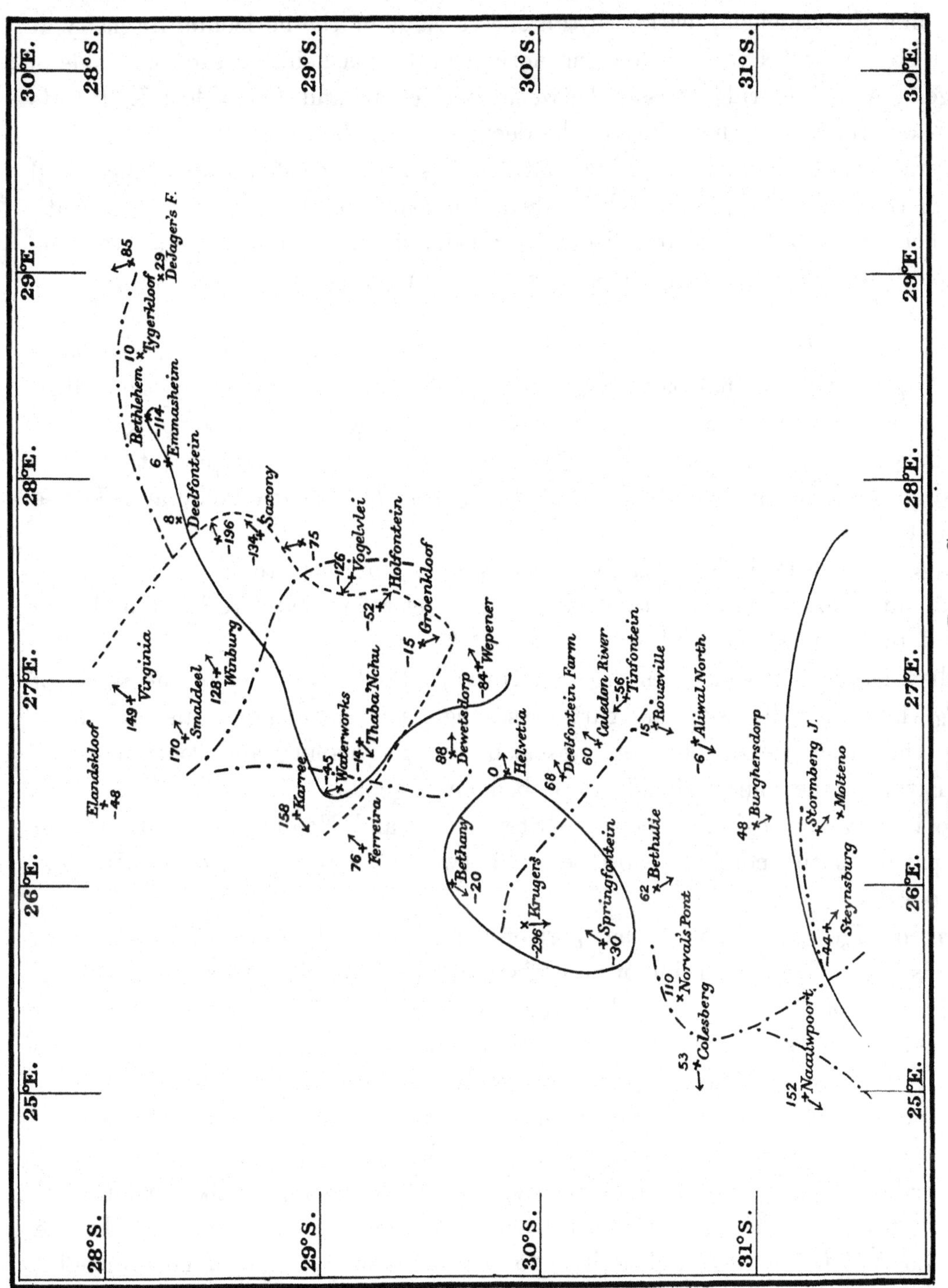

Map 19. Orange River Colony.

Showing the vertical intensity disturbance in γ, the ridges - - - -, the valleys .—.—., the lines of no vertical disturbance ———, and the directions of the disturbing horizontal forces ↑. A negative sign denotes that the actual vertical intensity is greater than the calculated.

Near Deelfontein a valley begins, passes in an easterly direction north of Bethlehem, south of Van Reenen, and probably intersects the Natal-Zululand ridge between Modder Spruit and Colenso. There are also two small ridges, one between Emmasheim and Bethlehem, marked by an excess value of the horizontal intensity at the latter station and a value below the normal at Emmasheim on the south side of this ridge. A second ridge passes between Bethlehem and Tygerkloof Drift; its position is determined by the values of the declination at these stations.

The directions of the disturbing horizontal forces at Emmasheim, Bethlehem and Tygerkloof Drift are all approximately towards the same point, a point to the south of Bethlehem, where a line of no horizontal intensity disturbance intersects one of no declination disturbance, and which is a peak. (See Maps 14 and 18.)

Ridges and valleys in the south of the Orange River Colony and the northern part of Cape Colony. The main ridge referred to in the previous section, after passing to the west of Vogelvlei, continues in a southerly direction, keeping east of Holfontein and of Groenkloof; a little to the south of the latter place this ridge turns and goes in a north-westerly direction to the south of Thaba 'Nchu and Karree (see Map 19).

A ridge goes from Krugers in a south-easterly direction to a point between Tinfontein and Rouxville. Another ridge runs between Waterworks and Thaba 'Nchu and to the west of Smaldeel.

A valley passes between Colesberg and Norval's Pont. A little to the south of Colesberg the valley divides; one part passes between Naauwpoort and Rosmead Junction, the other passes between Rosmead and Steynsburg and continues east of Fish River as the valley already described on page 95.

There is a valley beginning between Steynsburg and Naauwpoort and running east; it passes to the north of Stormberg. Between the latter place and Molteno there is a ridge.

The main ridge, together with that part of it passing to the south of Dewetsdorp, corresponds in position with a line of no horizontal force disturbance (see Map 20).

To the east of Vogelvlei and then west of Winburg and Smaldeel there is a valley which coincides with a line of no declination disturbance. The values of the declination at Saxony and Holfontein, stations to the west of the ridge, are greater than the normal, whereas at Vogelvlei, to the east, the value is less. The part of the ridge running towards Karree coincides with a line of no declination disturbance, and the values of the declination at Karree, Waterworks, Thaba 'Nchu and Groenkloof, places to the east of this ridge, are smaller than the normal. In the neighbourhood of Groenkloof and a little to the south of it a line of no horizontal force disturbance intersects one of no declination disturbance; the point of intersection is a peak, towards which the horizontal disturbing forces at Groenkloof and at Wepener are directed.

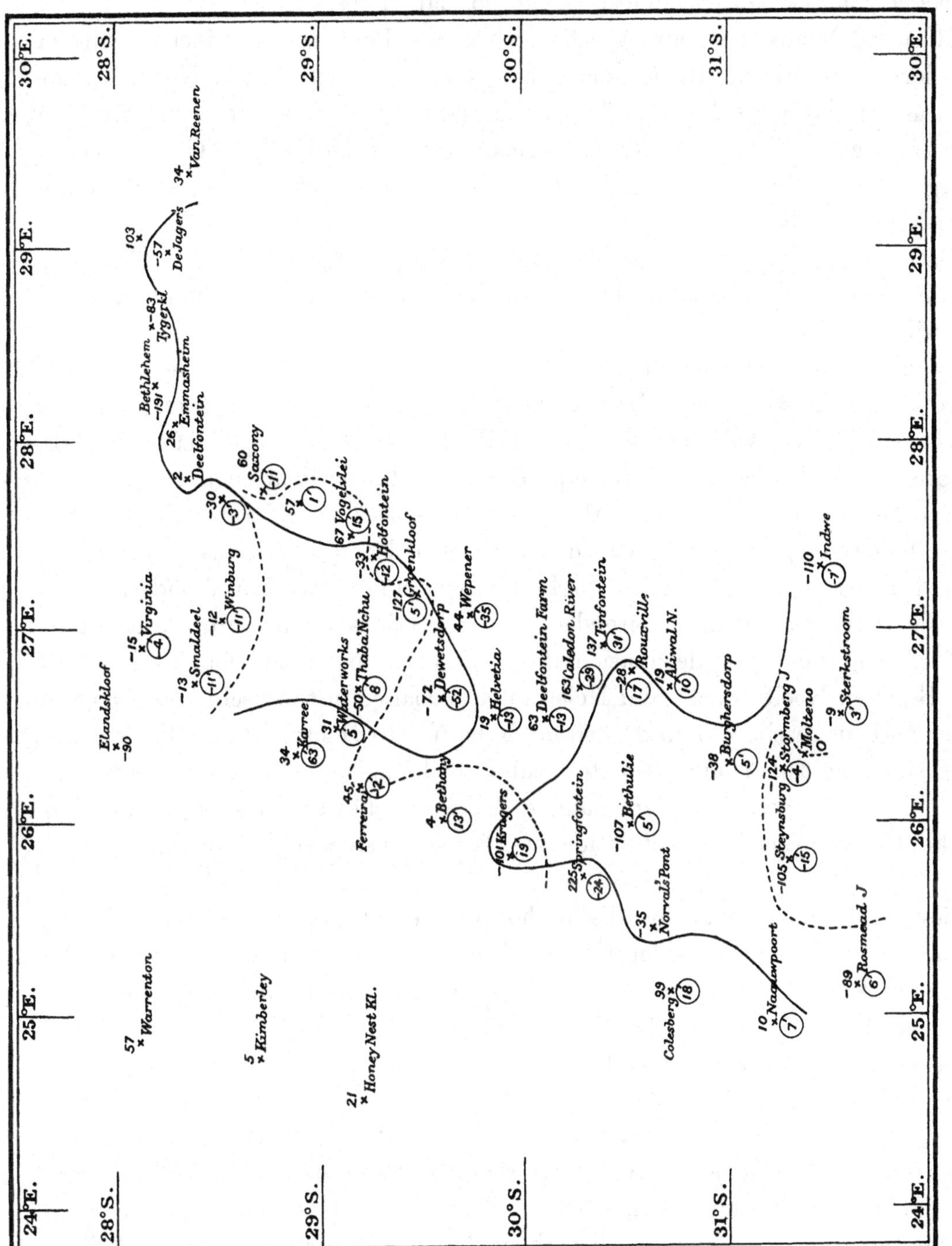

Map 20.

Showing the horizontal intensity and declination anomalies for the Orange River Colony. The horizontal intensity anomalies are expressed in terms of γ. The declination anomalies are in minutes of arc. A negative sign denotes that the observed is greater than the calculated value. The continuous lines are lines of no disturbance of horizontal intensity. The dotted lines are lines of no disturbance of declination.

The valley running from Krugers to Rouxville coincides with a line of no horizontal intensity disturbance.

The value of the declination at Deelfontein Farm, on the eastern side, is in excess, and at Krugers, on the west, in defect of the normal values.

The Rosmead-Naauwpoort and Colesberg-Norval's Pont valley coincides with a line of no horizontal intensity disturbance; the values of this element at Norval's Pont and Rosmead on the one side and at Naauwpoort, Colesberg, and Springfontein on the other are greater and less than the normal values respectively. The Rosmead-Steynsburg valley coincides with a line of no declination disturbance; so does the Stormberg-Molteno ridge.

The disturbances in parts of the Transvaal and of the Orange River Colony can be shown in another way, viz. by plotting the value of the declination for a given parallel against the longitude.

In Map 21 the stations with latitude 25° 45′ S. have the values of their declinations plotted against longitude. In the case of those stations not exactly on this parallel the declination values have been corrected to the proper latitude. It will be noticed that between Pretoria and Balmoral the declination decreases at a rate a little less than the normal. After that there is a considerable drop in the declination, followed by a region in which instead of a decrease there is a very sharp increase with increase of longitude, the highest value being reached at Wonderfontein. Between the latter place and Nooitgedacht—two stations distant from each other about 38′ of longitude—the declination drops in value by about four and a half degrees. Between Nooitgedacht and Nelspruit the change in the declination is again an increase with increasing longitude; from Nelspruit to Hector Spruit the decrease with increasing longitude is approximately equal to the mean decrease for this region, but after this point there is a very sudden decrease followed by such an increase as to make the value of the declination at Picene considerably above the normal value.

The view of the facts given by this method of presentation agrees with what we have already seen from a consideration of Map 14. For example between the station marked No. 391 and Wonderfontein is a centre of repulsion giving rise to an abnormally small value of the declination to the west of it and an abnormally great one to the east. A glance at Map 14 shows that the disturbing horizontal forces at No. 391 and Thirty-First on the west and at Botha's Berg and Wonderfontein on the east of this region of repulsion are away from it. Between Wonderfontein and Nooitgedacht is a strong centre of attraction which accounts for the abnormal values at Wonderfontein and Nooitgedacht. This region of attraction is shown in Map 14 by the directions of the disturbing horizontal forces at Machadodorp, Grobler's Bridge and Botha's Berg, which point to a place where a line of no declination disturbance intersects one of no disturbance of horizontal intensity.

A second Map—22—gives the change of declination with longitude for stations on the parallel 27° 20′ S. As in the other similar cases the value of the declination has been corrected to the proper latitude when necessary.

Between Abelsdam and Winkeldrift the declination, instead of decreasing, increases with increasing longitude; this increase is more than compensated by a very rapid decrease in the declination in passing from Winkeldrift to Vredefort.

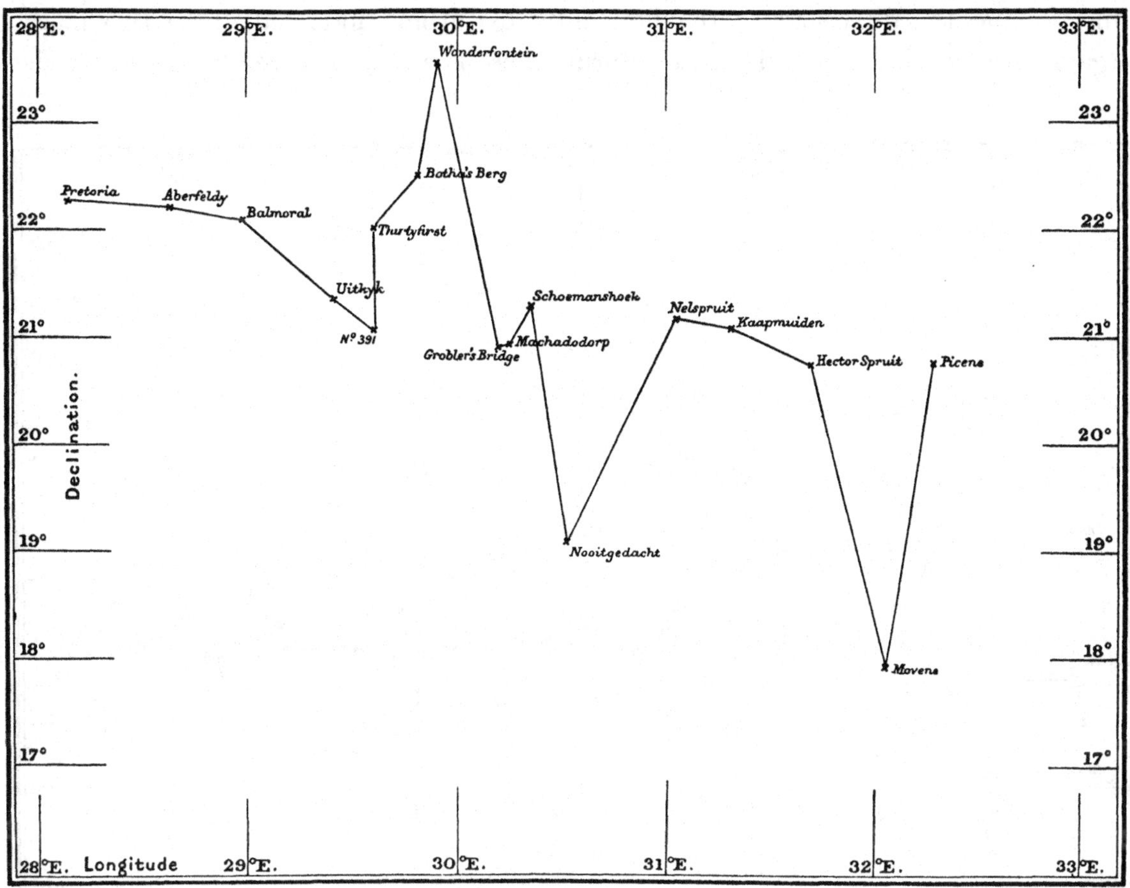

Map 21. Showing values of declination at different longitudes along parallel 25° 45′ S.

From the latter place there is again an increase with longitude, first slight, then more pronounced until Kromm River is reached; this is followed by a sharp fall in the declination between Kromm River and Groenplaats, from which place to No. 396 there is practically no change; another abnormal decrease takes place between No. 396 and Twelfelhoek; from this latter to Platrand there is an increase with increasing longitude. Between Platrand and Charlestown the change is a rapid decrease with increasing longitude; between Charlestown and Graskop there is a very sudden increase in the value of the declination amounting to over a degree. From Graskop to

Wakkerstroom there is a decrease with increasing longitude; this is followed by an increase which is first slight, and finally between Pivaan's Poort and Schikhoek rapid. Between Schikhoek and Driehoek there is a very pronounced decrease, while from the latter station to Alma there is an increase with increasing longitude. From Alma to Hluti the change is a decrease with increasing longitude. From Map 22 it is seen that there is a strong centre of attraction between Winkeldrift and Vredefort; a glance at Map 14 shows that the position of this centre is at the intersection of the lines giving the directions of the horizontal disturbing forces at these two places. Again between Heilbron and Honing Spruit there is a weak centre of repulsion which

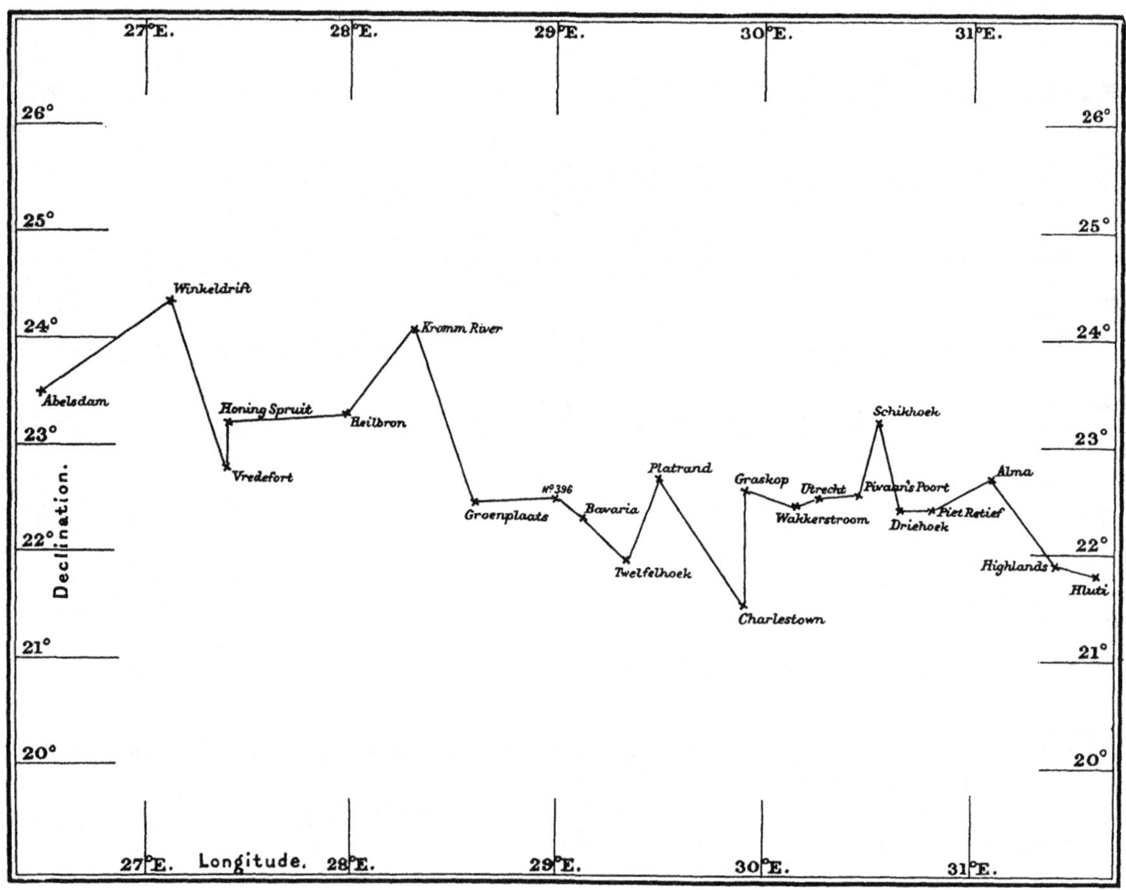

Map 22. Showing values of declination at different longitudes along parallel 27° 20′ S.

is in accordance also with the direction of the disturbing horizontal forces at these two stations (see Map 14). It will be seen from Map 18 that this repulsion centre is at the intersection of a line of no horizontal intensity disturbance with one of no declination disturbance.

There is another and strong centre of attraction between Heilbron and Kromm River; the horizontal disturbing forces at these two stations point towards this attraction region which is in the neighbourhood of the intersection of two lines of no disturbance (see Map 18). There is further a weak centre of attraction between Bavaria and station No. 396, and a weak repulsion centre between Platrand and Twelfelhoek. The position and nature of these two centres are also indicated by the direction of the disturbing horizontal forces (Map 14). Between Charlestown and Platrand there is a centre of attraction, and between Charlestown and Graskop

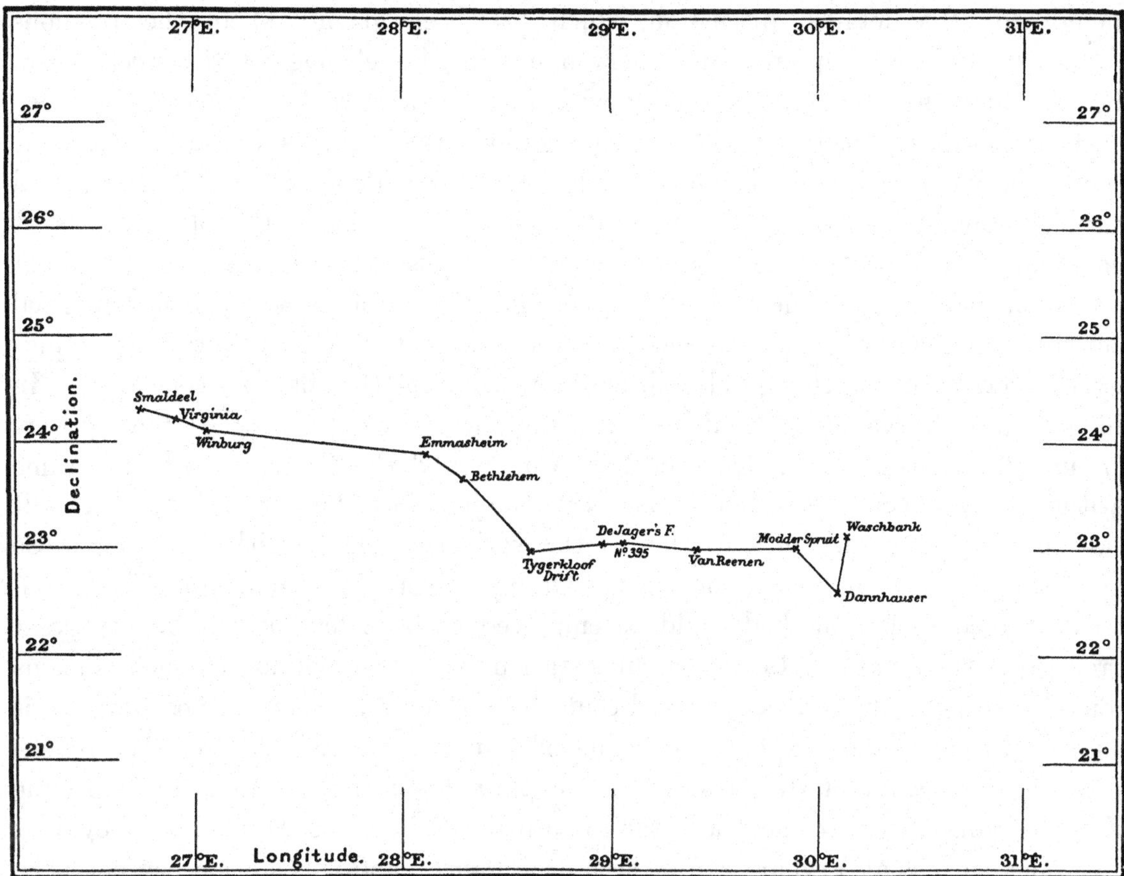

Map 23. Showing values of declination at different longitudes for parallel 28° 18′ S.

one of repulsion. Another centre of repulsion is seen between Pivaan's Poort and Schikhoek, and one of attraction between Driehoek and Schikhoek. Finally between Alma and Highlands there is a well marked centre of attraction which coincides with the intersection of a line of no declination disturbance with one of no horizontal intensity disturbance.

In Map 23 the change of declination with longitude is shown for the parallel 28° 18′ S. The most striking points in this are the centres of attraction between Bethlehem and Tygerkloof Drift, and between Modder Spruit and Dannhauser, and the centre of repulsion between Dannhauser and Waschbank.

The Bechuanaland and Rhodesian results.

It has not been thought advisable to combine in every case the results in these two regions according to districts. The stations are not well distributed for such treatment; in some cases, e.g. between Warrenton and Bulawayo, they are along a line which runs north with a slightly easterly tendency, and as a rule there is only one line of stations, which though of value for some purposes are not numerous enough for the detailed discussion which has been attempted in the previous paragraphs of this chapter. In order to obtain the change in the declination, the dip, and the horizontal intensity with latitude and with longitude, equations must be obtained which connect the changes in these various elements with the change in latitude and in longitude. This has been done for the Bechuanaland stations from Warrenton to Seruli in the following manner: the observations at places in approximately the same latitudes as those of the stations now under discussion and lying to the east in the Orange River Colony and in the Transvaal had the values of their magnetic elements reduced to the desired latitude by applying the correction for change of latitude deduced from the districts concerned. In this way probable values of the declination, the dip, and the horizontal intensity were obtained for positions in the Orange River Colony and the Transvaal with the same latitude as the Bechuanaland stations. By combining the former with the latter it was possible to obtain the variation of the various elements with longitude. So far as the results from the stations in the Orange River Colony and in the Transvaal are concerned such a method could be employed without fear of introducing great errors from abnormal results at particular stations, for the deviation at such stations from the normal value was known and could be allowed for. The method however is not so satisfactory when applied to a line of stations such as that from Warrenton to Seruli on account of the fact that there are not sufficient stations to determine with the same degree of accuracy the abnormalities in these cases. Very evident deviations such as at Vryburg can be seen at a glance; and this station has not been used to determine the variation with longitude.

The following table gives the variations per degree of longitude along a number of parallels. The negative sign denotes a decrease of the element concerned with increasing longitude.

Latitude for which variation holds	Change per degree increase of longitude for D	θ	H	Longitudes of stations used in calculation		
° ′	′	′		° ′		° ′
27 35 S.	−16	23	− 51 γ	26 30 E.	to	24 45 E.
26 25 ..	−19	28	−137 ..	27 30 ..	..	25 15 ..
25 52 ..	−11	24	−100 ..	28 0 ..	..	25 39 ..
25 14 ..	−16	19	−147 ..	28 18 ..	..	25 40 ..
24 47 ..	− 9	25	− 87 ..	28 18 ..	..	25 50 ..
23 35 ..	−34	0	100 ..	29 18 ..	..	26 34 ..
23 7 ..	− 8	41	−132 ..	29 8 ..	..	26 50 ..
22 33 ..	−17	28	−164 ..	29 6 ..	..	27 7 ..
21 56 ..	−17	26	−173 ..	29 10 ..	..	27 19 ..

A comparison of this table with table 23, page 53, shows that the variations with longitude for the parallels 27° 35′ S. and 26° 25′ S. differ only slightly from those of district XI; this district extends from 28° S. to 31° S. and from 24° E. to 27° E., and for it the change per degree increase of longitude is a decrease in declination of 18′, of dip an increase of 17′, and of horizontal intensity a decrease of 72 γ.

The variations for the parallel 27° 35′ S. may also be compared with those for district VIII through which this parallel passes, and which is bounded by the meridians 27° and 28° E. It will be seen that the decrease of the declination with longitude is less in the present case, the increase of dip is greater, and there is a decrease of horizontal intensity instead of an increase.

The parallels 26° 25′, 25° 52′, 25° 14′ S. pass also through district X. The variations with longitude of the various elements along these parallels—variations which have been obtained by combining those of district X with the results from the Bechuanaland stations—differ very considerably from those of X; the decrease in declination is considerably less, the increase of dip greater, and the horizontal intensity decreases with increasing longitude instead of increasing.

The changes along the parallels 24° 47′ S. to 21° 56′ S. may be compared with the changes with increase of longitude in district XV through which these parallels also pass.

The changes with latitude in Bechuanaland have been obtained by using the changes with longitude as given above to reduce the results at two successive stations to the same longitude, then taking the differences of latitude, of declination, of dip and of horizontal intensity and calculating from these the corresponding changes per degree change of latitude.

Change of magnetic elements of the Bechuanaland stations with latitude.

	Lat.	Change per degree increase of latitude for *D*	*θ*	*H*	Remarks
	° ′	′	′		
Taungs to Maribogo	27 35 S. 26 25 ..	33	51	− 462 γ	Results from observation reduced to 25° E.
Maribogo to Mafeking	26 25 .. 25 52 ..	43	31	− 308 ..	Reduced to 25° 27′ E.
Mafeking to Lobatsi	25 52 .. 25 14 ..	59	19	− 87 ..	Difference in longitude neglected
Lobatsi to Crocodile Pools	25 14 .. 24 47 ..	113	113	− 1035 ..	Reduced to longitude of Lobatsi by using longitude variation along 24° 47′ S.
Crocodile Pools to Shoshong Road	24 47 .. 23 35 ..	− 6	− 15	179 ..	Reduced to 26° 12′ E.
Shoshong Road to Magalapye	23 35 .. 23 7 ..	180	152	− 1632 ..	,, ,, 26° 42′ E.
Magalapye to Palapye	23 7 .. 22 33 ..	− 21	63	− 330 ..	,, ,, 26° 58′·5 E.
Palapye to Seruli	22 33 .. 21 56 ..	6	28	− 86 ..	,, ,, 27° 7′ E.
Lobatsi to Seruli		36	47	− 380 ..	

A comparison of the results given in the above table with the results for districts X and XV given on page 53 shows that for the first three changes, viz. from Taungs to Lobatsi the results are of the same sign as for district X but on the average greater in magnitude in the case of each of the three elements.

The same is true when the mean result from Lobatsi to Seruli is compared with the results for district XV. When the individual results are considered from Lobatsi to Seruli it is seen that there are great irregularities; in two cases the change of declination with decreasing latitude is a decrease and not an increase; in one case the change in dip is irregular and in another that of horizontal intensity.

The stations at which observations were made in Rhodesia were divided into two groups, one, XVII, containing twelve stations mostly in Matabeleland, the other, XVI, containing thirteen stations in Mashonaland. All these stations are situated on the railway lines from Bulawayo to Victoria Falls and from Bulawayo to Umtali.

To obtain the variations of the various elements with latitude and with longitude the procedure adopted was the same as that given on page 53. The results are contained in the following tables.

TABLE 34.

Table giving the latitude, longitude, and the magnetic elements at mean stations in district XVII, Matabeleland (12 stations), and in district XVI, Mashonaland (13 stations).

District	ϕ_m ° ′	λ_m ° ′	D_m ° ′	θ_m ° ′	H_m	X_m	Y_m	Z_m	T_m	
XVI	18 21 S.	30 58 E.	16 27 W.	52 20 S.	·22218	·21308	·06288	·28866	·36429	Mashonaland
XVII	19 12 ..	27 58 ..	18 8 ..	52 52 ..	·21821	·20736	·06787	·28841	·36173	Matabeleland

TABLE 35.

Table giving the changes per degree of the declination, dip, and horizontal intensity.

District	Change per degree increase of latitude			Change per degree increase of longitude		
	$\frac{\partial D}{\partial \phi}$	$\frac{\partial \theta}{\partial \phi}$	$\frac{\partial H}{\partial \phi}$	$\frac{\partial D}{\partial \lambda}$	$\frac{\partial \theta}{\partial \lambda}$	$\frac{\partial H}{\partial \lambda}$
XVI	37′	66′	− 400 γ	− 37′	4′	88 γ
XVII	45′	54′	− 318 ..	− 23′	23′	114 ..

TABLE 36.

Table giving the changes per degree of the northerly, westerly, vertical and total intensities.

District	Change per degree increase of latitude				Change per degree increase of longitude			
	$\frac{\partial X}{\partial \phi}$	$\frac{\partial Y}{\partial \phi}$	$\frac{\partial Z}{\partial \phi}$	$\frac{\partial T}{\partial \phi}$	$\frac{\partial X}{\partial \lambda}$	$\frac{\partial Y}{\partial \lambda}$	$\frac{\partial Z}{\partial \lambda}$	$\frac{\partial T}{\partial \lambda}$
XVI	− 420 γ	12 γ	720 γ	348 γ	26 γ	− 174 γ	42 γ	18 γ
XVII	− 384 ..	174 ..	528 ..	234 ..	150 ..	− 102 ..	564 ..	516 ..

The tables 34, 35 and 36 were used in calculating the values given in tables 32 and 33 for the Rhodesian stations where calculated values are given.

A further consideration of the magnetic state of these two districts shows that at Bulawayo, Lochard, Shangani and Gwelo the value of the declination is greater than that calculated from the above changes with latitude and longitude. The greatest difference is at Shangani; the differences at Bulawayo and at Lochard are so small as to be negligible, viz. 1′ and 2′ respectively. At Globe and Phoenix the value of the declination is about half a degree less than the calculated result; it is also less at Hartley, Forty-one mile Siding, Ayrshire Mine and Marandellas. The line of no declination disturbance runs between Gwelo and Globe and Phoenix, between the latter station and Battlefields—where the declination is again greater than the calculated value—, between Battlefields and Hartley, and then between Hartley and Forty-one mile Siding on the west and Makwiro and Salisbury on the east. This line corresponds to a valley, and the directions of the disturbing horizontal forces are away from the valley throughout its course.

The line of no disturbance of horizontal intensity passes between Hartley and Battlefields, south of Makwiro, between Makwiro and Ayrshire Mine on the west and Salisbury and Marandellas on the east; this coincides with a ridge.

Note on the Geological Map.

The rocks belonging to the Cretaceous and Karroo systems are only slightly folded, excepting that part of the Karroo system along the edge of the southern strip occupied by the Cape system.

The rocks of the Cape system are strongly folded in the south coast region, less so between the west coast and the Karroo; the axes of folding are parallel to the boundary lines against the Karroo system.

The Pre-Cape rocks are generally strongly folded, but this is not the case in the area stretching northwards through Griquatown and Kuruman nor along the 20th meridian north of the granite area.

North of the line which indicates the southern limit of the dolorite intrusions masses of that rock are very abundant in the form of dykes and sills, though they are less important at the present surface in the area of the Pre-Cape rocks than in the country which is still covered by the Karroo system.

The acid igneous rocks of Pre-Cape age are chiefly intrusive, those of basic and intermediate composition mainly volcanic.

ARTHUR W. ROGERS.

APPENDICES

APPENDIX A.

GINGINHLOVU, ZULULAND, 27 October 1903. Observer and Recorder, Beattie.

Observations for Latitude.

		Chron. time	Levels		Circle readings	Mean circle readings after applying level corrections	
		h m s			° ′ ″	° ′ ″	
Circle Left.	App. upper limb	11 29 6	12·8	9·6	A. 72 49 15 B. 52 30	72 51 14	
,,	,,	11 30 30	12·2	10·2	A. 72 51 30 B. 54 0	72 52 52	
,,	,,	11 32 46	12·8	9·2	A. 72 53 15 B. 56 30	72 55 16	
,,	,,	11 33 56	13·2	12·0	A. 72 54 0 B. 57 0	72 55 59	(1)′
Circle Right.	App. lower limb	11 35 38	7·0	15·0	B. 73 54 0 A. 52 0	73 53 55	(2)′
,,	,,	11 36 51	7·0	15·2	B. 73 54 0 A. 51 30	73 53 41	
,,	,,	11 38 6	6·2	15·8	B. 73 53 30 A. 51 0	73 53 21	
,,	,,	11 39 27	7·0	15·0	B. 73 53 0 A. 50 30	73 52 40	

Observations for Longitude (time).

	Sun's limb	Chron. time	Levels		Circle readings	
		h m s			° ′ ″	
Circle Left.	Preceding limb	2 27 51	6·6	16·4	A. 46 28 30 B. 31 0	(1)
,,	Following limb	2 29 23·5	7·8	15·2	A. 46 41 15 B. 43 30	(2)
,,	Following limb	2 30 37·0	6·4	16·6	A. 46 25 30 B. 28 30	(3)
,,	Preceding limb	2 31 44	6·0	17·0	A. 45 39 0 B. 42 0	(4)
Circle Right.	Preceding limb	2 33 6·5	17·0	6·0	B. 45 48 15 A. 46 0	(5)
,,	Following limb	2 34 13·5	14·8	8·2	B. 46 6 0 A. 4 0	(6)
,,	Following limb	2 35 13·0	16·0	7·0	B. 45 54 0 A. 52 0	(7)
,,	Preceding limb	2 36 24·0	16·6	6·4	B. 45 7 0 A. 4 30	(8)

	Chron. time	Circle reading corrected for level
	h m s	° ′ ″
Combination of (1) (2) (5) (6)	2 31 8·6	46 15 3
,, (3) (4) (7) (8)	2 33 32·0	45 43 39

Observations for Azimuth.

Circle reading of reference line magnetometer-theodolite :—

Circle Right. A. 292° 44′ 30″
B. 46 0

	Sun's limb	Chron. time h m s	Striding level	Circle reading ° ′ ″	
,,	Preceding	2 37 25	21·8 – 1·0	B. 248 38 0 A. 36 0	(1)
,,	Following	2 39 15	20·0 0·8	B. 249 6 30 A. 4 45	(2)
,,	Following	2 40 36·5	20·0 0·9	B. 248 52 45 A. 51 0	(3)
,,	Preceding	2 41 55·0	17·6 3·4	B. 247 55 15 A. 53 30	(4)
Circle Left.	Preceding	2 47 17	18·0 2·8	A. 247 6 15 B. 8 0	(5)
,,	Following	2 48 57	16·2 4·2	A. 247 35 30 B. 36 45	(6)
,,	Following	2 50 19·5	17·8 2·8	A. 247 22 30 B. 24 0	(7)
,,	Preceding	2 51 47	19·2 1·2	A. 246 25 30 B. 27 0	(8)

Circle reading of reference line magnetometer-theodolite :—

Circle left. A. 112° 44′ 15″
B. 46 0

	Chron. time h m s	Circle reading corrected for level ° ′ ″
Combination of (1) (2) (7) (8)	2 44 41·6	247 54 38·5
,, (3) (4) (5) (6)	2 44 41·4	247 53 55·0

Reduction of the astronomical observations taken at Ginginhlovu, Zululand.

Approximate chronometer correction.

Approximate altitude, h, from (1)′ and (2)′ of last page, 73° 25′.

$$z = 90 - h$$
$$= 16° \; 35'$$
$$\delta = +12 \; 29$$
$$\phi \fallingdotseq 29° \; 4'$$

This value of the latitude is now applied in the formula given on page 13 to calculate the approximate chronometer correction.

Data	Chron. time	Corrected altitude
$z = 90° - 46° \; 15'$	$2^h \; 31^m \; 8^s{\cdot}6$	46° 15′ 3″
$= 43 \; 45$		
$\delta = 12 \; 29$		

$$\tfrac{1}{2}\{z + \phi - \delta\} = 30° \; 10'$$
$$\phi - \delta = 16° \; 35' \qquad \tfrac{1}{2}\{z - (\phi - \delta)\} = 13 \; 35$$

$\log \sin 30° \; 10' = 9{\cdot}70115$ $\qquad$ $\log \cos 29° \; 4' = 9{\cdot}94154$

$\log \sin 13 \; 15 = 9{\cdot}36022$ $\qquad$ $\log \cos 12 \; 29 = 9{\cdot}98961$

$\log \dfrac{1}{\cos\phi \cos\delta} = {\cdot}07885$ $\qquad$ $9{\cdot}93115$

$9{\cdot}14022$

$\log \sin \tfrac{1}{2}\lambda \quad 9{\cdot}57011$

$\tfrac{1}{2}\lambda$ in arc $\quad 21° \; 49'$

λ ,, $\quad 43 \; 38$

λ in time $\quad 2^h \; 54^m \; 32^s$

Chronometer correction = λ (in time) minus the chronometer time = $+ 23^m \; 23^s$

Latitude.

27 *October* 1903.

	Obs. Time	Corrected time (App. solar)	Obs. altitude corrected for level	h.*	Corrected meridian altitudes	
	h m s	h m s	° ′ ″	′ ″	° ′ ″	
Circle Left	11 29 6	11 52 29	72 51 14	5 20	72 56 34	
	11 30 30	11 53 53	72 52 52	3 32	72 56 24	
	11 32 46	11 56 9	72 55 16	1 23	72 56 39	
	11 33 56	11 57 19	72 55 59	37	72 56 38	
						———72° 56′ 34″
Circle Right	11 35 38	11 59 1	73 53 55	3	73 53 58	
	11 36 51	12 0 14	73 53 41	0	73 53 41	
	11 38 6	12 1 29	73 53 21	15	73 53 36	
	11 39 27	12 2 50	73 52 40	56	73 53 36	
						———73 53 43

Mean 73 25 9

16 34 51 = z

16 = refraction

16 35 7

12 27 7 = δ

29 2 14 = ϕ

Longitude (time).

27 *October* 1903.

$\phi = 29°\ 2'\ 14''$ $\delta = 12°\ 29'\ 18''$

Obs. time	Altitude
h m s	° ′ ″
2 31 8·6	46 15 3

43 44 57

52 = refraction

43 45 49 = z

30 9 53 = $\frac{1}{2}\{z + (\phi - \delta)\}$

13 35 57 = $\frac{1}{2}\{z - (\phi - \delta)\}$

$\log \sin \frac{1}{2}\{z + (\phi - \delta)\} = 9{\cdot}701122$

$\log \sin \frac{1}{2}\{z - (\phi - \delta)\} = 9{\cdot}371304$

$\log \frac{1}{\cos \phi \cos \delta} = {\cdot}068736$

9·141162

$\log \sin \frac{1}{2}\lambda = 9{\cdot}570581$

h m s

$\frac{1}{2}\lambda$ in time = 1 27 21·9

λ in time = 2 54 43·8

Chronometer correction = 23 35·2

$\log \cos \delta = 9{\cdot}989601$

$\log \cos \phi = 9{\cdot}941663$

9·931264

Obs. time	Altitude
h m s	° ′ ″
2 33 32·0	45 44 39

44 15 21

52 = refraction

44 16 13 = z

30 25 04 = $\frac{1}{2}\{z + (\phi - \delta)\}$

13 51 08 = $\frac{1}{2}\{z - (\phi - \delta)\}$

$\log \sin \frac{1}{2}\{z + (\phi - \delta)\} = 9{\cdot}704409$

$\log \sin \frac{1}{2}\{z - (\phi - \delta)\} = 9{\cdot}379140$

$\log \frac{1}{\cos \phi \cos \delta} = {\cdot}068736$

9·152285

$\log \sin \frac{1}{2}\lambda = 9{\cdot}576143$

h m s

$\frac{1}{2}\lambda$ in time = 1 28 33·2

λ in time = 2 57 6·4

Chronometer correction = 23 34·4

* See p. 13.

$\phi = 29°\ 2'\ 14''$ $\qquad\qquad$ $\delta = 12°\ 29'\ 40''$

Time h m s	Altitude ° ′ ″
2 58 35·1	40 21 28
	49 38 32

refraction = 1 2

$z = 49\ 39\ 34$

$\frac{1}{2}\{z + (\phi - \delta)\} = 33\ \ 6\ 34$

$\frac{1}{2}\{z - (\phi - \delta)\} = 16\ 33\ 00$

$\log \sin \frac{1}{2}\{z + (\phi - \delta)\} = 9·737384$

$\log \sin \frac{1}{2}\{z - (\phi - \delta)\} = 9·454619$

$\log \frac{1}{\cos\phi \cos\delta} = ·068746$

9·260749

$\log \sin \frac{\lambda}{2} = 9·630375$

$\frac{\lambda}{2} = 1^h\ 41^m\ 5^s·8$

$\lambda = 3\ 22\ 11·6$

Chron. correction = 23 36·5

$\log \cos \delta = 9·989591$

$\log \cos \phi = 9·941663$

$\log \cos \delta \cos \phi = ·931254$

Time h m s	Altitude ° ′ ″
3 0 47·1	39 52 50
	50 7 10

refraction = 1 3

$z = 50\ 8\ 13$

$\frac{1}{2}\{z + (\phi - \delta)\} = 33\ 21\ 14$

$\frac{1}{2}\{z - (\phi - \delta)\} = 16\ 47\ 20$

$\log \sin \frac{1}{2}\{z + (\phi - \delta)\} = 9·740215$

$\log \sin \frac{1}{2}\{z - (\phi - \delta)\} = 9·460666$

$\log \frac{1}{\cos\phi \cos\delta} = ·068746$

9·269627

$\log \sin \frac{\lambda}{2} = 9·634814$

$\frac{\lambda}{2} = 1^h\ 42^m\ 12^s·5$

$\lambda = 3\ 24\ 25·0$

Chron. correction = 23 37·9

28 *October* 1903.

$\phi = 29°\ 2'\ 14''$ $\qquad\qquad$ $\delta = 12°\ 45'\ 12''$

Time h m s	Altitude ° ′ ″
9 12 19·3	52 52 18
	37 7 42

refraction = 42

$z = 37\ \ 8\ 24$

$\frac{1}{2}\{z + (\phi - \delta)\} = 26\ 42\ 43$

$\frac{1}{2}\{z - (\phi - \delta)\} = 10\ 25\ 41$

$\log \sin \frac{1}{2}\{z + (\phi - \delta)\} = 9·652735$

$\log \sin \frac{1}{2}\{z - (\phi - \delta)\} = 9·257681$

$\log \frac{1}{\cos\phi \cos\delta} = ·069186$

8·979602

$\frac{\lambda}{2} = 4^h\ 48^m\ 1^s·85$

$\lambda = 9\ 36\ \ 3·7$

Chron. correction = 23 44·4

$\log \cos \phi = 9·941663$

$\log \cos \delta = 9·989151$

$\log \cos \phi \cos \delta = 9·930814$

Time h m s	Altitude ° ′ ″
9 14 35·5	53 20 32
	36 39 28

refraction = 42

$z = 36\ 40\ 10$

$\frac{1}{2}\{z + (\phi - \delta)\} = 26\ 28\ 36$

$\frac{1}{2}\{z - (\phi - \delta)\} = 10\ 11\ 34$

$\log \sin \frac{1}{2}\{z + (\phi - \delta)\} = 9·649172$

$\log \sin \frac{1}{2}\{z - (\phi - \delta)\} = 9·247876$

$\log \frac{1}{\cos\phi \cos\delta} = ·069186$

8·966234

$\frac{\lambda}{2} = 4^h\ 49^m\ 10^s·0$

$\lambda = 9\ 38\ 20·0$

Chron. correction = 23 44·5

27 *October* 1903.

m s	
23 36·0	= mean chronometer correction
− 1 20·0	= error of chronometer
− 15 57·1	= equation of time
+ 6 18·9	E. of 30° E. of Greenwich

Longitude = 31° 34′·8 E. of Greenwich

28 *October* 1903.

m s	
23 44·4	= mean chronometer correction
− 1 22·5	= error of chronometer
− 16 1·5	= equation of time
+ 6 20·4	E. of 30° E. of Greenwich

31° 35′ E. of Greenwich

Mean adopted 31° 35′·0 E. of Greenwich

Azimuth.

27 *October* 1903.

$\delta = 12° \; 29' \; 27''$

Time	Circle reading	Circle reading of reference line magnetometer-theodolite
h m s	° ′ ″	° ′ ″
2 44 41·5	247 54 17	292 45 11
+ 23 37·5 = Chron. cor.		
3 8 19·0 = λ		
1 34 9·5 = $\frac{\lambda}{2}$		

$\delta = 12° \; 29' \; 51''$

Time	Circle reading	Circle reading of reference line magnetometer-theodolite
h m s	° ′ ″	° ′ ″
3 12 9·8	243 47 40	292 45 0
+ 23 37·5 = Chron. cor.		
3 35 47·3 = λ		
1 47 53·7 = $\frac{\lambda}{2}$		

69° 14′ 10″ = $\frac{1}{2}(\phi' + \delta')$
8 16 23 = $\frac{1}{2}(\phi' - \delta')$

69° 13′ 57″ = $\frac{1}{2}(\phi' + \delta')'$
8 16 12 = $\frac{1}{2}(\phi' - \delta')'$

9·995458 = log cos $\frac{1}{2}(\phi' - \delta')$	9·158034 = log sin $\frac{1}{2}(\phi' - \delta')$
9·549638 = log cos $\frac{1}{2}(\phi' + \delta')$	9·970835 = log sin $\frac{1}{2}(\phi' + \delta')$
0·445820	9·187199
0·360878 = log cot $\frac{\lambda}{2}$	0·360878 = log cot $\frac{\lambda}{2}$
0·806698 = log tan $\frac{1}{2}(A + B)$	9·548077 = log tan $\frac{1}{2}(A - B)$
$\frac{1}{2}(A + B)$ = 81° 7′ 47″	$\frac{1}{2}(A - B)$ = 19° 27′ 20″

	A = 100° 35′ 7″
Circle reading for sun	= 247 54 17
Circle reading for meridian	= 147 19 10
Circle reading for reference line magnetometer-theodolite	= 292 45 11
Bearing of reference line magnetometer-theodolite	= 145 26 1 E.N.

9·995460 = log cos $\frac{1}{2}(\phi' - \delta')'$	9·157874 = log sin $\frac{1}{2}(\phi' - \delta')'$
9·549711 = log cos $\frac{1}{2}(\phi' + \delta')'$	9·970824 = log sin $\frac{1}{2}(\phi' + \delta')'$
0·445749	9·187050
0·293325 = log cot $\frac{\lambda}{2}$	0·293325 = log cot $\frac{\lambda}{2}$
0·739074 = log tan $\frac{1}{2}(A + B)$	9·480375 = log tan $\frac{1}{2}(A - B)$
$\frac{1}{2}(A + B)$ = 79° 39′ 54″	$\frac{1}{2}(A - B)$ = 16° 49′ 4″

	A = 96° 28′ 58″
Circle reading for sun	= 243 47 40
Circle reading for meridian	= 147 18 42
Circle reading for reference line magnetometer-theodolite	= 292 45 0
Bearing of reference line magnetometer-theodolite	= 145 26 18 E.N.

Mean bearing of reference line magnetometer-theodolite = 145° 26′·2 E.N.

APPENDIX B.

UMTALI, 16 April 1903. Magnetometer 73. Observer and Recorder, Beattie.

Observations for Declination.

G.M.T.	Magnet	Scale reading	Circle reading	Torsion zero	Magnetic meridian	Circle reading of line magnetometer-theodolite on magnetometer
h m			° ′ ″		° ′	° ′ ″
2 20 p.m.	erect	+ 11·9		70		
	inverted	− 6·75	A. 115 52 30		115 53·7	A. 179 50 30
	inverted	− 6·75	B. 295 54 50		+ 4·6 = reduction to zero	B. 359 52 40
2 27 p.m.	erect	+ 12·1			0·0 = torsion correction	
					115 58·3	
				70		A. 179 50 30
						B. 359 52 40
2 53 p.m.	erect	+ 12·0				
	inverted	− 6·8	A. 115 52 50		115 53·9	
	inverted	− 6·9	B. 295 55 0		+ 4·7 = reduction to zero	
3 1 p.m.	erect	+ 12·5			− 0·8 = torsion correction	
				50	115 57·8	
						A. 179 50 30
						B. 359 52 40
					Mean	179 51·5

The true bearing of the line magnetometer-theodolite was 48° 1′·0 E.N.

This gives for the declination at $2^h\ 23^m{\cdot}5$ p.m., 15° 52′·2 W.

,, ,, 2 27 ,, 15 52·7 ,,

APPENDIX C.

UMTALI, 18 April 1903. Circle 142. Observer and Recorder, Beattie.

Dip Observations.

Meridian.

Needle 3.

		Upper end	Lower end	Mean	Final mean
		° ′	° ′	° ′	° ′
Facing N.	A dipping	11 26	11 23		
		11 39	11 37	$11\ 22\frac{1}{8}$	
” S.	A dipping	11 19	11 19		
		11 7	11 17		11 22·5
Backing S.	A dipping	10 53	10 56		
		10 43	10 46	$11\ 22\frac{7}{8}$	
” N.	A dipping	11 50	11 52		
		11 61	11 61		

Dip.

		Needle 3			Needle 4		
Time beginning		$9^h\ 30^m$ a.m. G.M.T.			$9^h\ 38^m$ a.m. G.M.T.		
		Upper end	Lower end	Mean	Upper end	Lower end	Mean
		° ′	° ′	° ′	° ′	° ′	° ′
Facing W.	A dipping	52 79	$52\ 70\frac{1}{2}$		$52\ 66\frac{1}{2}$	52 60	
		$19\frac{1}{2}$	11		$66\frac{1}{2}$	60	
Facing E.	A dipping	52 66	$52\ 47\frac{1}{2}$		$52\ 74\frac{1}{2}$	52 55	
		66	48	53 4·9	$74\frac{1}{2}$	56	$52\ 62\frac{3}{4}$
Backing E.	A dipping	$52\ 71\frac{1}{2}$	52 51		52 55	$52\ 37\frac{1}{2}$	
		71	52		56	39	
Backing W.	A dipping	$52\ 69\frac{1}{2}$	52 64		$52\ 79\frac{1}{2}$	52 72	
		69	64		80	72	
Backing W.	B dipping	52 57	52 48		52 60	$52\ 53\frac{1}{2}$	
		57	$48\frac{1}{2}$		$59\frac{1}{2}$	53	
Backing E.	B dipping	$52\ 61\frac{1}{2}$	52 45		52 65	$52\ 46\frac{1}{2}$	
		62	45	52 50·7	65	46	$52\ 55\frac{3}{8}$
Facing E.	B dipping	52 55	$52\ 35\frac{1}{2}$		52 54	52 36	
		56	$35\frac{1}{2}$		55	36	
Facing W.	B dipping	52 54	$52\ 48\frac{1}{2}$		52 68	$52\ 60\frac{1}{2}$	
		54	48		$67\frac{1}{2}$	$60\frac{1}{2}$	
Time ending		$10^h\ 30^m$ a.m. G.M.T.			$10^h\ 23^m$ a.m. G.M.T.		
Mean Dip		52° 57′·8			52° 59′·0		
G.M.T.		$10^h\ 0^m$ a.m.			$10^h\ 0·5^m$ a.m.		

Needle 1

Time beginning 9h 47m a.m. G.M.T.

		Upper end	Lower end	Mean
		° ′	° ′	° ′
Facing W.	A dipping	52 56½	52 48½	
		56½	49	
Facing E.	A dipping	52 73	52 53	
		73	53	
Backing E.	A dipping	52 63	52 46	52 57·9
		63½	46	
Backing W.	A dipping	52 66	52 57½	
		65½	57	
Backing W.	B dipping	52 62	52 55	
		62	55½	
Backing E.	B dipping	52 73½	52 53½	
		73½	54	
Facing E.	B dipping	52 57	52 41	52 59·3
		57½	42	
Facing W.	B dipping	52 69	52 61½	
		69	61	

Time ending 10h 15m a.m. G.M.T.

Mean Dip 52° 58′·6

G.M.T. 10h 1m a.m.

APPENDIX D.

UMTALI, 16 April 1903. Magnetometer 73. Magnet 73 A. Chron. Reid and Son, rate -1^s. Observer and Recorder, Beattie.

Observations for Horizontal Intensity.

First Vibration.

			Temp. C.	Temp. corrected
Semiarc at beginning	75′	At beginning	34°·4	33°·6 C.
,, end	24	At middle	34 ·5	
		At end	34 ·5	

Scale moving apparently to right					Scale moving apparently to left				
No.	Time of passing	No.	Time of passing	Time of 100	No.	Time of passing	No.	Time of passing	Time of 100
	h m s		h m s	m s		h m s		h m s	m s
0	1 12 49·0	100	1 20 8·6	7 19·5	5	1 13 11·2	105	1 20 30·45	7 19·25
10	1 13 33·05	110	1 20 52·55	7 19·5	15	1 13 55·05	115	1 21 14·5	7 19·45
20	1 14 17·0	120	1 21 36·3	7 19·3	25	1 14 38·85	125	1 21 58·2	7 19·35
30	1 15 1·0	130	1 22 20·25	7 19·25	35	1 15 22·8	135	1 22 42·05	7 19·25
40	1 15 44·8	140	1 23 4·1	7 19·3	45	1 16 6·8	145	1 23 26·2	7 19·4
50	1 16 28·9	150	1 23 48·15	7 19·25	55	1 16 50·75	155	1 24 10·0	7 19·25
60	1 17 12·8	160	1 24 32·05	7 19·25	65	1 17 34·8	165	1 24 54·0	7 19·2

2 55·8 = Time for 40

1 20 8·6 = Time of 100th

Mean Time for 100 I A 4^s·3940 I B 4^s·3931 at 11.18 a.m. G.M.T.

Second Vibration.

			Temp. C.	Temp. corrected
Semiarc at beginning	70′	At beginning	30°·4	30°·1 C.
,, end	24	At middle	31 ·0	
		At end	31 ·7	

Scale moving apparently to right					Scale moving apparently to left				
No.	Time of passing	No.	Time of passing	Time of 100	No.	Time of passing	No.	Time of passing	Time of 100
	h m s		h m s	m s		h m s		h m s	m s
0	2 52 38·85	100	2 59 58·15	7 19·3	5	2 53 0·85	105	3 0 20·15	7 19·3
10	2 53 22·8	110	3 0 42·1	7 19·3	15	2 53 44·75	115	3 1 4·1	7 19·35
20	2 54 6·8	120	3 1 26·1	7 19·3	25	2 54 28·75	125	3 1 48·15	7 19·4
30	2 54 50·65	130	3 2 10·0	7 19·35	35	2 55 12·55	135	3 2 32·0	7 19·45
40	2 55 34·55	140	3 2 54·0	7 19·45	45	2 55 56·55	145	3 3 15·9	7 19·35
50	2 56 18·55	150	3 3 37·9	7 19·35	55	2 56 40·35	155	3 3 59·8	7 19·45
60	2 57 2·35	160	3 4 21·75	7 19·4	65	2 57 24·35	165	3 4 43·7	7 19·35

2 55·7 = Time for 40

2 59 58·05 = Time of 100th

Mean Time for 100 II A 4^s·3934 II B 4^s·3938 at 12.58 p.m. G.M.T.

Final Means I A and II B 4·3939 } at 12.8 p.m. G.M.T. and temp. 31°·85 C.
,, II A and I B 4·3933 }

Deflections.

Deflecting magnet 73 A, deflected 73 C.

Time at beginning 11.42 a.m. G.M.T. Time at end 12.38 p.m.

Distance cms.	Deflecting Magnet North end	Deflecting Magnet Temp. C.	Circle reading	Scale reading	Corrected Circle reading
West		°	° ′ ″		° ′ ″
40	W.	31·5	109 40 0 37 10	200·1	109 38 41
40	E.	31·6	121 6 30 4 10	199·8	121 5 8
35	E.	32·0	123 56 10 53 50	199·95	123 54 57
35	W.	32·0	106 48 20 46 20	200·4	106 47 44
30	W.	32·0	101 37 20 35 10	200·7	101 36 57
30	E.	32·2	129 3 30 1 30	200·0	129 2 30
25	E.	32·5	139 31 40 28 30	199·75	139 29 50
25	W.	32·5	91 3 30 1 30	200·5	91 3 0
East					
25	W.	31·5	91 20 0 17 10	199·95	91 18 32
25	E.	29·2	139 47 10 44 20	200·3	139 46 3
30	E.	29·3	129 10 40 8 0	200·2	129 9 32
30	W.	29·8	101 46 0 43 40	200·7	101 45 32
35	W.	30·5	106 53 30 51 30	199·85	106 52 21
35	E.	29·5	124 0 30 3 57 40	200·35	123 59 26
40	E.	29·5	121 9 20 6 50	200·2	121 8 17
40	W.	30·1	109 42 30 0 40	200·6	109 42 11

	Means and differences	Temp. C.	Corrected Temp. C.	G.M.T.
	° ′ ″	°	°	h m
	121 6 42·5 109 40 26			
	11 26 16·5			
40 cm.	5 43 8·3 = θ	30·7	31·2	12 10 p.m.
	123 57 11·5 106 50 2·5			
	17 7 9			
35 cm.	8 33 34·5 = θ	30·8	31·3	12 10 p.m.

	Means and differences	Temp. C.	Corrected Temp. C.	G.M.T.
	° ′ ″	°	°	h m
	129 6 1 101 41 14·5			
	27 24 46·5			
30 cm.	13 42 23·3 = θ	30·8	31·3	12 10 p.m.
	139 37 56·5 91 10 46·0			
	48 27 10·5			
25 cm.	24 13 35·3 = θ	31·4	31·9	12 10 p.m.

Reduction of Vibration Experiments.

		IA and IIB		IIA and IB	
$1 + \frac{G}{F} =$	1·00129	log 4·3939	= ·64285	log 4·3933	= ·64278
$- qt - q_1t^2 = -$	·01015		·64285		·64279
$\frac{2\mu}{r^2 \sin \theta} =$	·00146		1·28570		1·28557
	0·99260	log 0·99260	$= \bar{1}·99677$	log 0·99260	$= \bar{1}·99677$
			1·28247		1·28234
		log $\pi^2 K$	= 3·47439	log $\pi^2 K$	= 3·47439
		log MH	= 2·19192	log MH	= 2·19205
		log (correction for rate and arc) =	0	log (correction for rate and arc) =	0

Reduction of Deflection Experiments.

Distance	25 cms.	30 cms.	35 cms.	40 cms.
$1 + \frac{2\mu}{r^3}$	= 1·00061	1·00035	1·00022	1·00015
$qt + q_1t^2$	= ·01016	·00996	·00996	·00992
(a)	1·01077	1·01031	1·01018	1·01007
$\log \frac{r^3}{2}$	= 3·89317	4·13076	4·33168	4·50569
$\log \sin \theta$	= 9·61315	9·37465	9·17271	8·99849
log (a)	= ·00465	·00445	·00440	·00435
$\log \frac{M}{H}$	= 3·51097	3·50986	3·50879	3·50853
log MH	= 2·19192	2·19192	2·19205	2·19205
log M^2	= 5·70289	5·70178	5·70084	5·70058
log M	= 2·85145	2·85089	2·85042	2·85029
log P. Correction from all observations	= −·00230	−·00177	−·00119	−·00099
log M. Corrected	= 2·84915	2·84912	2·84923	2·84930
M	= 706·55	706·52	706·69	706·82
log H^2	$= \bar{2}·68095$	$\bar{2}·68206$	$\bar{2}·68326$	$\bar{2}·68352$
log H	$= \bar{1}·34048$	$\bar{1}·34103$	$\bar{1}·34163$	$\bar{1}·34176$
log P. Correction	= +·00230	+·00177	+·00119	+·00099
	$\bar{1}·34278$	$\bar{1}·34280$	$\bar{1}·34282$	$\bar{1}·34275$
H	= ·22018	·22020	·22020	·22017

APPENDIX E.

The stations are numbered and arranged in alphabetical order and for each is given the latitude, and the longitude, followed by a short description of the position, the declination, the dip, and the horizontal intensity.

Under declination (D) is given in column 1 the date, in column 2 the Greenwich mean time of the observation, in 3 the actual observed value, in 4 the value—corrected for daily and for secular variation—at the epoch 1st July 1903. The last two columns give the name of the observer and the number of the instrument respectively.

Under dip (θ) is given in the first two columns the date and the Greenwich mean time of the observation, the number of the needle* in column 3, the actually observed dip in column 4, the dip—corrected for secular variation—at the epoch 1st July 1903 in column 5. The last two columns give the name of the observer and the number of the instrument respectively.

Under horizontal intensity (H) in the first column is given the date, in the second column the times of the different observations are given—the time of a vibration being followed by a V and of a deflection by a D, in the third column the actual values of the horizontal intensity—calculated for different distances at places where both vibration and deflection were observed—are given. The method of calculation has already been explained in Appendix D. At stations where only a vibration or a deflection was observed, the most probable value of the moment of the magnet was found at the time of the particular observation and used with it to calculate the horizontal intensity. In the fourth column the horizontal intensity—corrected for secular variation—is given at the epoch 1st July 1903. The two last columns give the names of the observer and the number of the instrument respectively.

* A needle number with an index number—e.g. 4_9—means Needle 4 of instrument 9. If no index number is used then the needle belongs to the instrument given in the last column.

1. ABELSDAM. Lat. 27° 37′·1 S.; Long. 26° 29′·6 E. Between farmhouse and Zandspruit drift, about 100 paces from house on road Bothaville to Hoopstad. Hoopstad side of Zandspruit.

Declination. D.

Date	G.M.T.	D (observed)	D	Observer	Instrument
1904 Feb. 14	7 50 a.m.	23° 34′·1 W.			
	12 10 p.m.	23 29 ·9	23° 38′·1 W.	Beattie	73
	12 18 p.m.	23 29 ·9			

Dip. θ.

Date	G.M.T.	Needle	θ (observed)	θ	Observer	Instrument
1904 Feb. 14	9 43 a.m.	1	59° 6′·2 S.	59° 2′·5 S.	Beattie	142
	9 43 a.m.	4_9	59 7 ·5			

Horizontal Intensity. H.

Date	G.M.T.	H (observed)	H	Observer	Instrument
1904 Feb. 14	8 43 a.m. V.	H_{30} ·18741			
	10 50 a.m. D.	H_{40} ·18739	·18790	Beattie	73
	11 56 a.m. V.				

2. ABERDEEN, C. C. Lat. 32° 29′·1 S.; Long. 24° 3′·0 E.

Declination. D.

Date	G.M.T.	D (observed)	D	Observer	Instrument
1900 July 11	7 44 a.m.	27° 44′·8 W.			
	7 54 a.m.	27 45 ·1			
	8 2 a.m.	27 45 ·4	27° 14′·3 W.	Beattie	31
	1 19 p.m.	27 43 ·0		Morrison	
	1 29 p.m.	27 42 ·9			
	1 36 p.m.	27 43 ·1			

Dip. θ.

Date	G.M.T.	Needle	θ (observed)	θ	Observer	Instrument
1900 July 11	9 25 a.m.	1	60° 0′·5 S.	60° 24′·7 S.	Beattie	9
	9 25 a.m.	2	60 0 ·9			

Horizontal Intensity. H.

Date	G.M.T.	H (observed)	H	Observer	Instrument
1900 July 11	12 24 p.m. V.	H_{30} ·18245			
	1 9 p.m. D.	H_{40} ·18242	·17991	Morrison	31
	1 42 p.m. V.				

3. ABERDEEN (TRANSVAAL). Lat. 26° 3′·8 S.; Long. 29° 33′·0 E. In hollow on Bethal side of farmhouse; about half a mile to the right of the pool on Bethal Road. 22 miles from Bethal.

Declination. D.

Date	G.M.T.	D (observed)	D	Observer	Instrument
1903 Aug. 3	12 50 p.m.	21° 14′·9 W.	21° 15′·6 W.	Beattie	73
	1 1 p.m.	21 14 ·4			

Dip. θ.

Date	G.M.T.	Needle	θ (observed)	θ	Observer	Instrument
1903 Aug. 3	9 22 a.m.	4	58° 43′·2 S.			
	9 22 a.m.	1	58 44 ·5	58° 43′·3 S.	Beattie	142
	9 22 a.m.	3	58 44 ·1			

Horizontal Intensity. H.

Date	G.M.T.	H (observed)	H	Observer	Instrument
1903 Aug. 3	11 26 a.m. V.	H_{30} ·19211	·19217	Beattie	73
	12 24 p.m. D.	H_{40} ·19210			
	1 25 p.m. V.				

4. ABERDEEN ROAD. Lat. 32° 46′·0 S.; Long. 24° 20′·0 E. About one and a half miles from station on right-hand side of road going towards Aberdeen.

Declination. D.

Date	G.M.T.	D (observed)	D	Observer	Instrument
1900 July 12	2 45 p.m.	27° 33′·1 W.			
	2 58 p.m.	27 32 ·7	27° 6′·6 W.	Beattie Morrison	31
	3 13 p.m.	27 34 ·2			
1904 July 6	2 20 p.m.	26 55 ·4	27 5 ·9	Morrison	31
	2 28 p.m.	26 55 ·9			
			27 6 ·2 (mean adopted)		

Dip. θ.

Date	G.M.T.	Needle	θ (observed)	θ	Observer	Instrument
1900 July 12	1 40 p.m.	1	59° 56′·7 S.	60° 19′·6 S.	Beattie	9
	1 40 p.m.	2	59 55 ·0			
1904 July 6	1 22 p.m.	1	60 31 ·4	60 22 ·0	Morrison	9
	1 20 p.m.	2	60 28 ·7			
				60 20 ·8 (mean adopted)		

Horizontal Intensity. H.

Date	G.M.T.	H (observed)	H	Observer	Instrument
1900 July 12	10 2 a.m. V.	H_{30} ·18441	·18187	Morrison	31
	10 56 a.m. D.	H_{40} ·18437			
	11 32 a.m. V.				
1904 July 6	7 a.m. V.	H_{30} ·18104	·18197	Morrison	31
	8 12 a.m. D.	H_{40} ·18109			
	8 30 a.m. V.				
			·18192 (mean adopted)		

5. Aberfeldy. Lat. 25° 45′·7 S.; Long. 28° 34′·5 E. Right-hand side of railway Middelburg to Pretoria. 66 paces from dead end towards Pretoria, then 220 paces at right angles to railway.

Declination. D.

Date	G.M.T.	D (observed)	D	Observer	Instrument
1903 Sept. 22	4 40 a.m.	22° 7′·2 W.			
	4 51 a.m.	22 7·9	22° 14′·4 W.	Beattie	73
	8 4 a.m.	22 17·4			
	8 12 a.m.	22 18·2			

Dip. θ.

Date	G.M.T.	Needle	θ (observed)	θ	Observer	Instrument
1903 Sept. 22	8 58 a.m.	1	58° 5′·1 S.	58° 3′·5 S.	Beattie	142
	8 59 a.m.	4_9	58 5·1			

Horizontal Intensity. H.

Date	G.M.T.	H (observed)	H	Observer	Instrument
1903 Sept. 22	6 38 a.m. V.	H_{30} ·19411			
	7 30 a.m. D.	H_{40} ·19409	·19425	Beattie	73
	7 54 a.m. V.				

6. Adelaide. Lat. 32° 43′·0 S.; Long. 26° 18′·0 E. To reach the magnetometer from the railway station pass close to and below an excavated dam with a pump at the lower end; the magnetometer was about 135 yards from this and about 100 yards from the graveyard.

Declination. D.

Date	G.M.T.	D (observed)	D	Observer	Instrument
1904 July 9	1 10 p.m.	26° 11′·6 W.			
	1 19 p.m.	26 10·6			
	1 27 p.m.	26 12·2			
1904 July 10	7 20 a.m.	26 11·1	26° 21′·5 W.	Morrison	31
	7 29 a.m.	26 10·2			
	7 38 a.m.	26 12·3			

Dip. θ.

Date	G.M.T.	Needle	θ (observed)	θ	Observer	Instrument
1904 July 9	12 23 a.m.	1	61° 30′·0 S.	61° 22′·0 S.	Morrison	9
	12 22 a.m.	2	61 30·1			

Horizontal Intensity. H.

Date	G.M.T.	H (observed)	H	Observer	Instrument
1904 July 9	7 46 a.m. V.	H_{30} ·17656			
	8 22 a.m. D.	H_{40} ·17650	·17743	Morrison	31
	9 7 a.m. V.				

7. ALBERT FALLS. Lat. 29° 26'·0 S.; Long. 30° 29'·0 E. On triangle side of railway. 120 paces from end of triangle farther from railway and at right angles to it.

Declination. D.

Date	G.M.T.	D (observed)	D	Observer	Instrument
1903 Oct. 14	6 4 a.m.	23° 43'·1 W.			
	6 27 a.m.	23 44 ·4			
	6 45 a.m.	23 45 ·9	23° 46'·5 W.	Beattie	73
	8 31 a.m.	23 45 ·4			
	8 39 a.m.	23 44 ·9			

Dip. θ.

Date	G.M.T.	Needle	θ (observed)	θ	Observer	Instrument
1903 Oct. 14	7 53 a.m.	1	60° 58'·3 S.			
	7 54 a.m.	4_9	60 55 ·6	60° 54'·7 S.	Beattie	142
	7 54 a.m.	4	60 55 ·5			

Horizontal Intensity. H.

Date	G.M.T.	H (observed)	H	Observer	Instrument
1903 Oct. 14	6 56 a.m. V.	·18320	·18338	Beattie	73
	8 53 a.m. V.	·18320			

8. ALICEDALE. Lat. 33° 18'·9 S.; Long. 26° 2'·5 E. In field adjoining Munro's Hotel. On side of hotel away from Station and about 200 yards from house.

Declination. D.

Date	G.M.T.	D (observed)	D	Observer	Instrument
1900 Jan. 20	2 15 p.m.	28° 0'·3 W.			
	2 30 p.m.	28 0 ·3	27° 27'·2 W.	Beattie Morrison	31
	2 45 p.m.	28 0 ·1			

Dip. θ.

Date	G.M.T.	Needle	θ (observed)	θ	Observer	Instrument
1900 Jan. 20	9 10 a.m.	1	61° 9'·3 S.	61° 36'·5 S.	Beattie	9
	9 10 a.m.	2	61 8 ·3			

Horizontal Intensity. H.

Date	G.M.T.	H (observed)	H	Observer	Instrument
1900 Jan. 20	8 51 a.m. V.				
	9 33 a.m. D.	H_{30} ·17853 H_{40} ·17862	·17546	Morrison	31
	10 5 a.m. V.				

9. ALIWAL NORTH. Lat. 30° 41′·7 S.; Long. 26° 42′·0 E. In field on north side of the recreation ground.

Declination. D.

Date		G.M.T.	D (observed)	D	Observer	Instrument
1901	Dec. 26	7 8 a.m.	25° 40′·8 W.			
		7 16 a.m.	25 41 ·5			
		7 25 a.m.	25 40 ·8	25° 21′·0 W.	Beattie Morrison	31
		3 50 p.m.	25 33 ·0			
		4 2 p.m.	25 33 ·0			
		4 12 p.m.	25 33 ·4			

Dip. θ.

Date		G.M.T.	Needle	θ (observed)	θ	Observer	Instrument
1901	Dec. 26	9 0 a.m.	2	60° 11′·7 S.	60° 22′·1 S.	Beattie	142
		9 12 a.m.	1	60 8 ·5			

Horizontal Intensity. H.

Date		G.M.T.	H (observed)	H	Observer	Instrument
1901	Dec. 26	1 51 p.m. V.	H_{30} ·18337	·18205	Morrison	31
		2 34 p.m. D.	H_{40} ·18342			
		3 9 p.m. V.				

10. ALMA. Lat. 27° 7′·6 S.; Long. 31° 5′·5 E. Right-hand side of road coming from Piet Retief. About six miles from sulphur springs.

Declination. D.

Date		G.M.T.	D (observed)	D	Observer	Instrument
1903	Aug. 18	1 18 p.m.	22° 34′·6 W.			
		1 30 p.m.	22 36 ·3	22° 37′·1 W.	Beattie	73
		1 41 p.m.	22 36 ·3			

Dip. θ.

Date		G.M.T.	Needle	θ (observed)	θ	Observer	Instrument
1903	Aug. 18	11 54 a.m.	4	59° 21′·4 S.			
		11 57 a.m.	1	59 23 ·0	59° 21′·8 S.	Beattie	142
		11 56 a.m.	4_9	59 20 ·9			

Horizontal Intensity. H.

Date		G.M.T.	H (observed)	H	Observer	Instrument
1903	Aug. 18	1 55 p.m. V.	·18851	·18857	Beattie	73

11. AMABELE JUNCTION. Lat. 32° 43′·1 S.; Long. 27° 19′·2 E. At a place in a field south of the railway station and about 300 yards from it.

Declination. D.

Date	G.M.T.	D (observed)	D	Observer	Instrument
1906 Jan. 11	5 58 a.m.	25° 40′·3 W.			
	6 10 a.m.	25 39 ·7	26° 2′·9 W.	Brown Morrison	31
	6 15 a.m.	25 39 ·9			

Dip. θ.

Date	G.M.T.	Needle	θ (observed)	θ	Observer	Instrument
1906 Jan. 11	9 59 a.m.	1	61° 41′·5 S.	61° 22′·1 S.	Morrison	9
	9 59 a.m.	2	61 43 ·1			

Horizontal Intensity. H.

Date	G.M.T.	H (observed)	H	Observer	Instrument
1906 Jan. 11	7 49 a.m. V.				
	8 23 a.m. D.	H_{30} ·17576 H_{40} ·17568	·17815	Morrison	31
	9 0 a.m. V.				

12. AMARANJA. Lat. 31° 14′·7 S.; Long. 29° 30′·0 E. Nine miles from Flagstaff.

Dip. θ.

Date	G.M.T.	Needle	θ (observed)	θ	Observer	Instrument
1906 Jan. 26	4 15 p.m.	1	61° 20′·5 S.	61° 0′·2 S.	Morrison	9
	4 17 p.m.	2	61 21 ·2			

Horizontal Intensity. H.

Date	G.M.T.	H (observed)	H	Observer	Instrument
1906 Jan. 26	9 9 a.m. V.				
	9 45 a.m. D.	H_{30} ·17826	·18073	Morrison	31
	11 6 a.m. V.				

13. AMATONGAS. Lat. 19° 11′·2 S.; Long. 33° 45′·0 E. In corner of garden distant from store, about 100 yards from railway, and 60 from store. Right-hand side of railway Umtali to Beira.

Declination. D.

Date	G.M.T.	D (observed)	D	Observer	Instrument
1903 April 21	7 55 a.m.	16° 6′·0 W.	16° 4′·0 W.	Beattie	73

Dip. θ.

Date	G.M.T.	Needle	θ (observed)	θ	Observer	Instrument
1903 April 21	9 6 a.m.	3	53° 34′·8 S.	53° 38′·0 S.	Beattie	142
	9 6 a.m.	4	53 38 ·3			

Horizontal Intensity. H.

Date	G.M.T.	H (observed)	H	Observer	Instrument
1903 April 21	8 14 a.m. V.	·21851	·21839	Beattie	73

14. Ashton. Lat. 33° 50′·0 S.; Long. 20° 4′·0 E. In field opposite front of boarding-house.

Dip. θ.

Date	G.M.T.	Needle	θ (observed)	θ	Observer	Instrument
1901 Feb. 6	1 32 p.m.	1	59° 17′·9 S.	59° 40′·8 S.	Beattie	142
	1 32 p.m.	2	59 20 ·2			

Horizontal Intensity. H.

Date	G.M.T.	H (observed)	H	Observer	Instrument
1901 Feb. 6	12 15 p.m. V.	·18344	·18135	Morrison	31

15. Assegai Bosch. Lat. 33° 56′·7 S.; Long. 24° 20′·5 E. In field adjoining hotel, 200 yards west of hotel.

Declination. D.

Date	G.M.T.	D (observed)	D	Observer	Instrument
1903 Feb. 10	7 17 a.m.	27° 42′·3 W.	27° 33′·4 W.	Beattie	31
	8 6 a.m.	27 39 ·7			
	8 17 a.m.	27 38 ·6			

Horizontal Intensity. H.

Date	G.M.T.	H (observed)	H	Observer	Instrument
1903 Feb. 10	9 6 a.m. V.	H_{30} ·17778	·17746	Beattie	31
	9 57 a.m. D.	H_{40} ·17779			
	11 23 a.m. V.				

16. Avontuur. Lat. 33° 44′·2 S.; Long. 23° 13′·0 E. In field on right-hand side of road going to Knysna from Avontuur; about 300 yards from post office.

Declination. D.

Date	G.M.T.	D (observed)	D	Observer	Instrument
1903 Feb. 13	4 30 p.m.	27° 51′·4 W.	27° 48′·7 W.	Beattie	31
1903 Feb. 14	7 58 a.m.	27 56 ·7			
	8 13 a.m.	27 56 ·0			
	8 28 a.m.	27 55 ·8			

Dip. θ.

Date	G.M.T.	Needle	θ (observed)	θ	Observer	Instrument
1903 Feb. 13	2 20 p.m.	3	60° 39′·0 S.	60° 42′·1 S.	Beattie	142
	2 21 p.m.	4	60 39 ·3			

Horizontal Intensity. H.

Date	G.M.T.	H (observed)	H	Observer	Instrument
1903 Feb. 13	8 47 a.m. V.	H_{30} ·17876	·17842	Beattie	31
	9 33 a.m. D.	H_{40} ·17884			

17. Ayrshire Mine (Lo Maghonda District). Lat. 17° 11′·5 S.; Long. 30° 23′·0 E. In kloof between manager's house and kopje on which granite was being quarried. On line joining house and kopje and half-way between them.

Declination. D.

Date	G.M.T.	D (observed)	D	Observer	Instrument
1903 April 29	7 25 a.m.	16° 11′·8 W.			
	7 40 a.m.	16 10 ·2	16° 5′·0 W.	Beattie	73
	2 16 p.m.	16 2 ·5			
	2 32 p.m.	16 2 ·4			

Dip. θ.

Date	G.M.T.	Needle	θ (observed)	θ	Observer	Instrument
1903 April 29	12 11 p.m.	3	50° 51′·9 S.			
	12 11 p.m.	4	50 53 ·7	50° 54′·9 S.	Beattie	142
	12 10 p.m.	1	50 55 ·0			

Horizontal Intensity. H.

Date	G.M.T.	H (observed)	H	Observer	Instrument
1903 April 29	8 2 a.m. V.	H_{30} ·22813			
	8 48 a.m. D.	H_{40} ·22822	·22805	Beattie	73
	9 29 a.m. V.	H_{25} ·22813			
		H_{35} ·22811			

17 a. Balmoral. Lat. 25° 51′·3 S.; Long. 28° 58′·0 E. Right-hand side of railway Middelburg to Pretoria. Along railway 54 paces from dead end towards Middelburg, then 263 paces at right angles to the railway.

Declination. D.

Date	G.M.T.	D (observed)	D	Observer	Instrument
1903 Sept. 21	4 44 a.m.	22° 3′·0 W.			
	4 56 a.m.	22 2 ·3			
	6 30 a.m.	22 5 ·0			
	8 30 a.m.	22 9 ·2	22° 7′·0 W.	Beattie	73
	8 41 a.m.	22 7 ·9			
	8 53 a.m.	22 9 ·6			

Dip. θ.

Date	G.M.T.	Needle	θ (observed)	θ	Observer	Instrument
1903 Sept. 21	9 41 a.m.	1	58° 12′·6 S.	58° 9′·9 S.	Beattie	142
	9 40 a.m.	4_9	58 10 ·3			

Horizontal Intensity. H.

Date	G.M.T.	H (observed)	H	Observer	Instrument
1903 Sept. 21	7 4 a.m. V.	H_{30} ·19345			
	7 38 a.m. D.	H_{40} ·19348	·19362	Beattie	73
	8 16 a.m. V.				

18. BAMBOO CREEK. Lat. 19° 16′·5 S.; Long. 34° 12′·0 E. Right-hand side of railway Umtali to Beira. On Beira side of engine shed.

Declination. D.

Date	G.M.T.	D (observed)	D	Observer	Instrument
1903 April 22	7 44 a.m. 7 59 a.m.	15° 7′·4 W. 15 6·7	15° 5′·1 W.	Beattie	73

Dip. θ.

Date	G.M.T.	Needle	θ (observed)	θ	Observer	Instrument
1903 April 22	11 21 a.m. 11 22 a.m. 11 23 a.m.	3 4 1	53° 44′·3 S. 53 44·8 53 44·3	53° 46′·0 S.	Beattie	142

Horizontal Intensity. H.

Date	G.M.T.	H (observed)	H	Observer	Instrument
1903 April 22	8 17 a.m. V. 8 58 a.m. D. 10 2 a.m. V.	H_{30} ·21943 H_{40} ·21937 H_{25} ·21947 H_{35} ·21946	·21930	Beattie	73

19. BANKPAN. Lat. 26° 18′·0 S.; Long. 29° 35′·0 E. On right-hand side of road coming from Middelburg. Farmhouse about three-quarters of a mile off to right.

Dip. θ.

Date	G.M.T.	Needle	θ (observed)	θ	Observer	Instrument
1903 Aug. 4	10 14 a.m. 10 13 a.m.	1 4	59° 43′·9 S. 59 43′·6	59° 43′·2 S.	Beattie	142

Horizontal Intensity. H.

Date	G.M.T.	H (observed)	H	Observer	Instrument
1903 Aug. 4	11 6 a.m. V. 11 55 a.m. D. 12 32 p.m. D.	H_{30} ·18482 H_{40} ·18476	·18485	Beattie	73

20. **Barberton.** Lat. 25° 47′·3 S.; Long. 31° 0′·0 E. Place of observation on left-hand side of railway coming from Kaapmuiden; to reach it go from railway station towards the town till the road running parallel to the railway line is crossed, continue till the trench on the town side of the road. Magnetometer alongside trench on side away from town.

Declination. D.

Date	G.M.T.	D (observed)	D	Observer	Instrument
1903 Sept. 12	7 38 a.m. 8 5 a.m. 12 9 p.m. 12 16 p.m.	21° 14′·3 W. 21 14 ·0 21 7 ·8 21 7 ·3	21° 12′·8 W.	Beattie	73

Dip. θ.

Date	G.M.T.	Needle	θ (observed)	θ	Observer	Instrument
1903 Sept. 12	11 19 a.m. 11 18 a.m.	1 4	58° 38′·1 S. 58 37 ·2	58° 36′·2 S.	Beattie	142

Horizontal Intensity. H.

Date	G.M.T.	H (observed)	H	Observer	Instrument
1903 Sept. 12	8 22 a.m. V. 9 0 a.m. D. 9 27 a.m. D. 10 8 a.m. V.	H_{30} ·19276 H_{40} ·19274	·19287	Beattie	73

21. **Barrington (Rooi Kraal).** Lat. 33° 55′·2 S.; Long. 22° 52′·0 E. On public outspan, 200 paces south west from Wallace's store.

Declination. D.

Date	G.M.T.	D (observed)	D	Observer	Instrument
1903 Jan. 27	5 19 a.m. 5 29 a.m. 5 39 a.m.	26° 42′·6 W. 26 43 ·4 26 44 ·4	26° 37′·7 W.	Beattie	31

Dip. θ.

Date	G.M.T.	Needle	θ (observed)	θ	Observer	Instrument
1903 Jan. 26	4 11 p.m. 4 11 p.m.	3 4	60° 5′·1 S. 60 3 ·7	60° 7′·7 S.	Beattie	142

Horizontal Intensity. H.

Date	G.M.T.	H (observed)	H	Observer	Instrument
1903 Jan. 27	8 15 a.m. V. 9 47 a.m. V.	·18353	·18308	Beattie	31

22. BATTLEFIELDS. Lat. 18° 36′·4 S.; Long. 30° 0′·0 E. Left-hand side of railway Gwelo to Salisbury. 144 paces from points at Gwelo end of siding along railway towards Salisbury, then 192 paces at right angles to railway.

Declination. D.

Date	G.M.T.	D (observed)	D	Observer	Instrument
1903 April 11	8 8 a.m.	17° 26′·9 W.			
	8 25 a.m.	17 26 ·5	17° 23′·0 W.	Beattie	73
	2 22 p.m.	17 23 ·4			
	2 37 p.m.	17 22 ·8			

Dip. θ.

Date	G.M.T.	Needle	θ (observed)	θ	Observer	Instrument
1903 April 11	12 30 p.m.	3	52° 28′·5 S.			
	12 32 p.m.	4	52 30 ·4	52° 31′·0 S.	Beattie	142
	12 33 p.m.	1	52 28 ·9			

Horizontal Intensity. H.

Date	G.M.T.	H (observed)	H	Observer	Instrument
1903 April 11	8 47 a.m. V.	H_{30} ·21890			
	9 33 a.m. D.	H_{40} ·21887	·21872	Beattie	73
	10 48 a.m. V.	H_{25} ·21894			
		H_{35} ·21888			

23. BAVARIA. Lat. 27° 30′·7 S.; Long. 29° 8′·4 E. About 3 miles from Vrede. Right-hand side of road Harrismith to Vrede. On Vrede side of spruit with four trees growing; about 150 paces from the two largest trees.

Declination. D.

Date	G.M.T.	D (observed)	D	Observer	Instrument
1904 Feb. 3	8 15 a.m.	22° 31′·3 W.			
	8 24 a.m.	22 30 ·1	22° 29′·3 W.	Beattie	73
	12 37 p.m.	22 16 ·0			
	12 48 p.m.	22 16 ·5			

Dip. θ.

Date	G.M.T.	Needle	θ (observed)	θ	Observer	Instrument
1904 Feb. 3	11 43 a.m.	1	59° 18′·0 S.	59° 15′·6 S.	Beattie	142
	11 44 a.m.	4_9	59 21 ·6			

Horizontal Intensity. H.

Date	G.M.T.	H (observed)	H	Observer	Instrument
1904 Feb. 3	9 5 a.m. V.	H_{30} ·18798			
	9 39 a.m. D.	H_{40} ·18795	·18843	Beattie	73
	1 35 p.m. V.				

24. Baviaanskrantz Farm. Lat. 27° 23′·0 S.; Long. 26° 47′·0 E. The farm was formerly part of the farm Doornkuil. Left-hand side of road to Bothaville coming from east just opposite dam with big fig tree. About 13 miles from Bothaville.

Declination. D.

Date	G.M.T.	D (observed)	D	Observer	Instrument
1904 Feb. 12	1 58 p.m.	26° 43′·2 W.	26° 50′·0 W.	Beattie	73

Dip. θ.

Date	G.M.T.	Needle	θ (observed)	θ	Observer	Instrument
1904 Feb. 12	11 30 a.m.	1	58° 45′·7 S.	58° 41′·5 S.	Beattie	142
	11 37 a.m.	4_9	58 45 ·9			

Horizontal Intensity. H.

Date	G.M.T.	H (observed)	H	Observer	Instrument
1904 Feb. 12	12 22 p.m. V.	·18857	·18903	Beattie	73

24 a. Beaconsfield. Lat. 28° 45′·0 S.; Long. 24° 44′·0 E. On right-hand side of road, and 200 yards from it between Kimberley and Beaconsfield. Just past large brick building, 2 miles from Kimberley.

Dip. θ.

Date	G.M.T.	Needle	θ (observed)	θ	Observer	Instrument
1898 Jan. 25	5 30 a.m.	2	57° 58′·5 S.	58° 39′·8 S.	Beattie	9

25. Beaufort West. Lat. 32° 20′·9 S.; Long. 22° 34′·1 E. In a field on left-hand side of road to Lemoenfontein, on west side of railway.

Declination. D.

Date	G.M.T.	D (observed)	D	Observer	Instrument
1900 Jan. 9	7 40 a.m.	27° 29′·8 W.	27° 1′·7 W.	Beattie Morrison	31
	8 0 a.m.	27 31 ·3			
	8 20 a.m.	27 31 ·6			
1905 Jan. 14	7 44 a.m.	27 8 ·5	27 16 ·2	Beattie Brown	73
	7 52 a.m.	27 8 ·2			
			27 8 ·9 (mean adopted)		

Dip. θ.

Date	G.M.T.	Needle	θ (observed)	θ	Observer	Instrument
1899 July 13	10 35 a.m.	1	59° 15′·2 S.	59° 46′·6 S.	Beattie	9
1899 July 12	10 41 a.m.	2	59 11 ·6			
	11 40 a.m.	1	59 17 ·5			
	11 50 a.m.	2	59 14 ·0			
1900 Jan. 10	—	1	59 12 ·8	59 42 ·0	Beattie	9
		2	59 15 ·1			
1905 Jan. 14	10 12 a.m.	3_9	59 55 ·4	59 43 ·4	Beattie	142
	10 12 a.m.	4_9	59 55 ·9			
				59 43 ·0 (mean adopted)		

Horizontal Intensity. H.

Date	G.M.T.	H (observed)	H	Observer	Instrument
1899 July 12	12 36 p.m. V.	H_{30} ·18499	·18182		
	1 26 p.m. D.	H_{40} ·18505			
1899 July 13	10 41 a.m. V.	H_{30} ·18513		Morrison	31
	11 25 a.m. D.	H_{40} ·18501	·18187		
	12 0 noon V.				
1900 Jan. 10	7 6 a.m. V.				
	7 45 a.m. D.	H_{30} ·18507	·18227	Morrison	31
	9 50 a.m. V.	H_{40} ·18507			
1905 Jan. 14	8 12 a.m. V.	H_{30} ·18070	·18193	Beattie	73
	8 55 a.m. D.	H_{40} ·18070			
			·18205 (mean adopted)		

26. Beira. Lat. 19° 49′·2 S.; Long. 34° 50′·0 E. On low ground adjoining the station. Left-hand side of railway Beira to Umtali. 600 paces along railway from end of platform towards Umtali, then 400 paces at right angles to railway.

Declination. D.

Date	G.M.T.	D (observed)	D	Observer	Instrument
1903 April 23	7 29 a.m.	16° 0′·2 W.			
	8 12 a.m.	15 56 ·0			
	8 20 a.m.	15 56 ·0	15° 55′·9 W.	Beattie	73
	1 38 p.m.	15 56 ·3			
	1 53 p.m.	15 55 ·5			

Dip. θ.

Date	G.M.T.	Needle	θ (observed)	θ	Observer	Instrument
1903 April 23	12 8 p.m.	3	54° 32′·7 S.			
	12 8 p.m.	4	54 33 ·7	54° 34′·4 S.	Beattie	142
	12 8 p.m.	1	54 32 ·2			

Horizontal Intensity. H.

Date	G.M.T.	H (observed)	H	Observer	Instrument
1903 April 23	8 46 a.m. V.	H_{30} ·21685			
	9 56 a.m. D.	H_{40} ·21680	·21673	Beattie	73
	10 58 a.m. V.	H_{25} ·21688			
		H_{35} ·21687			

27. Belleville. Lat. 33° 49′·0 S.; Long. 18° 39′·0 E. (from Cape Meteorological Commission Report). Right-hand side of road Belleville to Durban Road, about half a mile from the station.

Dip. θ.

Date	G.M.T.	Needle	θ (observed)	θ	Observer	Instrument
1901 Aug. 31	11 38 a.m.	2	58° 56′·1 S.	59° 11′·3 S.	Beattie	9
	11 38 a.m.	1	58 53 ·5			

Horizontal Intensity. H.

Date	G.M.T.	H (observed)	H	Observer	Instrument
1901 Aug. 31	1 19 p.m. V.				
	2 6 p.m. D.	H_{30} ·18432	·18262	Morrison	31
	3 28 p.m. V.	H_{40} ·18427			

28. Berg River Mouth. Lat. 32° 46′·5 S.; Long. 18° 10′·0 E. At mouth of the Berg River, near the landing place, about half a mile north west of Stephan's house.

Declination. D.

Date	G.M.T.	D (observed)	D	Observer	Instrument
1901 July 15	3 12 p.m.	28° 27′·3 W.	28° 18′·3 W.	Beattie Morrison	31
	3 18 p.m.	28 27 ·8			
	3 26 p.m.	28 27 ·8			
1901 July 16	8 23 a.m.	28 25 ·8			
	8 31 a.m.	28 25 ·7			
	8 45 a.m.	28 24 ·9			

Dip. θ.

Date	G.M.T.	Needle	θ (observed)	θ	Observer	Instrument
1901 July 16	10 0 a.m.	2	58° 13′·1 S.	58° 29′·4 S.	Beattie	142
	10 0 a.m.	1	58 10 ·3			

Horizontal Intensity. H.

Date	G.M.T.	H (observed)	H	Observer	Instrument
1901 July 15	1 37 p.m. V.	H_{30} ·18623	·18435	Morrison	31
	2 17 p.m. D.	H_{40} ·18623			
	3 41 p.m. V.				

29. Bethal. Lat. 26° 28′·1 S.; Long. 29° 27′·5 E. Half-way between post office and hotel.

Declination. D.

Date	G.M.T.	D (observed)	D	Observer	Instrument
1903 Aug. 5	7 33 a.m.	22° 41′·2 W.	22° 44′·5 W.	Beattie	73
	7 46 a.m.	22 42 ·7			
	9 5 a.m.	22 47 ·8			
	9 16 a.m.	22 49 ·3			
	12 31 p.m.	22 40 ·3			
	12 42 p.m.	22 40 ·7			

Dip. θ.

Date	G.M.T.	Needle	θ (observed)	θ	Observer	Instrument
1903 Aug. 5	1 38 p.m.	1	58° 14′·1 S.	58° 13′·0 S.	Beattie	142
	1 38 p.m.	4	58 13 ·3			

Horizontal Intensity. H.

Date	G.M.T.	H (observed)	H	Observer	Instrument
1903 Aug. 5	9 38 a.m. V.	H_{30} ·19407	·19416	Beattie	73
	11 17 a.m. D.	H_{40} ·19413			
	11 56 a.m. V.				

30. BETHANY. Lat. 29° 37′·0 S.; Long. 26° 2′·0 E. Right-hand side of railway going towards Bloemfontein, 250 paces at right angles to railway starting from Bloemfontein end of platform.

Declination. D.

Date	G.M.T.	D (observed)	D	Observer	Instrument
1903 May 31	7 10 a.m.	24° 51′·6 W.			
	7 42 a.m.	24 52 ·1	24° 50′·0 W.	Beattie	73
	12 53 p.m.	24 49 ·3			
	1 10 p.m.	24 49 ·9			

Dip. θ.

Date	G.M.T.	Needle	θ (observed)	θ	Observer	Instrument
1903 May 31	8 53 a.m.	3	59° 41′·8 S.			
	8 53 a.m.	4	59 41 ·6	59° 42′·4 S.	Beattie	142
	8 54 a.m.	1	59 42 ·1			

Horizontal Intensity. H.

Date	G.M.T.	H (observed)	H	Observer	Instrument
1903 May 31	9 45 a.m. V.	H_{30} ·18444			
	11 51 a.m. D.	H_{40} ·18442	·18443	Beattie	73
	12 35 p.m. V.				

31. BETHESDA ROAD. Lat. 31° 55′·3 S.; Long. 24° 38′·0 E. Behind Sutton's house, about two hundred yards from it.

Declination. D.

Date	G.M.T.	D (observed)	D	Observer	Instrument
1900 July 4	1 39 p.m.	26° 48′·4 W.			
	1 47 p.m.	26 49 ·2			
	1 54 p.m.	26 48 ·3	26° 21′·4 W.	Beattie Morrison	31
1900 July 6	8 11 a.m.	26 54 ·2			
	8 23 a.m.	26 54 ·7			
	8 32 a.m.	26 54 ·4			

Dip. θ.

Date	G.M.T.	Needle	θ (observed)	θ	Observer	Instrument
1900 July 5	10 0 a.m.	2	60° 3′·7 S.			
1900 July 6	10 20 a.m.	1	60 3 ·9			
	10 20 a.m.	2	60 3 ·8	60° 28′·9 S.	Beattie	31
	11 10 a.m.	1	60 7 ·9			
	11 10 a.m.	2	60 6 ·4			

Horizontal Intensity. H.

Date	G.M.T.	H (observed)	H	Observer	Instrument
1900 July 4	2 17 p.m. V.	H_{30} ·18309			
	3 8 p.m. D.	H_{40} ·18316			
1900 July 5	1 0 p.m. V.	H_{30} ·18315	·18062	Morrison	31
	1 44 p.m. D.	H_{40} ·18318			
	2 12 p.m. V.				

32. BETHLEHEM. Lat. 28° 13′·8 S.; Long. 28° 17′·3 E. Outside village on left-hand side of road coming from Senekal. In hollow about 1½ miles from village.

Declination. D.

Date		G.M.T.	D (observed)	D	Observer	Instrument
1904	Jan. 28	1 33 p.m. 1 41 p.m.	23° 31′·6 W. 23 31 ·7	23° 38′·9 W.	Beattie	73

Dip. θ.

Date		G.M.T.	Needle	θ (observed)	θ	Observer	Instrument
1904	Jan. 28	8 40 a.m. 8 41 a.m.	1 4_9	59° 19′·1 S. 59 23 ·2	59° 17′·0 S.	Beattie	142

Horizontal Intensity. H.

Date		G.M.T.	H (observed)	H	Observer	Instrument
1904	Jan. 28	10 49 a.m. V. 11 29 a.m. D.	H_{30} ·18887 H_{40} ·18889	·18931	Beattie	73

33. BETHULIE. Lat. 30° 30′·5 S.; Long. 25° 59′·0 E. Left-hand side of railway, Bethulie to Springfontein. 200 paces at right angles to railway starting from end of platform away from Bloemfontein.

Declination. D.

Date		G.M.T.	D (observed)	D	Observer	Instrument
1903	May 29	7 32 a.m. 7 46 a.m. 2 36 p.m. 2 54 p.m.	25° 32′·3 W. 25 30 ·8 25 29 ·8 25 28 ·6	25° 30′·1 W.	Beattie	73

Dip. θ.

Date		G.M.T.	Needle	θ (observed)	θ	Observer	Instrument
1903	May 29	9 15 a.m. 9 16 a.m. 9 17 a.m.	3 4 1	59° 52′·5 S. 59 48 ·5 59 49 ·7	59° 50′·8 S.	Beattie	142

Horizontal Intensity. H.

Date		G.M.T.	H (observed)	H	Observer	Instrument
1905	May 29	10 56 a.m. V. 11 59 a.m. D. 12 35 p.m. D. 1 27 p.m. V.	H_{30} ·18398 H_{40} ·18403	·18400	Beattie	73

34. BIESJESPOORT. Lat. 31° 43′·8 S.; Long. 23° 12′·0 E. Left-hand side of railway coming from Cape Town. On Cape Town side of store alongside the road, and 150 paces from the railway.

Declination. D.

Date	G.M.T.	D (observed)	D	Observer	Instrument
1902 Feb. 10	3 5 p.m. 4 46 p.m.	26° 10′·3 W. 26 13 ·5	26° 1′·5 W.	Beattie	31

Dip. θ.

Date	G.M.T.	Needle	θ (observed)	θ	Observer	Instrument
1902 Feb. 11	5 52 a.m. 5 54 a.m.	2 1	59° 28′·5 S. 59 24 ·4	59° 37′·5 S.	Beattie	142

Horizontal Intensity. H.

Date	G.M.T.	H (observed)	H	Observer	Instrument
1902 Feb. 11	7 50 a.m. V.	·18564	·18445	Beattie	31

35. BIRTHDAY. Lat. 23° 19′·5 S.; Long. 30° 46′·0 E. Left-hand side of road, Pietersburg to Birthday. 150 paces perpendicular to road starting from a point on road 250 paces from house and on Pietersburg side of it.

Declination. D.

Date	G.M.T.	D (observed)	D	Observer	Instrument
1903 July 17	1 51 p.m. 2 4 p.m.	16° 6′·7 W. 16 7 ·6	16° 7′·5 W.	Beattie Löwinger	73

Dip. θ.

Date	G.M.T.	Needle	θ (observed)	θ	Observer	Instrument
1903 July 17	3 15 p.m. 3 15 p.m.	1 4	59° 7′·8 S. 59 8 ·5	59° 7′·0 S.	Beattie	142

Horizontal Intensity. H.

Date	G.M.T.	H (observed)	H	Observer	Instrument
1903 July 17	12 20 p.m. V. 1 12 p.m. D. 1 39 p.m. V.	H_{30} ·18566	·18566	Beattie	73

36. Blaauwbosch. Lat. 30° 38′·9 S.; Long. 22° 14′·1 E. Right-hand side of road coming from Carnarvon and passing the house. About 150 paces from the house.

Declination. D.

Date	G.M.T.	D (observed)	D	Observer	Instrument
1904 Dec. 10	6 8 a.m. 6 20 a.m.	26° 22′·2 W. 26 23 ·8	26° 33′·9 W.	Beattie Hough	73

Dip. θ.

Date	G.M.T.	Needle	θ (observed)	θ	Observer	Instrument
1904 Dec. 10	5 1 a.m. 5 1 a.m.	3_9 4_9	59° 3′·2 S. 59 4 ·6	58° 52′·2 S.	Beattie	142

Horizontal Intensity. H.

Date	G.M.T.	H (observed)	H	Observer	Instrument
1904 Dec. 10	6 40 a.m. V. 7 25 a.m. D.	H_{40} ·18381	·18496	Beattie	73

37. Blaauwkrantz. Lat. 33° 57′·0 S.; Long. 23° 35′·0 E. In middle of cleared ground immediately in front of the post office.

Declination. D.

Date	G.M.T.	D (observed)	D	Observer	Instrument
1903 Feb. 1	5 22 a.m. 7 12 a.m.	27° 50′·8 W. 27 50 ·2	27° 43′·9 W.	Beattie	31

Dip. θ.

Date	G.M.T.	Needle	θ (observed)	θ	Observer	Instrument
1903 Feb. 1	1 43 p.m. 1 44 p.m. 1 44 p.m.	3 4 1	60° 52′·4 S. 60 54 ·3 60 56 ·8	60° 57′·8 S.	Beattie	142

Horizontal Intensity. H.

Date	G.M.T.	H (observed)	H	Observer	Instrument
1903 Feb. 1	8 39 a.m. V. 9 45 a.m. D. 11 20 a.m. V.	H_{30} ·17853 H_{40} ·17849	·17813	Beattie	31

38. The Bluff. Lat. 29° 52′·5 S.; Long. 31° 4′·0 E. On west side of lighthouse on border of grass plot.

Declination. D.

Date	G.M.T.	D (observed)	D	Observer	Instrument
1903 Oct. 30	5 50 a.m.	23° 39′·0 W.	23° 40′·1 W.	Beattie	73
	5 59 a.m.	23 39 ·1			
	7 57 a.m.	23 36 ·6			
	8 5 a.m.	23 36 ·7			

Dip. θ.

Date	G.M.T.	Needle	θ (observed)	θ	Observer	Instrument
1903 Oct. 30	9 38 a.m.	1	61° 24′·6 S.	61° 22′·1 S.	Beattie	142
	9 43 a.m.	4_9	61 24 ·1			

Horizontal Intensity. H.

Date	G.M.T.	H (observed)	H	Observer	Instrument
1903 Oct. 30	6 37 a.m. V.	H_{30} ·17901	·17924	Beattie	73
	7 25 a.m. D.	H_{40} ·17903			
	8 21 a.m. V.				

39. Boschkopjes. Lat. 23° 11′·5 S.; Long. 29° 55′·0 E. Right-hand side of road, Pietersburg to Loveday, opposite farmhouse, 100 paces from road.

Declination. D.

Date	G.M.T.	D (observed)	D	Observer	Instrument
1903 July 14	6 28 a.m.	19° 50′·6 W.	19° 50′·9 W.	Beattie	73
	6 41 a.m.	19 50 ·6		Löwinger	

Dip. θ.

Date	G.M.T.	Needle	θ (observed)	θ	Observer	Instrument
1903 July 14	5 38 a.m.	1	56° 49′·4 S.	56° 49′·2 S.	Beattie	142
	5 30 a.m.	4	56 49 ·6			

Horizontal Intensity. H.

Date	G.M.T.	H (observed)	H	Observer	Instrument
1903 July 14	7 47 a.m. V.	H_{30} ·19958	·19957	Beattie	73
	8 27 a.m. D.	H_{40} ·19957			
	9 9 a.m. V.				

40. Boschrand. Lat. 27° 45′·8 S.; Long. 27° 12′·0 E. Right-hand side of railway going to Kroonstad. 225 paces at right angles to railway, starting from a point 64 paces from dead end towards Kroonstad.

Declination. D.

Date	G.M.T.	D (observed)	D	Observer	Instrument
1903 June 10	8 5 a.m.	23° 38′·6 W.			
	8 22 a.m.	23 39·6			
	12 20 p.m.	23 39·1	23° 38′·7 W.	Beattie	73
	12 33 p.m.	23 40·0			
	1 56 p.m.	23 39·6			

Dip. θ.

Date	G.M.T.	Needle	θ (observed)	θ	Observer	Instrument
1903 June 10	9 29 a.m.	3	59° 2′·4 S.	59° 2′·7 S.	Beattie	142
	9 35 a.m.	1	59 2·2			

Horizontal Intensity. H.

Date	G.M.T.	H (observed)	H	Observer	Instrument
1903 June 10	10 35 a.m. V.				
	11 8 a.m. D.	H_{30} ·18844	·18842	Beattie	73
	12 8 p.m. V.	H_{40} ·18840			

41. Boston. Lat. 29° 41′·0 S.; Long. 30° 1′·0 E. At a point in corner between roads to Howick and to Bulwer. On bank of river near hotel.

Declination. D.

Date	G.M.T.	D (observed)	D	Observer	Instrument
1903 Nov. 11	6 9 a.m.	23° 45′·9 W.			
	7 59 a.m.	23 42·8	23° 44′·9 W.	Beattie	73
	10 9 a.m.	23 39·6			

Dip. θ.

Date	G.M.T.	Needle	θ (observed)	θ	Observer	Instrument
1903 Nov. 11	11 24 a.m.	1	60° 57′·1 S.	60° 54′·1 S.	Beattie	142
	11 18 a.m.	4_9	60 56·0			

Horizontal Intensity. H.

Date	G.M.T.	H (observed)	H	Observer	Instrument
1903 Nov. 11	8 48 a.m. V.				
	9 25 a.m. D.	H_{30} ·18225			
	10 12 a.m. V.	H_{40} ·18219	·18242	Beattie	73
	12 22 p.m. V.	·18213			

42. Botha's Berg. Lat. 25° 25′·0 S.; Long. 29° 49′·0 E. Left-hand side of road, Pietersburg to Middelburg. 70 paces from road, starting from point where Roos Senekal and Pietersburg roads join.

Declination. D.

Date	G.M.T.	D (observed)	D	Observer	Instrument
1903 July 30	1 23 p.m.	22° 21′·5 W.			
	1 34 p.m.	22 22 ·7	22° 23′·5 W.	Beattie Löwinger	73
	1 46 p.m.	22 24 ·7			

Dip. θ.

Date	G.M.T.	Needle	θ (observed)	θ	Observer	Instrument
1903 July 30	10 54 a.m.	1	57° 59′·1 S.	57° 57′·4 S.	Beattie	142
	10 54 a.m.	4	57 56 ·8			

Horizontal Intensity. H.

Date	G.M.T.	H (observed)	H	Observer	Instrument
1903 July 30	12 0 noon V.				
	12 39 p.m. D.	H_{30} ·20342	·20347	Beattie	73
	1 12 p.m. V.	H_{40} ·20340			

43. Brak River. Lat. 22° 52′·2 S.; Long. 29° 13′·0 E. Right-hand side of road, Pietersburg to Cream of Tartarfontein. 233 paces from stables along road away from Pietersburg, then 53 paces from road.

Declination. D.

Date	G.M.T.	D (observed)	D	Observer	Instrument
1903 July 5	6 18 a.m.	19° 31′·5 W.	19° 31′·6 W.	Beattie Löwinger	73
	6 27 a.m.	19 31 ·6			

Dip. θ.

Date	G.M.T.	Needle	θ (observed)	θ	Observer	Instrument
1903 July 5	5 39 a.m.	1	55° 56′·2 S.	55° 55′·5 S.	Beattie	142
	5 41 a.m.	4	55 54 ·8			

Horizontal Intensity. H.

Date	G.M.T.	H (observed)	H	Observer	Instrument
1903 July 5	6 55 a.m. V.	·20452	·20452	Beattie	73

44. Brandboontjes. Lat. 23° 28′·0 S.; Long. 30° 16′·0 E. Right-hand side of road, Birthday to Pietersburg. 300 paces along road from river towards Pietersburg, then 80 paces from road.

Declination. D.

Date	G.M.T.	D (observed)	D	Observer	Instrument
1903 July 19	7 25 a.m.	19° 51′·9 W.	19° 52′·3 W.	Beattie	73
	7 39 a.m.	19 52 ·1		Löwinger	

Dip. θ.

Date	G.M.T.	Needle	θ (observed)	θ	Observer	Instrument
1903 July 19	10 43 a.m.	1	56° 40′·2 S.	56° 40′·2 S.	Beattie	142
	10 45 a.m.	4	56 40 ·7			

Horizontal Intensity. H.

Date	G.M.T.	H (observed)	H	Observer	Instrument
1903 July 19	8 45 a.m. V.	H_{30} ·20063	·20065	Beattie	73
	9 27 a.m. D.	H_{40} ·20067			
	9 56 a.m. V.				

45. Bredasdorp. Lat. 34° 32′·2 S.; Long. 20° 3′·0 E. In middle of agricultural showyard.

Declination. D.

Date	G.M.T.	D (observed)	D	Observer	Instrument
1901 Jan. 26	6 42 a.m.	28° 54′·6 W.			
	6 52 a.m.	28 54 ·4			
	7 0 a.m.	28 53 ·6			
1901 Jan. 27	3 20 p.m.	28 51 ·9	28° 39′·2 W.	Beattie Morrison	31
	3 30 p.m.	28 50 ·5			
	3 39 p.m.	28 49 ·8			
	4 9 p.m.	28 51 ·5			

Dip. θ.

Date	G.M.T.	Needle	θ (observed)	θ	Observer	Instrument
1901 Jan. 27	9 24 a.m.	1	59° 33′·7 S.	59° 56′·3 S.	Beattie	142
	9 24 a.m.	2	59 35 ·0			

Horizontal Intensity. H.

Date	G.M.T.	H (observed)	H	Observer	Instrument
1901 Jan. 27	12 50 p.m. V.	H_{30} ·18248	·18015	Morrison	31
	1 33 p.m. D.	H_{40} ·18242			
	2 8 p.m. V.				

46. Breekkerrie. Lat. 30° 6′·7 S.; Long. 21° 35′·0 E. Right-hand side of road, Van Wyk's Vlei to Breekkerrie, just past windmill on right-hand side of dam. Farm about a mile farther on.

Declination. D.

Date	G.M.T.	D (observed)	D	Observer	Instrument
1904 Dec. 13	7 0 a.m.	26° 52′·0 W.	27° 3′·1 W.	Beattie Hough	73

Dip. θ.

Date	G.M.T.	Needle	θ (observed)	θ	Observer	Instrument
1904 Dec. 13	6 24 a.m. 6 24 a.m.	3_9 4_9	58° 31′·8 S. 58 33 ·5	58° 20′·9 S.	Beattie	142

Horizontal Intensity. H.

Date	G.M.T.	H (observed)	H	Observer	Instrument
1904 Dec. 13	6 55 a.m. V. 7 33 a.m. D.	H_{30} ·18414 H_{40} ·18417	·18531	Beattie	73

47. Britstown. Lat. 30° 35′·0 S.; Long. 23° 33′·0 E. Right-hand side of road coming from Prieska. In field adjoining first houses of village.

Declination. D.

Date	G.M.T.	D (observed)	D	Observer	Instrument
1904 Dec. 27	6 37 a.m. 6 57 a.m.	26° 16′·9 W. 26 17 ·7	26° 28′·7 W.	Beattie Hough	73

Dip. θ.

Date	G.M.T.	Needle	θ (observed)	θ	Observer	Instrument
1904 Dec. 27	5 30 a.m. 5 30 a.m.	3_9 4_9	59° 19′·6 S. 59 17 ·0	59° 6′·3 S.	Beattie	142

Horizontal Intensity. H.

Date	G.M.T.	H (observed)	H	Observer	Instrument
1904 Dec. 27	7 13 a.m. V. 7 54 a.m. D. 8 24 a.m. V.	H_{30} ·17982 H_{40} ·17973	·18097	Beattie	73

48. BUFFELSBERG. Lat. 23° 36′·7 S.; Long. 30° 1′·0 E. Right-hand side of road, Birthday to Pietersburg. 150 paces along road from store towards Pietersburg, then 60 paces from road.

Declination. D.

Date	G.M.T.	D (observed)	D	Observer	Instrument
1903 July 20	1 33 p.m.	20° 54′·9 W.	20° 55′·3 W.	Beattie Löwinger	73

Dip. θ.

Date	G.M.T.	Needle	θ (observed)	θ	Observer	Instrument
1903 July 20	10 49 a.m. 10 51 a.m	1 4	57° 11′·1 S. 57 12·6	57° 11′·4 S.	Beattie	142

Horizontal Intensity. H.

Date	G.M.T.	H (observed)	H	Observer	Instrument
1903 July 20	11 56 a.m. V. 12 33 p.m. D. 1 8 p.m. V.	H_{30} ·19482 H_{40} ·19485	·19484	Beattie	73

49. BUFFELSHOEK (Blaauwberg Police Camp). Lat. 23° 8′·3 S.; Long. 28° 55′·0 E. On Buffelshoek farm. Right-hand side of road from Kalkbank to Camp. 30 yards from road on Kalkbank side of residency.

Declination. D.

Date	G.M.T.	D (observed)	D	Observer	Instrument
1903 July 3	7 19 a.m. 7 40 a.m.	20° 7′·2 W. 20 7·3	20° 7′·2 W.	Beattie Löwinger	73

Dip. θ.

Date	G.M.T.	Needle	θ (observed)	θ	Observer	Instrument
1903 July 3	9 6 a.m. 9 6 a.m.	1 4	57° 5′·2 S. 57 5·6	57° 5′·4 S.	Beattie	142

Horizontal Intensity. H.

Date	G.M.T.	H (observed)	H	Observer	Instrument
1903 July 3	9 51 a.m. V. 10 36 a.m. D. 11 14 a.m. V.	H_{30} ·19808 H_{40} ·19809	·19809	Beattie	73

50. BUFFELSKLIP. Lat. 33° 31′·7 S.; Long. 22° 52′·5 E. In field 200 yards from store on Zwart's Farm.

Declination. D.

Date	G.M.T.	D (observed)	D	Observer	Instrument
1903 Feb. 16	4 24 p.m.	27° 49′·4 W.			
	4 29 p.m.	27 49 ·6			
1903 Feb. 17	5 10 a.m.	27 50 ·5	27° 46′·6 W.	Beattie	31
	5 27 a.m.	27 51 ·2			
	7 7 a.m.	27 53 ·3			
	7 22 a.m.	27 55 ·2			

Dip. θ.

Date	G.M.T.	Needle	θ (observed)	θ	Observer	Instrument
1903 Feb. 17	11 14 a.m.	3	60° 35′·5 S.	60° 38′·1 S.	Beattie	142
	11 15 a.m.	4	60 34 ·9			

Horizontal Intensity. H.

Date	G.M.T.	H (observed)	H	Observer	Instrument
1903 Feb. 17	8 18 a.m. V.	H_{30} ·17903	·17867	Beattie	31
	9 0 a.m. D.	H_{40} ·17907			
	9 44 a.m. V.				

51. BULAWAYO. Lat. 20° 9′·1 S.; Long. 28° 36′·3 E. The place of observation in 1898 was in the public gardens near Lundie's monument. That in 1903 was in the grounds of the meteorological observatory conducted by the Jesuits.

Declination. D.

Date	G.M.T.	D (observed)	D	Observer	Instrument
1904 July 16	7 11 a.m.	18° 26′·6 W.			
	7 18 a.m.	18 27 ·4	18° 37′·3 W.	Beattie	73
	9 24 a.m.	18 27 ·4			

Dip. θ.

Date	G.M.T.	Needle	θ (observed)	θ	Observer	Instrument
1898 Jan. 14	8 45 a.m.	1	52° 50′·1 S.	53° 27′·5 S.	Morrison	9
	10 0 a.m.	2	52 48 ·8			
1904 July 15	9 35 a.m.	1	53 36 ·1	53 28 ·3	Beattie	142
	9 35 a.m.	4	53 35 ·0			
				53 27 ·9 (mean adopted)		

Horizontal Intensity. H.

Date	G.M.T.	H (observed)	H	Observer	Instrument
1898 Feb. 15	9 40 a.m. V.	H_{30} ·21816	·21473	Morrison	31
	11 0 a.m. D.				
1898 Jan. 16	11 10 a.m. V.	H_{30} ·21791		Beattie	31
	9 13 a.m. D.	H_{40} ·21788			
1904 July 15	1 16 p.m. V.	H_{30} ·21372			
	3 12 p.m. D.	H_{40} ·21372			
1904 July 16	7 44 a.m. V.		·21450	Beattie	73
	8 16 a.m. D.	H_{30} ·21409			
	9 11 a.m. V.	H_{40} ·21404			
			·21461 (mean adopted)		

52. BULT AND BAATJES. Lat. 26° 8′·0 S.; Long. 30° 16′·0 E. About 200 yards S.E. of Trigonometrical Survey beacon, marked 3rd June 1903.

Dip. θ.

Date	G.M.T.	Needle	θ (observed)	θ	Observer	Instrument
1903 Aug. 27	11 24 a.m.	1	59° 13′·9 S.			
	11 28 a.m.	4	59 13 ·9	59° 12′·4 S.	Beattie	142
	11 27 a.m.	4_9	59 13 ·1			

Horizontal Intensity. H.

Date	G.M.T.	H (observed)	H	Observer	Instrument
1903 Aug. 27	12 21 p.m. V.				
	1 10 p.m. D.	H_{30} ·19217 H_{40} ·19215	·19223	Beattie	73
	1 30 p.m. D.				

52 A. BULWER. Lat. 29° 48′·4 S.; Long. 29° 41′·0 E. On line joining courthouse and hotel, and alongside second road from hotel on side opposite to courthouse.

Declination. D.

Date	G.M.T.	D (observed)	D	Observer	Instrument
1903 Nov. 12	7 56 a.m.	24° 21′·0 W.			
	8 10 a.m.	24 19 ·5	24° 21′·5 W.	Beattie	73
	8 30 a.m.	24 19 ·4			

Horizontal Intensity. H.

Date	G.M.T.	H (observed)	H	Observer	Instrument
1903 Nov. 12	10 48 a.m. V.				
	11 27 a.m. D.	H_{30} ·18057			
	11 48 a.m. D.	H_{40} ·18054	·18080	Beattie	73
	12 38 p.m. V.				

53. BURGHERSDORP. Lat. 31° 0′·0 S.; Long. 26° 18′·0 E. In field between police barracks and the native location.

Declination. D.

Date	G.M.T.	D (observed)	D	Observer	Instrument
1901 Dec. 25	6 39 a.m.	25° 59′·6 W.			
	6 42 a.m.	25 59 ·5	25° 42′·2 W.	Beattie Morrison	31
	6 53 a.m.	25 59 ·5			
	6 57 a.m.	25 59 ·5			

Dip. θ.

Date	G.M.T.	Needle	θ (observed)	θ	Observer	Instrument
1901 Dec. 25	5 26 a.m.	2	60° 6′·3 S.	60° 17′·7 S.	Beattie	142
	5 25 a.m.	1	60° 5 ·1			

Horizontal Intensity. H.

Date	G.M.T.	H (observed)	H	Observer	Instrument
1901 Dec. 25	8 53 a.m. V.				
	9 29 a.m. D.	H_{30} ·18348 H_{40} ·18346	·18212	Morrison	31
	9 55 a.m. V.				

54. BUSHMANSKOP. Lat. 32° 20′·8 S.; Long. 22° 14′·5 E. At a place 3½ hours by cart from Beaufort West; on left-hand side of road to Fraserburg. 500 paces from farmhouse, and on Fraserburg side of it.

Declination. D.

Date	G.M.T.	D (observed)	D	Observer	Instrument
1905 Jan. 15	6 59 a.m. 7 6 a.m.	27° 9′·4 W. 27 8·7	27° 16′·8 W.	Beattie Brown	73

Dip. θ.

Date	G.M.T.	Needle	θ (observed)	θ	Observer	Instrument
1905 Jan. 15	4 58 a.m. 4 54 a.m.	3_9 4_9	59° 54′·0 S. 59 52·0	59° 40′·7 S.	Beattie	142

Horizontal Intensity. H.

Date	G.M.T.	H (observed)	H	Observer	Instrument
1905 Jan. 15	7 52 a.m. V. 8 39 a.m. D.	H_{30} ·18110 H_{40} ·18105	·18231	Beattie	73

55. BUTTERWORTH ROAD. Lat. 32° 21′·3 S.; Long. 28° 4′·0 E. On outspan on road to Butterworth from Komgha.

Declination. D.

Date	G.M.T.	D (observed)	D	Observer	Instrument
1906 Jan. 15	5 51 a.m. 6 0 a.m. 6 8 a.m.	25° 27′·3 W. 25 28·4 25 28·1	25° 51′·1 W.	Brown Morrison	31

Dip. θ.

Date	G.M.T.	Needle	θ (observed)	θ	Observer	Instrument
1906 Jan. 15	9 36 a.m. 9 37 a.m.	1 2	61° 39′·8 S. 61 40·9	61° 20′·1 S.	Morrison	9

Horizontal Intensity. H.

Date	G.M.T.	H (observed)	H	Observer	Instrument
1906 Jan. 15	7 19 a.m. V. 7 58 a.m. D. 8 30 a.m. V.	H_{30} ·17595 H_{40} ·17599	·17841	Morrison	31

56. Caledon River. Lat. 30° 16′·8 S.; Long. 26° 41′·7 E. About 3 miles on the Rouxville side of the Caledon River Bridge, on right-hand side of road to Smithfield. Smithfield side of a small river just on top of rise after crossing the drift.

Declination. D.

Date	G.M.T.	D (observed)	D	Observer	Instrument
1904 Jan. 11	1 7 p.m.	25° 36′·7 W.	25° 43′·8 W.	Beattie	73
	1 14 p.m.	25 36 ·7			

Dip. θ.

Date	G.M.T.	Needle	θ (observed)	θ	Observer	Instrument
1904 Jan. 11	11 59 a.m.	1	60° 26′·7 S.	60° 22′·9 S.	Beattie	142
	11 59 a.m.	4_9	60 26 ·2			

Horizontal Intensity. H.

Date	G.M.T.	H (observed)	H	Observer	Instrument
1904 Jan. 11	12 20 p.m. V.	·18049	·18089	Beattie	73

57. Calitzdorp. Lat. 33° 32′·1 S.; Long. 21° 41′·0 E.

Declination. D.

Date	G.M.T.	D (observed)	D	Observer	Instrument
1903 Jan. 12	6 37 a.m.	28° 24′·3 W.	28° 17′·9 W.	Beattie	31
	6 50 a.m.	28 24 ·3		Morrison	
	6 57 a.m.	28 23 ·7			

Dip. θ.

Date	G.M.T.	Needle	θ (observed)	θ	Observer	Instrument
1903 Jan. 12	5 55 a.m.	3	60° 14′·8 S.	60° 18′·8 S.	Beattie	142
	5 52 a.m.	4	60 15 ·1			

Horizontal Intensity. H.

Date	G.M.T.	H (observed)	H	Observer	Instrument
1903 Jan. 12	8 32 a.m. V.	H_{30} ·17964	·17918	Morrison	31
	9 26 a.m. D.	H_{40} ·17962			
	10 54 a.m. V.				

58. CAMPERDOWN. Lat. 29° 44′·0 S.; Long. 30° 37′·0 E. Left-hand side of railway, Pietermaritzburg to Durban. 150 paces at right angles to railway, starting from end of Pietermaritzburg dead end.

Dip. θ.

Date	G.M.T.	Needle	θ (observed)	θ	Observer	Instrument
1903 Oct. 16	12 24 p.m.	1	61° 31′·3 S.			
	12 24 p.m.	4_9	61 28 ·9	61° 28′·8 S.	Beattie	142
	12 24 p.m.	4	61 31 ·5			

Horizontal Intensity. H.

Date	G.M.T.	H (observed)	H	Observer	Instrument
1903 Oct. 16	8 57 a.m. V.				
	9 40 a.m. D.	H_{30} ·17811	·17833	Beattie	73
	10 38 a.m. D.	H_{40} ·17813			
	11 27 a.m. V.				

59. CANGO. Lat. 33° 24′·8 S.; Long. 22° 14′·5 E. On hillside to the west of the hotel, and on opposite side of sluit to hotel. 200 yards from smithy.

Declination. D.

Date	G.M.T.	D (observed)	D	Observer	Instrument
1903 Jan. 10	8 6 a.m.	28° 22′·0 W.			
	8 15 a.m.	28 22 ·0			
	8 25 a.m.	28 21 ·7	28° 18′·6 W.	Beattie	31
	2 32 p.m.	28 25 ·0		Morrison	
	3 0 p.m.	28 24 ·8			
	3 18 p.m.	28 24 ·6			

Dip. θ.

Date	G.M.T.	Needle	θ (observed)	θ	Observer	Instrument
1903 Jan. 10	1 17 p.m.	3	60° 40′·2 S.	60° 43′·7 S.	Beattie	142
	1 17 p.m.	4	60 39 ·6			

Horizontal Intensity. H.

Date	G.M.T.	H (observed)	H	Observer	Instrument
1903 Jan. 10	11 3 a.m. V.	H_{30} ·17827			
	12 25 p.m. D.		·17780	Morrison	31
	1 21 p.m. V.	H_{40} ·17822			

60. Cape Town. Royal Observatory, Observatory Road. Lat. 33° 56′·1 S.; Long. 18° 28′·7 E. Under large tree on Cape Town side of new transit house.

Declination. D.

Date	G.M.T.	D (observed)	D	Observer	Instrument
1900 Dec. 28	8 8 a.m.	28° 56′·0 W.			
	8 20 a.m.	28 54·6			
	8 30 a.m.	28 54·6			
1900 Dec. 29	2 34 p.m.	28 54·4			
	2 50 p.m.	28 55·2	28° 44′·0 W.	Beattie	31
	3 1 p.m.	28 55·2			
	3 37 p.m.	28 55·7			
	3 46 p.m.	28 55·7			
	3 53 p.m.	28 54·9			
1900 Dec. 30	8 28 a.m.	28 55·7			
	8 42 a.m.	28 55·3			
	8 54 a.m.	28 54·0	28 43·3	Beattie	31
1900 Dec. 31	1 17 p.m.	28 48·9			
	1 29 p.m.	28 49·1			
	1 42 p.m.	28 49·3			
1901 Jan. 1	1 42 p.m.	28 53·4			
	1 52 p.m.	28 53·1			
1901 Jan 2	1 14 p.m.	28 52·5	28 45·0	Beattie	31
	1 26 p.m.	28 52·8			
	1 35 p.m.	28 53·0			
1902 March 28	9 30 a.m.	28 52·3			
	2 40 p.m.	28 46·8	28 44·1	Beattie	31
	4 7 p.m.	28 47·9			
1902 July 1	3 10 p.m.	28 47·6	28 44·8	Beattie	31
	3 18 p.m.	28 47·4			
1902 July 2	9 52 a.m.	28 47·1			
	9 58 a.m.	28 47·3	28 42·0	Beattie	31
	2 48 p.m.	28 44·7			
	2 55 p.m.	28 45·6			
1902 Nov. 10	8 10 a.m.	28 46·7			
	8 26 a.m.	28 45·7	28 40·4	Beattie	31
	8 40 a.m.	28 45·2			
1902 Nov. 16	3 46 p.m.	28 45·0	28 43·3	Beattie	31
	3 58 p.m.	28 44·8			
1902 Nov. 19	7 40 a.m.	28 53·8	28 48·2	Beattie	31
	7 55 a.m.	28 54·5			
			28 43·9	(mean adopted)	

Dip. θ.

Date		G.M.T.	Needle	θ (observed)	θ	Observer	Instrument
1898	March 27	6 30 a.m.	1	58° 20′·8 S.	59° 8′·1 S.	Beattie	9
		6 30 a.m.	2	58 19·4			
1898	May 7	10 30 a.m.	1	58 21·4	59 7·1	Beattie	9
		11 0 a.m.	2	58 19·9			
1899	Oct. 2	11 0 a.m.	1	58 31·6	59 4·9	Beattie	9
		11 0 a.m.	2	58 30·1			
1900	Aug. 19	10 30 a.m.	1	58 40·4	59 3·9	Beattie	9
		10 30 a.m.	2	58 37·4			
1900	Dec. 26	10 6 a.m.	1	58 40·3			
		10 4 a.m.	2	58 38·9		Beattie	9
1900	Dec. 27	7 58 a.m.	1	58 41·1			
		8 3 a.m.	2	58 38·5			
1900	Dec. 1	10 48 a.m.	1	58 40·6			
		10 48 a.m.	2	58 44·1			
1900	Dec. 2	9 53 a.m.	1	58 42·6			
		9 55 a.m.	2	58 44·9			
1900	Dec. 19	9 34 a.m.	1	58 39·4	59 4·8	Beattie	142
		9 34 a.m.	2	58 42·6			
		9 30 a.m.	1_9	58 41·7			
		9 28 a.m.	2_9	58 39·2			
1900	Dec. 21	2 16 p.m.	1	58 40·2			
		2 11 p.m.	2	58 42·8			
1900	Dec. 26	8 7 a.m.	1	58 43·0			
		8 7 a.m.	2	58 44·5			
1900	Dec. 27	10 0 a.m.	1	58 40·2		Beattie	142
		10 0 a.m.	2	58 42·4			
1900	Dec. 28	8 30 a.m.	1	58 42·7			
		8 30 a.m.	2	58 45·4			
1902	March 30	1 47 p.m.	1	58 53·2			
1902	March 31	8 32 a.m.	1	58 54·2	59 5·2	Beattie	142
		10 25 a.m.	4	58 51·8			
		10 26 a.m.	3	58 56·5			
1902	Nov. 8	8 26 a.m.	4	58 58·1			
		8 26 a.m.	3	58 57·1			
		8 26 a.m.	2	59 3·1			
		8 26 a.m.	1	58 58·2	59 5·6	Beattie	142
1902	Nov. 23	10 19 a.m.	4	59 2·2			
		10 19 a.m.	3	58 59·0			
		10 20 a.m.	2	58 58·7			
		10 18 a.m.	1	59 1·3			
1904	April 24	9 8 a.m.	1	59 14·3			142
		9 8 a.m.	4_9	59 16·7	59 7·4	Beattie	
		9 40 a.m.	1_{142}	59 12·7			9
		9 40 a.m.	4	59 15·9			
1905	April 27	9 12 a.m.	4_9	59 18·9			
		9 14 a.m.	3_9	59 18·2	59 7·2	Beattie	142
		9 14 a.m.	5	59 19·1			
					59 6·0 (mean adopted)		

Horizontal Intensity. H.

Date	G.M.T.	H (observed)	H	Observer	Instrument
1898 March 27	12 28 p.m. V. 2 20 p.m. D.	H_{30} ·18683 H_{40} ·18685	·18180	Morrison	31
1900 Dec. 29	10 5 a.m. V. 11 1 a.m. D. 11 45 a.m. V.	H_{30} ·18512 H_{40} ·18512	·18272	Morrison	31
1900 Dec. 31	9 59 a.m. V. 10 40 a.m. D. 11 33 a.m. V.	H_{30} ·18496 H_{40} ·18499	·18258	Morrison	31
1901 Jan. 1	8 51 a.m. V. 9 30 a.m. D. 10 7 a.m. V.	H_{30} ·18501 H_{40} ·18503	·18262	Morrison	31
1901 Jan. 2	8 43 a.m. V. 9 21 a.m. D. 9 57 a.m. V.	H_{30} ·18527 H_{40} ·18535	·18291	Morrison	31
1902 March 28	9 43 a.m. V. 12 53 p.m. V.	·18417	·18297	Beattie	31
1902 March 30	4 30 p.m. V.	·18426	·18306	Beattie	31
1902 July 2	8 50 a.m. V.	·18381	·18285	Beattie	31
1903 March 6	10 21 a.m. V.	·18292	·18260	Beattie	73
1903 March 9	8 6 a.m. V.	·18318	·18296	Beattie	73
	1 29 p.m. V.	·18276	·18244	Beattie	73
	12 12 p.m. V. 1 40 p.m. D.	H_{30} ·18313 H_{40} ·18317	·18283	Beattie	31
1903 March 10	9 50 a.m. V. 10 34 a.m. D. 11 20 a.m. V.	H_{30} ·18287 H_{40} ·18286	·18255	Beattie	73
	7 55 a.m. V. 10 13 a.m. D. 12 26 p.m. V.	H_{30} ·18307 H_{40} ·18301	·18272	Beattie	31
1903 May 19	9 13 a.m. V. 10 16 a.m. D. 12 19 p.m. V.	H_{30} ·18280 H_{40} ·18281 H_{25} ·18270 H_{35} ·18271	·18268	Beattie	73 with vibration magnet of 31
	12 19 p.m. V. 1 49 p.m. D.	H_{30} ·18288 H_{40} ·18292	·18282	Beattie	73 with vibration magnet of 31
	1 20 p.m. D. 12 47 p.m. V.	H_{30} ·18267 H_{40} ·18269	·18260	Beattie	73
	8 43 a.m. V. 11 0 a.m. D. 12 47 p.m. V.	H_{30} ·18288 H_{40} ·18287 H_{25} ·18292 H_{35} ·18287	·18281	Beattie	73
1903 May 20	10 7 a.m. V. 12 0 noon D.	H_{30} ·18281 H_{40} ·18279	·18272	Beattie	31 with vibration magnet of 73
	9 38 a.m. V. 12 36 p.m. D. 1 14 p.m. D.	H_{30} ·18320 H_{40} ·18320	·18312	Beattie	31
1904 April 1	11 20 a.m. V. 12 17 p.m. D. 12 46 p.m. D. 1 18 p.m. V.	H_{30} ·18181 H_{40} ·18181	·18250	Beattie	31
1904 April 12	8 30 a.m. V. 9 11 a.m. D. 9 46 a.m. V.	H_{30} ·18239 H_{40} ·18238	·18308	Beattie	31
	12 50 p.m. V. 1 24 p.m. D. 2 0 p.m. V.	H_{30} ·18162 H_{40} ·18165	·18234	Beattie	31
1904 May 1	9 7 a.m. V. 9 59 a.m. D. 10 32 a.m. V.	H_{30} ·18223 H_{40} ·18216	·18290	Beattie	73
	1 58 p.m. V. 2 32 p.m. D. 3 3 p.m. V.	H_{30} ·18183 H_{40} ·18191	·18257	Beattie	73

Date	G.M.T.	H (observed)	H	Observer	Instrument
1904 April 4	9 52 a.m. V.				
	10 26 a.m. D.	H_{30} ·18161	·18231	Beattie	73
	11 55 a.m. V.				
1904 May 12	11 54 a.m. V.				
	10 54 a.m. D.	H_{30} ·18185 H_{40} ·18187	·18256	Beattie	73
	11 26 a.m. D.				
1905 April 22	8 51 a.m. V.	H_{30} ·18117	·18268		
	10 0 a.m. D.	H_{40} ·18111			
1905 April 24	9 12 a.m. V.	H_{30} ·18127		Beattie	73
	10 17 a.m. D.	H_{40} ·18121	·18280		
	3 27 p.m. V.	·18128			
			·18271 (mean adopted)		

61. Cathcart. Lat. 32° 18′·0 S.; Long. 27° 9′·0 E. In field below railway station, about 300 paces above Railway Hotel.

Dip. θ.

Date	G.M.T.	Needle	θ (observed)	θ	Observer	Instrument
1902 Jan. 9	11 45 a.m.	2	61° 32′·1 S.	61° 46′·0 S.	Morrison	142
	11 49 a.m.	1	61 36 ·2			

Horizontal Intensity. H.

Date	G.M.T.	H (observed)	H	Observer	Instrument
1902 Jan. 9	2 13 p.m. V.				
	2 55 p.m. D.	H_{30} ·17610 H_{40} ·17613	·17477	Morrison	31
	3 29 p.m. V.				

62. Ceres Road. Lat. 33° 25′·6 S.; Long. 19° 19′·0 E. The road along the railway from the station to Ceres divides into two, one to Ceres the other to the right. The right-hand road again breaks into two a short distance from the railway. The magnetometer was placed in the fork of this branch.

Declination. D.

Date	G.M.T.	D (observed)	D	Observer	Instrument
1900 May 24	10 20 a.m.	28° 45′·3 W.			
	10 30 a.m.	28 45 ·2	28° 31′·7 W.	Beattie Morrison	31
	10 50 a.m.	28 44 ·7			

Dip. θ.

Date	G.M.T.	Needle	θ (observed)	θ	Observer	Instrument
1898 March 5	3 0 p.m.	1	58° 24′·0 S.	59° 12′·4 S.	Morrison	9
	3 0 p.m.	2	58 24 ·7			
1900 May 25	10 30 a.m.	1	58 41 ·1	59 8 ·9	Beattie	9
	10 30 a.m.	2	58 42 ·6			
				59 10 ·7 (mean adopted)		

Horizontal Intensity. H.

Date	G.M.T.	H (observed)	H	Observer	Instrument
1898 March 6	8 51 a.m. V.	H_{30} ·18720	·18274		
	10 12 a.m. D.	H_{40} ·18723			
1900 May 24	2 3 p.m. V.			Morrison	31
	2 53 p.m. D.	H_{30} ·18557 H_{40} ·18546	·18293		
	3 32 p.m. V.				
			·18284 (mean adopted)		

63. CHARLESTOWN. Lat. 27° 24′·9 S.; Long. 29° 54′·0 E. Right-hand side of railway, Charlestown to Newcastle. Along railway, 118 paces from dead end towards Newcastle, then 220 paces at right angles to railway.

Declination. D.

Date	G.M.T.	D (observed)	D	Observer	Instrument
1903 Sept. 30	4 49 a.m.	21° 32′·1 W.			
	5 0 a.m.	21 32 ·2			
	8 26 a.m.	21 38 ·4	21° 33′·4 W.	Beattie	73
	8 37 a.m.	21 37 ·5			
	12 5 p.m.	21 22 ·3			
	12 14 p.m.	21 23 ·0			

Dip. θ.

Date	G.M.T.	Needle	θ (observed)	θ	Observer	Instrument
1903 Sept. 30	10 52 a.m.	1	59° 25′·4 S.			
	10 52 a.m.	4_9	59 25 ·7	59° 23′·7 S.	Beattie	142
	10 49 a.m.	4	59 25 ·4			

Horizontal Intensity. H.

Date	G.M.T.	H (observed)	H	Observer	Instrument
1903 Sept. 30	5 18 a.m. V.				
	7 10 a.m. D.	H_{30} ·18845 H_{40} ·18847	·18861	Beattie	73
	8 58 a.m. V.				

64. CLARKSON. Lat. 34° 1′·0 S.; Long. 24° 10′·0 E. On right-hand side of path from missionary's house going up the hill, about 200 paces from house towards the cattle kraal.

Dip. θ.

Date	G.M.T.	Needle	θ (observed)	θ	Observer	Instrument
1903 Feb. 4	3 50 p.m.	3	61° 4′·3 S.	61° 8′·6 S.	Beattie	142
	3 51 p.m.	4	61 6 ·5			

Horizontal Intensity. H.

Date	G.M.T.	H (observed)	H	Observer	Instrument
1903 Feb. 4	4 50 p.m. V.	·17838	·17800	Beattie	31

65. Coerney. Lat. 33° 27′·6 S.; Long. 25° 44′·0 E. About 300 yards to west of railway station, and north of hotel.

Declination. D.

Date	G.M.T.	D (observed)	D	Observer	Instrument
1902 July 14	6 24 a.m.	27° 27′·4 W.			
	1 20 p.m.	27 25 ·6			
	1 26 p.m.	27 25 ·9	27° 17′·5 W.	Beattie Morrison	31
	1 41 p.m.	27 25 ·7			
	1 51 p.m.	27 25 ·7			
	2 2 p.m.	27 24 ·6			

Dip. θ.

Date	G.M.T.	Needle	θ (observed)	θ	Observer	Instrument
1902 July 14	8 55 a.m.	1	61° 24′·7 S.			
	8 55 a.m.	2	61 24 ·6	61° 30′·2 S.	Beattie	142
	8 55 a.m.	3	61 21 ·1			
	8 55 a.m.	4	61 19 ·2			

Horizontal Intensity. H.

Date	G.M.T.	H (observed)	H	Observer	Instrument
1902 July 14	9 56 a.m. V.	H_{30} ·17715	·17626	Morrison	31
	10 39 a.m. D.	H_{40} ·17717			
	11 17 a.m. V.				

66. Colenso. Lat. 28° 44′·0 S.; Long. 29° 50′·0 E. Right-hand side of railway, Pietermaritzburg to Ladysmith. 200 paces at right angles to railway, starting from stile on Pietermaritzburg side of railway station.

Declination. D.

Date	G.M.T.	D (observed)	D	Observer	Instrument
1903 Nov. 9	4 34 a.m.	22° 45′·3 W.			
	4 44 a.m.	22 46 ·1			
	5 59 a.m.	22 48 ·1			
	6 5 a.m.	22 47 ·9	22° 48′·3 W.	Beattie	73
	8 38 a.m.	22 44 ·2			
	8 48 a.m.	22 45 ·2			
	10 48 a.m.	22 44 ·7			

Dip. θ.

Date	G.M.T.	Needle	θ (observed)	θ	Observer	Instrument
1903 Nov. 9	10 25 a.m.	4_9	60° 11′·1 S.	60° 9′·3 S.	Beattie	142
	10 25 a.m.	4	60 12 ·5			

Horizontal Intensity. H.

Date	G.M.T.	H (observed)	H	Observer	Instrument
1903 Nov. 9	6 59 a.m. V.	H_{30} ·18522	·18548	Beattie	73
	7 38 a.m. D.	H_{40} ·18526			
	8 22 a.m. V.				

67. Colesberg. Lat. 30° 42′·8 S.; Long. 25° 8′·0 E. Field below railway station, about 300 yards in a north easterly direction from it.

Declination. D.

Date	G.M.T.	D (observed)	D	Observer	Instrument
1902 July 5	3 4 p.m.	25° 43′·0 W.			
	3 17 p.m.	25 42 ·1			
1902 July 6	8 37 a.m.	25 47 ·7			
	8 45 a.m.	25 48 ·1			
	8 55 a.m.	25 48 ·6	25° 34′·2 W.	Beattie Morrison	31
	9 2 a.m.	25 48 ·2			
	1 42 p.m.	25 41 ·6			
	1 52 p.m.	25 41 ·9			
	2 52 p.m.	25 43 ·1			
	3 2 p.m.	25 43 ·6			
1904 July 1	8 19 a.m.	25 31 ·8			
	8 26 a.m.	25 31 ·7			
	8 35 a.m.	25 31 ·8	25 38 ·8	Morrison	31
	2 17 p.m.	25 26 ·4			
	2 25 p.m.	25 26 ·4			
	2 36 p.m.	25 26 ·5			
			25 36 ·5 (mean adopted)		

Dip. θ.

Date	G.M.T.	Needle	θ (observed)	θ	Observer	Instrument
1902 July 5	2 17 p.m.	1	59° 56′·6 S.			
	2 19 p.m.	3	59 52 ·3			
1902 July 6	9 45 a.m.	1	59 51 ·8	60° 0′·7 S.	Beattie	142
	9 5 a.m.	2	59 56 ·1			
	9 4 a.m.	3	59 51 ·1			
	9 4 a.m.	4	59 52 ·2			
1904 July 1	1 30 p.m.	1	60 10 ·0	60 0 ·6	Morrison	9
	1 26 p.m.	2	60 5 ·2			
				60 0 ·6 (mean adopted)		

Horizontal Intensity. H.

Date	G.M.T.	H (observed)	H	Observer	Instrument
1902 July 6	1 13 p.m. V.				
	2 27 p.m. D.	H_{30} ·18270 H_{40} ·18268	·18179	Morrison	31
	3 18 p.m. V.				
1904 July 1	9 0 a.m. V.	·18095	·18186	Morrison	31
			·18183 (mean adopted)		

68. Connan's Farm. Lat. 28° 58′·4 S.; Long. 21° 19′·3 E. Between Kenhardt and Upington, 1½ hours by cart from Smit's farm. Pools on right-hand side and left-hand side of road. On right-hand side of road coming from Kenhardt.

Declination. D.

Date		G.M.T.	D (observed)	D	Observer	Instrument
1904	Dec. 18.	5 34 a.m.	24° 42′·9 W.	24° 54′·8 W.	Beattie Hough	73

Dip. θ.

Date		G.M.T.	Needle	θ (observed)	θ	Observer	Instrument
1904	Dec. 18	8 28 a.m. 8 28 a.m.	3_9 4_9	56° 28′·5 S. 56 27 ·6	56° 16′·2 S.	Beattie	142

Horizontal Intensity. H.

Date		G.M.T.	H (observed)	H	Observer	Instrument
1904	Dec. 18	6 47 a.m. V. 7 28 a.m. D.	H_{30} ·19480 H_{40} ·19489	·19602	Beattie	73

69. Cookhouse. Lat. 32° 44′·0 S.; Long. 25° 48′·0 E. Place of observation 100 yards to west of intersection of Somerset East Road and a second road, and sixty yards south of Somerset East Road.

Declination. D.

Date		G.M.T.	D (observed)	D	Observer	Instrument
1900	Jan. 18	4 26 p.m. 4 34 p.m.	27° 15′·1 W. 27 15 ·1	26° 42′·2 W.	Beattie Morrison	31
1904	July 8	2 10 p.m. 2 18 p.m.	26 33 ·6 26 33 ·6	26 43 ·9	Morrison	31
				26 43 ·0 (mean adopted)		

Dip. θ.

Date		G.M.T.	Needle	θ (observed)	θ	Observer	Instrument
1900	Jan. 19	9 15 a.m. 9 15 a.m.	1 2	60° 46′·7 S. 60 45 ·9	61° 14′·3 S.	Beattie	9
1904	July 8	1 42 p.m. 1 43 p.m.	1 2	61 22 ·5 61 22 ·5	61 14 ·5	Morrison	9
					61 14 ·4 (mean adopted)		

Horizontal Intensity. H.

Date		G.M.T.	H (observed)	H	Observer	Instrument
1900	Jan. 19	1 54 p.m. V. 2 35 p.m. D. 3 13 p.m. V.	H_{30} ·18039 H_{40} ·18041	·17729	Beattie	31
904	July 8	8 49 a.m. V. 9 29 a.m. D. 9 45 a.m. V.	H_{40} ·17645	·17735	Morrison	31
				·17732 (mean adopted)		

70. Cotswold Hotel. Lat. 32° 42′·7 S.; Long. 29° 53′·4 E. Road bifurcates just before reaching hotel. Observations taken on right-hand side road coming from Harding and in the bifurcation.

Declination. D.

Date	G.M.T.	D (observed)	D	Observer	Instrument
1903 Nov. 21	7 36 a.m. 7 45 a.m.	24° 21′·3 W. 24 21 ·6	24° 23′·2 W.	Beattie	73

Dip. θ.

Date	G.M.T.	Needle	θ (observed)	θ	Observer	Instrument
1903 Nov. 21	12 7 p.m. 12 7 p.m.	1 4_9	61° 30′·3 S. 61 30 ·4	61° 27′·7 S.	Beattie	142

Horizontal Intensity. H.

Date	G.M.T.	H (observed)	H	Observer	Instrument
1903 Nov. 21	10 2 a.m. V. 10 32 a.m. D. 11 5 a.m. V.	H_{30} ·17741 H_{40} ·17742	·17769	Beattie	73

71. Cradock. Lat. 32° 9′·6 S.; Long. 25° 38′·0 E. In field adjoining football ground. Right-hand distant corner coming into field from Cradock town.

Declination. D.

Date	G.M.T.	D (observed)	D	Observer	Instrument
1900 Jan. 17	1 20 p.m. 1 30 p.m. 1 40 p.m.	26° 59′·4 W. 26 59 ·9 27 0 ·0	26° 27′·2 W.	Beattie Morrison	31

Dip. θ.

Date	G.M.T.	Needle	θ (observed)	θ	Observer	Instrument
1898 April 14	9 32 a.m. 10 5 a.m. 11 12 a.m.	2 1 2	60° 4′·1 S. 60 6 ·3 60 6 ·5	60° 46′·6 S.	Beattie	9
1900 Jan. 15 1900 Jan. 17	9 45 a.m. 9 45 a.m. Forenoon Forenoon	1 2 1 2	60 24 ·9 60 23 ·9 60 23 ·1 60 22 ·7	60 51 ·7	Beattie	9
				60 49 ·1 (mean adopted)		

Horizontal Intensity. H.

Date	G.M.T.	H (observed)	H	Observer	Instrument
1898 April 15	8 45 a.m. V. 10 10 a.m. D. 11 9 a.m. D. 11 53 a.m. V.	H_{30} ·18449 H_{40} ·18450	·17981	Beattie	31
1900 Jan. 16	10 17 a.m. V. 10 50 a.m. D. 11 41 a.m. V.	H_{30} ·18294 H_{40} ·18295	·17980	Beattie	31
			·17981 (mean adopted)		

72. Cream of Tartar Fontein. Lat. 22° 35′·3 S.; Long. 29° 1′·0 E. Left-hand side of road going to Pont drift, opposite stables, 40 paces from road.

Declination. D.

Date	G.M.T.	D (observed)	D	Observer	Instrument
1903 July 6	6 25 a.m. 6 40 a.m. 6 56 a.m.	19° 40′·8 W. 19 42 ·1 19 41 ·5	19° 41′·6 W.	Beattie Löwinger	73

Dip. θ.

Date	G.M.T.	Needle	θ (observed)	θ	Observer	Instrument
1903 July 6	5 33 a.m. 5 33 a.m.	1 4	55° 45′·2 S. 55 43 ·7	55° 44′·5 S.	Beattie	142

Horizontal Intensity. H.

Date	G.M.T.	H (observed)	H	Observer	Instrument
1903 July 6	7 53 a.m. V. 8 31 a.m. D. 9 3 a.m. D. 9 31 a.m. V.	H_{30} ·20404 H_{40} ·20408	·20406	Beattie	73

73. Crocodile Pools. Lat. 24° 46′·7 S.; Long. 25° 50′·0 E. Right-hand side of railway, Mafeking to Bulawayo. 270 paces from water tank in direction of Mafeking, then 80 paces at right angles to railway line.

Declination. D.

Date	G.M.T.	D (observed)	D	Observer	Instrument
1903 March 22	2 6 p.m. 2 44 p.m.	21° 40′·6 W. 21 39 ·0	21° 37′·6 W.	Beattie	73

Dip. θ.

Date	G.M.T.	Needle	θ (observed)	θ	Observer	Instrument
1903 March 22	12 52 p.m. 12 52 p.m. 12 52 p.m.	3 4 1	56° 14′·0 S. 56 12 ·3 56 11 ·2	56° 14′·3 S.	Beattie	142

Horizontal Intensity. H.

Date	G.M.T.	H (observed)	H	Observer	Instrument
1903 March 24	5 22 a.m. V. 6 3 a.m. D. 6 39 a.m. V.	H_{30} ·20014 H_{40} ·20018	·19995	Beattie	73

74. DALTON. Lat. 29° 16′·0 S.; Long. 30° 42′·0 E. On left-hand side of railway, Pietermaritzburg to Greytown. About 120 paces from dead end of triangle in a direction making 45° with continuation of triangle.

Dip. θ.

Date	G.M.T.	Needle	θ (observed)	θ	Observer	Instrument
1903 Oct. 11	11 26 a.m.	1	61° 27′·3 S.	61° 23′·6 S.	Beattie	142
	11 26 a.m.	4_9	61 24 ·0			

Horizontal Intensity. H.

Date	G.M.T.	H (observed)	H	Observer	Instrument
1903 Oct. 11	8 1 a.m. V.	H_{30} ·17954	·17974	Beattie	73
	8 40 a.m. D.	H_{40} ·17957			
	9 19 a.m. V.				

75. DAMBIESFONTEIN. Lat. 31° 24′·2 S.; Long. 21° 17′·6 E. Right-hand side of road, Fraserburg to Williston. Half-way between road and pool on farm.

Declination. D.

Date	G.M.T.	D (observed)	D	Observer	Instrument
1905 Jan. 20	6 58 a.m.	26° 59′·4 W.	27° 7′·4 W.	Beattie Brown	73

Dip. θ.

Date	G.M.T.	Needle	θ (observed)	θ	Observer	Instrument
1905 Jan. 20	5 7 a.m.	3_9	59° 27′·5 S.	59° 16′·1 S.	Beattie	142
	5 7 a.m.	4_9	59 29 ·4			

Horizontal Intensity. H.

Date	G.M.T.	H (observed)	H	Observer	Instrument
1905 Jan. 20	7 17 a.m. V.	H_{30} ·18233	·18355	Beattie	73
	8 7 a.m. D.	H_{40} ·18229			

76. Dannhauser. Lat. 28° 1′·2 S.; Long. 30° 5′·0 E. Right-hand side of railway, Newcastle to Glencoe. 23 paces from Newcastle end of platform towards Newcastle, then 230 from that point at right angles to the railway.

Declination. D.

Date	G.M.T.	D (observed)	D	Observer	Instrument
1903 Oct. 2	4 56 a.m.	22° 26′·1 W.			
	5 6 a.m.	22 25 ·6	22° 29′·8 W.	Beattie	73
	7 45 a.m.	22 30 ·5			
	7 51 a.m.	22 31 ·5			

Dip. θ.

Date	G.M.T.	Needle	θ (observed)	θ	Observer	Instrument
1903 Oct. 2	9 8 a.m.	1	59° 44′·1 S.	59° 43′·7 S.	Beattie	142
	9 7 a.m.	4_9	59 47 ·0			

Horizontal Intensity. H.

Date	G.M.T.	H (observed)	H	Observer	Instrument
1903 Oct. 2	6 52 a.m. V.				
	7 25 a.m. D.	H_{30} ·18579 H_{40} ·18579	·18597	Beattie	73
	9 5 a.m. V.				

77. Dargle Road. Lat. 29° 29′·1 S.; Long. 30° 11′·0 E. Right-hand side of railway, Ladysmith to Durban. 290 paces at right angles to railway starting from goods station. Alongside bend of Lion River.

Declination. D.

Date	G.M.T.	D (observed)	D	Observer	Instrument
1903 Oct. 8	12 47 p.m.	24° 1′·4 W.			
	12 54 p.m.	24 1 ·1	24° 5′·7 W.	Beattie	73
	2 12 p.m.	24 2 ·9			
	2 17 p.m.	24 2 ·9			

Dip. θ.

Date	G.M.T.	Needle	θ (observed)	θ	Observer	Instrument
1903 Oct. 8	9 10 a.m.	1	60° 39′·5 S.			
	9 12 a.m.	4_9	60 40 ·1	60° 38′·8 S.	Beattie	142
	9 10 a.m.	4	60 42 ·6			

Horizontal Intensity. H.

Date	G.M.T.	H (observed)	H	Observer	Instrument
1903 Oct. 8	10 42 a.m. V.				
	12 3 p.m. D.	H_{30} ·18451 H_{40} ·18448	·18468	Beattie	73
	12 33 p.m. V.				

78. DARLING. Lat. 33° 22′·1 S.; Long. 18° 22′·0 E. In field in front of houses right-hand side of road coming from Malmesbury.

Declination. D.

Date	G.M.T.	D (observed)	D	Observer	Instrument
1901 July 10	8 52 a.m.	28° 39′·6 W.	28° 30′·6 W.	Beattie Morrison	31
	9 9 a.m.	28 39 ·4			
	9 22 a.m.	28 40 ·0			
	9 31 a.m.	28 40 ·3			
	2 49 p.m.	28 38 ·2			
	3 0 p.m.	28 39 ·1			
	3 10 p.m.	28 39 ·2			

Dip. θ.

Date	G.M.T.	Needle	θ (observed)	θ	Observer	Instrument
1901 July 10	10 47 a.m.	2	58° 38′·9 S.	58° 55′·5 S.	Beattie	142
	10 53 a.m.	1	58 36 ·0			

Horizontal Intensity. H.

Date	G.M.T.	H (observed)	H	Observer	Instrument
1901 July 10	12 53 p.m. V.	H_{30} ·18542	·18351	Morrison	31
	1 36 p.m. D.	H_{40} ·18543			
	3 16 p.m. V.				

79. DE AAR. Lat. 30° 40′·0 S.; Long. 24° 2′·0 E. In fork between Cape Town and Naauwpoort railways. On left-hand side of road going from De Aar.

Declination. D.

Date	G.M.T.	D (observed)	D	Observer	Instrument
1900 Jan. 12	1 13 p.m.	26° 12′·5 W.	25° 42′·0 W.	Beattie Morrison	31
	1 23 p.m.	26 12 ·9			
	1 32 p.m.	26 13 ·1			
1900 Jan. 13	5 50 a.m.	26 22 ·2			
	6 0 a.m.	26 20 ·8			
	6 10 a.m.	26 21 ·3			

Dip. θ.

Date	G.M.T.	Needle	θ (observed)	θ	Observer	Instrument
1900 Jan. 13	2 0 p.m.	1	58° 53′·4 S.	59° 52′·0 S.	Beattie	9
	2 0 p.m.	2	58 54 ·7			

Horizontal Intensity. H.

Date	G.M.T.	H (observed)	H	Observer	Instrument
1900 Jan. 13	9 56 a.m. V.	H_{30} ·18729	·18447	Morrison	31
	10 40 a.m. D.	H_{40} ·18725			
	11 17 a.m. V.				

80. De Doorns. Lat. 33° 29′·0 S.; Long. 19° 36′·0 E. In middle of Burger's vineyard.

Dip. θ.

Date	G.M.T.	Needle	θ (observed)	θ	Observer	Instrument
1899 Oct. 13	9 20 a.m.	1	58° 50′·9 S.	59° 20′·4 S.	Beattie	9
	9 20 a.m.	2	58 49 ·9			

Horizontal Intensity. H.

Date	G.M.T.	H (observed)	H	Observer	Instrument
1899 Oct. 13	3 0 p.m. V.	H_{30} ·18516	·18205	Beattie	31
	2 14 p.m. D.	H_{40} ·18525			

81. Deelfontein. Lat. 30° 5′·8 S.; Long. 26° 31′·7 E. On left-hand side of the road from Smithfield to Dewetsdorp, just opposite small dam near the garden.

Declination. D.

Date	G.M.T.	D (observed)	D	Observer	Instrument
1904 Jan. 12	1 30 p.m.	25° 15′·2 W.	25° 22′·0 W.	Beattie	73
	1 37 p.m.	25 15 ·1			

Dip. θ.

Date	G.M.T.	Needle	θ (observed)	θ	Observer	Instrument
1904 Jan. 12	10 49 a.m.	1	60° 8′·9 S.	60° 6′·4 S.	Beattie	142
	10 49 a.m.	4_9	60 11 ·5			

Horizontal Intensity. H.

Date	G.M.T.	H (observed)	H	Observer	Instrument
1904 Jan. 12	11 59 a.m. V.				
	12 36 p.m. D.	H_{30} ·18208	·18243	Beattie	73
	1 5 p.m. V.				

82. Deelfontein Farm. Lat. 28° 20′·0 S.; Long. 27° 48′·0 E. 8 miles from Senekal. Left-hand side of road, Senekal to Bethlehem, just over the spruit.

Dip. θ.

Date	G.M.T.	Needle	θ (observed)	θ	Observer	Instrument
1904 Jan. 26	12 42 p.m.	1	59° 28′·2 S.	59° 25′·0 S.	Beattie	142
	12 42 p.m.	4_9	59 29 ·9			

Horizontal Intensity. H.

Date	G.M.T.	H (observed)	H	Observer	Instrument
1904 Jan. 26	9 18 a.m. V.				
	9 53 a.m. D.	H_{30} ·18663	·18706	Beattie	73
	10 48 a.m. D.	H_{40} ·18663			
	11 15 a.m. V.				

83. De Jager's Farm. Lat. 28° 15′·9 S.; Long. 28° 57′·9 E. 9 miles from Harrismith. On right-hand side of road going to Harrismith from Bethlehem. Dam on other side of railway.

Declination. D.

Date	G.M.T.	D (observed)	D	Observer	Instrument
1904 Jan. 31	8 16 a.m.	23° 1′·5 W.	23° 4′·4 W.	Beattie	73
	8 26 a.m.	23 2 ·1			
	1 16 p.m.	22 55 ·5			
	1 23 p.m.	22 55 ·5			

Dip. θ.

Date	G.M.T.	Needle	θ (observed)	θ	Observer	Instrument
1904 Jan. 31	12 38 p.m.	1	59° 38′·2 S.	59° 33′·7 S.	Beattie	142
	12 38 p.m.	4_9	59 37 ·3			

Horizontal Intensity. H.

Date	G.M.T.	H (observed)	H	Observer	Instrument
1904 Jan. 31	8 55 a.m. V.		·18799	Beattie	73
	11 11 a.m. D.	H_{30} ·18754			
	11 47 a.m. D.	H_{40} ·18753			
	1 36 p.m. V.				

84. Dewetsdorp. Lat. 29° 26′·1 S.; Long. 26° 39′·3 E. On left-hand side of road from Helvetia to Dewetsdorp. Between farm Bultfontein and Dewetsdorp on town commonage.

Declination. D.

Date	G.M.T.	D (observed)	D	Observer	Instrument
1904 Jan. 14	9 18 a.m.	25° 38′·7 W.	25° 43′·3 W.	Beattie	73
	9 26 a.m.	25 38 ·1			
	1 35 p.m.	25 37 ·1			
	1 44 p.m.	25 37 ·4			

Dip. θ.

Date	G.M.T.	Needle	θ (observed)	θ	Observer	Instrument
1904 Jan. 14	10 11 a.m.	1	59° 40′·0 S.	59° 37′·1 S.	Beattie	142
	10 11 a.m.	4_9	59 41 ·8			

Horizontal Intensity. H.

Date	G.M.T.	H (observed)	H	Observer	Instrument
1904 Jan. 14	12 16 p.m. V.	·18488	·18528	Beattie	73

85. Draghoender. Lat. 29° 22′·3 S.; Long. 22° 7′·4 E. West side of hotel, and about 300 paces from it. Hotel, point of observation, and north end of gap in dam all in a line.

Declination. D.

Date		G.M.T.	D (observed)	D	Observer	Instrument
1904	Dec. 23	5 48 a.m. 6 3 a.m.	27° 1′·9 W. 27 3 ·9	27° 14′·4 W.	Beattie Hough	73

Dip. θ.

Date		G.M.T.	Needle	θ (observed)	θ	Observer	Instrument
1904	Dec. 23	5 21 a.m. 5 21 a.m.	3_9 4_9	58° 25′·8 S. 58 22 ·3	58° 16′·1 S.	Beattie	142

Horizontal Intensity. H.

Date		G.M.T.	H (observed)	H	Observer	Instrument
1904	Dec. 23	6 56 a.m. V. 7 46 a.m. D. 8 15 a.m. V.	H_{30} ·18606 H_{40} ·18603	·18723	Beattie	73

86. Drew. Lat. 33° 59′·5 S.; Long. 20° 13′·0 E. On the veld opposite the railway station, and on the other side of line to it.

Declination. D.

Date		G.M.T.	D (observed)	D	Observer	Instrument
1901	Feb. 4	7 22 a.m. 7 30 a.m.	28° 41′·1 W. 28 41 ·2	28° 25′·1 W.	Beattie Morrison	31

Dip. θ.

Date		G.M.T.	Needle	θ (observed)	θ	Observer	Instrument
1901	Feb. 5	11 17 a.m. 11 17 a.m.	1 2	59° 28′·5 S. 59 30 ·2	59° 51′·1 S.	Beattie	142

Horizontal Intensity. H.

Date		G.M.T.	H (observed)	H	Observer	Instrument
1901	Feb. 5	9 15 a.m. V. 9 51 a.m. D. 10 17 a.m. V.	H_{30} ·18339 H_{40} ·18341	·18130	Morrison	31

87. DRIEFONTEIN. Lat. 26° 29′·4 S.; Long. 29° 13′·0 E. Near farmhouse, about 3 miles from Trichardsfontein on Johannesburg side of Trichardsfontein.

Declination. D.

Date	G.M.T.	D (observed)	D	Observer	Instrument
1903 Aug. 6	12 20 p.m. 12 31 p.m. 12 43 p.m.	22° 21′·5 W. 22 23 ·3 22 21 ·8	22° 23′·4 W.	Beattie	73

Dip. θ.

Date	G.M.T.	Needle	θ (observed)	θ	Observer	Instrument
1903 Aug. 6	11 1 a.m. 11 1 a.m.	1 4	59° 7′·6 S. 59 5 ·8	59° 6′·0 S.	Beattie	142

Horizontal Intensity. H.

Date	G.M.T.	H (observed)	H	Observer	Instrument
1903 Aug. 6	11 52 a.m. V.	·18776	·18788	Beattie	73

88. DRIEHOEK. Lat. 27° 11′·6 S.; Long. 30° 41′·0 E. Left-hand side of road, Utrecht to Piet Retief. Alongside spruit, about a mile on Piet Retief side of farmhouse Driehoek.

Declination. D.

Date	G.M.T.	D (observed)	D	Observer	Instrument
1903 Aug. 15	7 38 a.m. 7 52 a.m.	22° 17′·4 W. 22 17 ·4	22° 18′·1 W.	Beattie	73

Dip. θ.

Date	G.M.T.	Needle	θ (observed)	θ	Observer	Instrument
1903 Aug. 15	9 29 a.m. 9 29 a.m. 9 31 a.m.	1 4 4_9	59° 27′·0 S. 59 27 ·4 59 24 ·4	59° 25′·4 S.	Beattie	142

Horizontal Intensity. H.

Date	G.M.T.	H (observed)	H	Observer	Instrument
1903 Aug. 15	8 21 a.m. V.	·18892	·18898	Beattie	73

89. East London. Lat. 33° 0′·0 S.; Long. 27° 56′·0 E. On grassy field behind Beach Hotel.

Declination. D.

Date	G.M.T.	D (observed)	D	Observer	Instrument
1902 Jan. 7	3 17 p.m. 4 21 p.m.	26° 45′·5 W. 26 45 ·4	26° 31′·4 W.	Beattie Morrison	31

Dip. θ.

Date	G.M.T.	Needle	θ (observed)	θ	Observer	Instrument
1902 Jan. 7	1 55 p.m. 1 55 p.m.	2 1	61° 36′·9 S. 61 35 ·7	61° 48′·3 S.	Beattie	142

Horizontal Intensity. H.

Date	G.M.T.	H (observed)	H	Observer	Instrument
1902 Jan. 8	10 36 a.m. V. 11 19 a.m. D. 1 20 p.m. V. 2 8 p.m. D. 2 44 p.m. V.	H_{30} ·17743 H_{40} ·17743 H_{30} ·17725 H_{40} ·17731	·17600	Morrison	31

90. Elandshoek. Lat. 25° 30′·0 S.; Long. 30° 41′·0 E. Right-hand side of railway, Waterval Onder to Komati Poort. Left-hand side of path going from station to house on hill.

Dip. θ.

Date	G.M.T.	Needle	θ (observed)	θ	Observer	Instrument
1903 Sept. 10	9 57 a.m. 9 57 a.m.	1 4_9	59° 54′·3 S. 59 52 ·4	59° 52′·0 S.	Beattie	142

Horizontal Intensity. H.

Date	G.M.T.	H (observed)	H	Observer	Instrument
1903 Sept. 10	5 42 a.m. V. 7 9 a.m. D. 7 47 a.m. D. 8 55 a.m. V.	H_{30} ·18857 H_{40} ·18854	·18867	Beattie	73

91. ELANDSKLOOF FARM. Lat. 28° 0′·0 S.; Long. 26° 24′·0 E. About 2 miles from boundary of farm Ganspan, on left-hand side of road from Ganspan to Vet River drift.

Dip. θ.

Date		G.M.T.	Needle	θ (observed)	θ	Observer	Instrument
1904	Feb. 15	11 28 a.m.	1	59° 0′·8 S.	58° 56′·6 S.	Beattie	142
		11 28 a.m.	4_9	59 1·2			

Horizontal Intensity. H.

Date		G.M.T.	H (observed)	H	Observer	Instrument
1904	Feb. 15	12 8 p.m. V.	·18780	·18830	Beattie	73

92. ELIM. Lat. 34° 35′·8 S.; Long. 19° 46′·0 E. In a field, about 200 yards south west of church.

Declination. D.

Date		G.M.T.	D (observed)	D	Observer	Instrument
1901	Jan. 22	5 47 a.m.	28° 56′·1 W.	28° 42′·0 W.	Beattie Morrison	31
		6 0 a.m.	28 55·9			
		6 13 a.m.	28 56·9			
		6 23 a.m.	28 57·7			
		6 32 a.m.	28 58·6			
1901	Jan. 23	5 40 a.m.	28 55·0			
		5 49 a.m.	28 55·8			
		6 0 a.m.	28 56·1			
		6 11 a.m.	28 56·7			
		6 21 a.m.	28 57·0			

Dip. θ.

Date		G.M.T.	Needle	θ (observed)	θ	Observer	Instrument
1901	Jan. 23	1 37 p.m.	1	59° 34′·5 S.	59° 55′·1 S.	Beattie	142
		1 37 p.m.	2	59 35·4			
1901	Jan. 24	9 12 a.m.	1	59.31·0			
		9 12 a.m.	2	59 31·8			

Horizontal Intensity. H.

Date		G.M.T.	H (observed)	H	Observer	Instrument
1901	Jan. 23	8 13 a.m. V. 8 57 a.m. D. 9 55 a.m. V.	H_{30} ·18301 H_{40} ·18300	·18065	Morrison	31

93. Ellerton. Lat. 23° 19′·0 S.; Long. 30° 30′·0 E. On right-hand side of road, Birthday to Pietersburg. On Pietersburg side of cross road to Spelonken.

Dip. θ.

Date	G.M.T.	Needle	θ (observed)	θ	Observer	Instrument
1903 July 18	11 39 a.m. 11 39 a.m.	1 4	56° 53′·2 S. 56 54 ·4	56° 53′·5 S.	Beattie	142

Horizontal Intensity. H.

Date	G.M.T.	H (observed)	H	Observer	Instrument
1903 July 18	12 30 p.m. V. 1 26 p.m. D.	H_{30} ·19819 H_{40} ·19819	·19819	Beattie	73

94. Elliot. Lat. 31° 18′·0 S.; Long. 27° 54′·0 E.

Declination. D.

Date	G.M.T.	D (observed)	D	Observer	Instrument
1906 Feb. 7	5 52 a.m. 6 0 a.m. 6 8 a.m.	24° 32′·2 W. 24 30 ·5 24 30 ·4	24° 54′·7 W.	Brown Morrison	31

Dip. θ.

Date	G.M.T.	Needle	θ (observed)	θ	Observer	Instrument
1906 Feb. 7	10 30 a.m. 10 32 a.m.	1 2	61° 24′·3 S. 61 26 ·6	61° 4′·7 S.	Morrison	9

Horizontal Intensity. H.

Date	G.M.T.	H (observed)	H	Observer	Instrument
1906 Feb. 7	8 33 a.m. V. 9 0 a.m. D. 9 27 a.m. V.	H_{30} ·17749 H_{40} ·17746	·17998	Morrison	31

95. ELSBURG. Lat. 26° 15′·0 S.; Long. 28° 11′·0 E. Left-hand side of railway going towards Pretoria. 143 paces at right angles to railway line reckoned from dead end on Pretoria side of platform.

Declination. D.

Date	G.M.T.	D (observed)	D	Observer	Instrument
1903 June 15	6 8 a.m.	22° 28′·0 W.			
	6 27 a.m.	22 29 ·7			
	6 40 a.m.	22 29 ·3	22° 28′·8 W.	Beattie Löwinger	73
	12 2 p.m.	22 29 ·0			
	12 15 p.m.	22 29 ·5			
	12 28 p.m.	22 29 ·2			

Dip. θ.

Date	G.M.T.	Needle	θ (observed)	θ	Observer	Instrument
1903 June 15	8 21 a.m.	3	58° 27′·1 S.			
	8 20 a.m.	4	58 24 ·4	58° 26′·1 S.	Beattie	142
	8 21 a.m.	1	58 26 ·0			

Horizontal Intensity. H.

Date	G.M.T.	H (observed)	H	Observer	Instrument
1903 June 15	9 26 a.m. V.				
	10 8 a.m. D.	H_{30} ·19142 H_{40} ·19138	·19140	Beattie	73
	10 46 a.m. V.				

96. EMMASHEIM. Lat. 28° 17′·2 S.; Long. 28° 7′·3 E. Between the two drifts over the Zand River on Senekal Bethlehem Road. Alongside destroyed farm-house on left-hand side of road to Bethlehem.

Declination. D.

Date	G.M.T.	D (observed)	D	Observer	Instrument
1904 Jan. 27	12 36 p.m.	23° 48′·2 W.	23° 55′·2 W.	Beattie	73
	12 44 p.m.	23 47 ·2			

Dip. θ.

Date	G.M.T.	Needle	θ (observed)	θ	Observer	Instrument
1904 Jan. 27	11 28 a.m.	1	59° 32′·1 S.	59° 28′·7 S.	Beattie	142
	11 28 a.m.	4_9	59 33 ·6			

Horizontal Intensity. H.

Date	G.M.T.	H (observed)	H	Observer	Instrument
1904 Jan. 27	12 12 p.m. V.	·18654	·18697	Beattie	73

97. Estcourt. Lat. 29° 0′·9 S.; Long. 29° 54′·0 E. Right-hand side of railway, Ladysmith to Durban. 200 paces at right angles to railway line starting from station.

Declination. D.

Date	G.M.T.	D (observed)	D	Observer	Instrument
1903 Oct. 7	7 36 a.m.	23° 15′·8 W.	23° 17′·2 W.	Beattie	73
	7 43 a.m.	23 15 ·6			

Dip. θ.

Date	G.M.T.	Needle	θ (observed)	θ	Observer	Instrument
1903 Oct. 7	10 38 a.m.	1	60° 13′·9 S.	60° 12′·0 S.	Beattie	142
	10 38 a.m.	4_9	60 13 ·9			

Horizontal Intensity. H.

Date	G.M.T.	H (observed)	H	Observer	Instrument
1903 Oct. 7	8 3 a.m. V.	H_{30} ·18358	·18376	Beattie	73
	8 40 a.m. D.	H_{40} ·18358			
	9 16 a.m. V.				

98. Ferreira. Lat. 29° 12′·0 S.; Long. 26° 11′·0 E. Left-hand side of railway Springfontein to Bloemfontein. 212 paces at right angles to railway line, starting from dead end and going 120 paces along railway towards Bloemfontein.

Declination. D.

Date	G.M.T.	D (observed)	D	Observer	Instrument
1903 June 1	6 59 a.m.	24° 46′·7 W.	24° 47′·3 W.	Beattie	73
	7 9 a.m.	24 47 ·7			
	12 54 p.m.	24 48 ·6			
	1 8 p.m.	24 48 ·2			

Dip. θ.

Date	G.M.T.	Needle	θ (observed)	θ	Observer	Instrument
1903 June 1	8 43 a.m.	3	59° 29′·3 S.	59° 30′·7 S.	Beattie	142
	8 42 a.m.	4	59 30 ·3			
	8 43 a.m.	1	59 30 ·6			

Horizontal Intensity. H.

Date	G.M.T.	H (observed)	H	Observer	Instrument
1903 June 1	9 37 a.m. V.	H_{30} ·18486	·18480	Beattie	73
	11 0 a.m. D.	H_{40} ·18480			
	11 48 a.m. D.	H_{25} ·18482			
	12 31 p.m. V.	H_{35} ·18480			

99. Fish River. Lat. 31° 55′·3 S.; Long. 25° 27′·0 E. On left-hand side of railway line, about 100 yards north of grave-yard.

Declination. D.

Date	G.M.T.	D (observed)	D	Observer	Instrument
1902 July 16	2 15 p.m.	26° 28′·4 W.			
	2 27 p.m.	26 27 ·9			
	2 40 p.m.	26 29 ·5			
1902 July 17	6 32 a.m.	26 26 ·1	26° 18′·5 W.	Beattie Morrison	31
	6 40 a.m.	26 25 ·9			
	6 54 a.m.	26 27 ·5			
	7 7 a.m.	26 27 ·5			

Dip. θ.

Date	G.M.T.	Needle	θ (observed)	θ	Observer	Instrument
1902 July 16	1 27 p.m.	1	60° 37′·5 S.			
	1 27 p.m.	3	60 35 ·6	60° 44′·0 S.	Beattie	142
	1 27 p.m.	4	60 36 ·1			

Horizontal Intensity. H.

Date	G.M.T.	H (observed)	H	Observer	Instrument
1902 July 17	7 46 a.m. V.				
	8 23 a.m. D.	H_{30} ·18110 H_{40} ·18106	·18018	Morrison	31
	9 52 a.m. V.				

100. Forty-one mile Siding. Lat. 17° 43′·0 S.; Long. 30° 33′·0 E. Left-hand side of railway, Salisbury to Ayrshire mine. 153 paces at right angles to railway line starting from point at Ayrshire end of loop.

Declination. D.

Date	G.M.T.	D (observed)	D	Observer	Instrument
1903 May 1	3 24 p.m.	15° 56′·2 W.			
	3 30 p.m.	15 55 ·7			
1903 May 2	5 44 a.m.	15 57 ·6	15° 55′·2 W.	Beattie	73
	5 57 a.m.	15 58 ·4			

Dip. θ.

Date	G.M.T.	Needle	θ (observed)	θ	Observer	Instrument
1903 May 2	8 26 a.m.	3	51° 45′·3 S.			
	8 26 a.m.	4	51 46 ·1	51° 47′·3 S.	Beattie	142
	8 26 a.m.	1	51 46 ·8			

Horizontal Intensity. H.

Date	G.M.T.	H (observed)	H	Observer	Instrument
1903 May 1	12 8 p.m. V. 1 45 p.m. D.	H_{40} ·22477 H_{25} ·22476 H_{35} ·22472	·22465	Beattie	73

101. FOUNTAIN HALL. Lat. 29° 15′·8 S.; Long. 29° 59′·0 E. Left-hand side of railway, Ladysmith to Pietermaritzburg. 150 paces at right angles to railway, starting from Pietermaritzburg dead end.

Declination. D.

Date	G.M.T.	D (observed)	D	Observer	Instrument
1903 Nov. 7	1 13 p.m.	23° 14′·6 W.	23° 20′·4 W.	Beattie	73
	1 24 p.m.	23 15 ·6			

Dip. θ.

Date	G.M.T.	Needle	θ (observed)	θ	Observer	Instrument
1903 Nov. 7	10 30 a.m.	1	60° 22′·4 S.	60° 19′·6 S.	Beattie	142
	10 30 a.m.	4_9	60 23 ·2			
	10 30 a.m.	4	60 20 ·7			

Horizontal Intensity. H.

Date	G.M.T.	H (observed)	H	Observer	Instrument
1903 Nov. 7	12 11 p.m. V.	·18476	·18500	Beattie	73

102. FRANCISTOWN (Tati Concession). Lat. 21° 4′·0 S.; Long. 27° 32′·0 E. The observations on the 18th were made at a place on the same side of the railway as the hotel, about 4 minutes walk from the railway. That on the 20th was at a place on opposite side of river to hotel on the left-hand side of road going to Monarch mine.

Dip. θ.

Date	G.M.T.	Needle	θ (observed)	θ	Observer	Instrument
1898 Jan. 18	9 10 a.m.	1	55° 52′·3 S.	56° 29′·8 S.	Beattie	9
	10 30 a.m.	2	55 49 ·3			
1898 Jan. 20	2 55 p.m.	2	53 6 ·0	53 45 ·0	Beattie	9

Horizontal Intensity. H.

Date	G.M.T.	H (observed)	H	Observer	Instrument
1898 Jan. 19	9 10 a.m. V.	H_{30} ·20704	·20380	Beattie	31
	10 23 a.m. D.	H_{40} ·20715			

102A. 10 MILES SOUTH OF FRANCISTOWN, ON ROAD TO PALAPYE.

Dip. θ.

Date	G.M.T.	Needle	θ (observed)	θ	Observer	Instrument
1898 Jan. 19	3 30 p.m.	2	53° 27′·0 S.	54° 6′·0 S.	Beattie	9

103. FRASERBURG. Lat. 31° 55′·2 S.; Long. 21° 31′·3 E. Left-hand side of road, Fraserburg to Fraserburg Road. 300 paces at right angles to road, and 230 paces from corner of Episcopal Church.

Declination. D.

Date	G.M.T.	D (observed)	D	Observer	Instrument
1905 Jan. 18	6 20 a.m.	27° 24′·8 W.	27° 33′·1 W.	Beattie Brown	73

Dip. θ.

Date	G.M.T.	Needle	θ (observed)	θ	Observer	Instrument
1905 Jan. 18	9 40 a.m. 9 42 a.m.	3_9 4_9	59° 28′·3 S. 59 27 ·5	59° 15′·5 S.	Beattie	142

Horizontal Intensity. H.

Date	G.M.T.	H (observed)	H	Observer	Instrument
1905 Jan. 18	7 32 a.m. V. 8 44 a.m. D.	H_{30} ·18170 H_{40} ·18161	·18290	Beattie	73

104. FRASERBURG ROAD. Lat. 32° 46′·0 S.; Long. 22° 0′·0 E. In bed of river behind the hotel.

Dip. θ.

Date	G.M.T.	Needle	θ (observed)	θ	Observer	Instrument
1897 Dec. 28 1897 Dec. 29 1897 Dec. 30	6 27 a.m. 8 51 a.m. 9 30 a.m.	3 1 2	59° 14′·6 S. 59 14 ·1 59 14 ·2	59° 58′·6 S.	Beattie	9
1902 Feb. 5	9 17 a.m. 9 17 a.m.	2 1	59 50 ·0 59 46 ·0	59 59 ·3	Beattie	142
				59 59 ·0 (mean adopted)		

Horizontal Intensity. H.

Date	G.M.T.	H (observed)	H	Observer	Instrument
1902 Feb. 5	2 0 p.m. V. 2 54 p.m. D. 3 38 p.m. V.	H_{30} ·18369 H_{40} ·18371	·18251	Beattie	31

105. Gamtoos River Bridge. Lat. 33° 15′·2 S.; Long. 25° 2′·5 E. Observations taken on flat land east of hotel, at the base of the hill.

Declination. D.

Date	G.M.T.	D (observed)	D	Observer	Instrument
1903 Feb. 8	2 18 p.m.	27° 28′·5 W.			
	4 6 p.m.	27 28 ·7	27° 25′·4 W.	Beattie	31
1903 Feb. 9	5 36 a.m.	27 31 ·1			
	5 53 a.m.	27 30 ·9			

Dip. θ.

Date	G.M.T.	Needle	θ (observed)	θ	Observer	Instrument
1903 Feb. 8	8 57 a.m.	3	61° 11′·5 S.	61° 14′·4 S.	Beattie	142
	8 59 a.m.	4	61 10 ·9			

Horizontal Intensity. H.

Date	G.M.T.	H (observed)	H	Observer	Instrument
1903 Feb. 8	11 10 a.m. V.	H_{30} ·17775	·17736	Beattie	31
	1 16 p.m. D.	H_{40} ·17773			
	2 39 p.m. V.				

106. George Town. Lat. 33° 57′·0 S.; Long. 22° 29′·0 E. In field behind rectory.

Dip. θ.

Date	G.M.T.	Needle	θ (observed)	θ	Observer	Instrument
1903 Jan. 24	6 41 a.m.	3	60° 24′·4 S.			
	2 10 p.m.	3	60 24 ·5	60° 29′·0 S.	Beattie	142
	2 10 p.m.	4	60 25 ·6			
	2 10 p.m.	1	60 27 ·9			

Horizontal Intensity. H.

Date	G.M.T.	H (observed)	H	Observer	Instrument
1903 Jan. 24	9 36 a.m. V.	H_{30} ·18017	·17970	Beattie	31
	10 28 a.m. D.	H_{40} ·18013			
	11 7 a.m. V.				

107. GEMSBOKFONTEIN. Lat. 31° 22′·8 S.; Long. 22° 57′·5 E. Right-hand side of road, about 350 paces from big tree on Victoria West side of brick house.

Declination. D.

Date	G.M.T.	D (observed)	D	Observer	Instrument
1904 Dec. 6	8 25 a.m.	27° 36′·9 W.	27° 47′·3 W.	Beattie Hough	73

Dip. θ.

Date	G.M.T.	Needle	θ (observed)	θ	Observer	Instrument
1904 Dec. 6	12 6 p.m.	4_9	59° 40′·8 S.	59° 27′·1 S.	Beattie	142
	12 6 p.m.	3_9	59 36 ·9			

Horizontal Intensity. H.

Date	G.M.T.	H (observed)	H	Observer	Instrument
1904 Dec. 6	8 47 a.m. V.	H_{30} ·18217	·18326	Beattie	73
	9 30 a.m. D.	H_{40} ·18207			

108. GINGINHLOVU. Lat. 29° 1′·7 S.; Long. 31° 35′·0 E. Right-hand side of railway, Durban to Hlabisa. 220 paces at right angles to railway reckoned from point 100 paces towards Tugela from Tugela end of platform.

Declination. D.

Date	G.M.T.	D (observed)	D	Observer	Instrument
1903 Oct. 27	1 37 p.m.	23° 0′·1 W.	23° 6′·5 W.	Beattie	73
	1 47 p.m.	22 59 ·9			
1903 Oct. 28	6 45 a.m.	23 6 ·8			
	6 55 a.m.	23 6 ·1			

Dip. θ.

Date	G.M.T.	Needle	θ (observed)	θ	Observer	Instrument
1903 Oct. 28	5 8 a.m.	1	61° 4′·4 S.	61° 3′·1 S.	Beattie	142
	5 11 a.m.	4_9	61 4 ·9			
	5 11 a.m.	4	61 6 ·5			

Horizontal Intensity. H.

Date	G.M.T.	H (observed)	H	Observer	Instrument
1903 Oct. 27	10 57 a.m. V.	H_{30} ·18232	·18251	Beattie	73
	11 37 a.m. D.	H_{40} ·18229			
	12 9 p.m. V.				

109. Glenallen. Lat. 29° 39′·0 S.; Long. 22° 36′·0 E.

Declination. D.

Date	G.M.T.	D (observed)	D	Observer	Instrument
1904 Dec. 24	6 0 a.m.	47° 46′·1 W.	47° 56′·5 W.	Beattie Hough	73

Dip. θ.

Date	G.M.T.	Needle	θ (observed)	θ	Observer	Instrument
1904 Dec. 24	5 58 a.m.	3_9	42° 6′·6 S.	41° 53′·0 S.	Beattie	142
	6 1 a.m.	4_9	42 6 ·2			

Horizontal Intensity. H.

Date	G.M.T.	H (observed)	H	Observer	Instrument
1904 Dec. 24	6 40 a.m. V.	·23339	·23457	Beattie	73

110. Glenconnor. Lat. 33° 25′·0 S.; Long. 25° 10′·0 E. In field opposite railway station on opposite side of railway to schoolhouse.

Declination. D.

Date	G.M.T.	D (observed)	D	Observer	Instrument
1900 July 23	1 41 p.m.	27° 57′·5 W.	27° 28′·8 W.	Beattie Morrison	31
	2 0 p.m.	27 57 ·6			
	2 13 p.m.	27 57 ·9			
	2 26 p.m.	27 58 ·4			
	2 36 p.m.	27 58 ·3			

Dip. θ.

Date	G.M.T.	Needle	θ (observed)	θ	Observer	Instrument
1900 July 23	9 30 a.m.	1	60° 56′·1 S.	61° 19′·0 S.	Beattie	9
	9 30 a.m.	2	60 54 ·6			
	11 0 a.m.	1	60 56 ·6			
	11 0 a.m.	2	60 55 ·0			

Horizontal Intensity. H.

Date	G.M.T.	H (observed)	H	Observer	Instrument
1900 July 23	10 59 a.m. V. 11 41 a.m. D. 12 14 p.m. V.	H_{30} ·17884 H_{40} ·17886	·17637	Morrison	31

111. Globe and Phœnix. Lat. 18° 56′·0 S.; Long. 29° 48′·0 E. Right-hand side of railway, Gwelo to Salisbury. 70 paces from dead end of triangle siding, 295 paces from main line.

Declination. D.

Date	G.M.T.	D (observed)	D	Observer	Instrument
1903 April 8	3 0 p.m.	16° 55′·7 W.	16° 53′·5 W.	Beattie	73
	3 10 p.m.	16 55 ·7			

Dip. θ.

Date	G.M.T.	Needle	θ (observed)	θ	Observer	Instrument
1903 April 8	10 54 a.m.	3	52° 40′·0 S.	52° 43′·6 S.	Beattie	142
	10 55 a.m.	4	52 42 ·7			
	10 54 a.m.	1	52 43 ·2			

Horizontal Intensity. H.

Date	G.M.T.	H (observed)	H	Observer	Instrument
1903 April 8	11 57 a.m. V.	H_{25} ·21965	·21945	Beattie	73
	12 49 p.m. D.	H_{35} ·21960			
	1 30 p.m. V.				

112. Goedgedacht. Lat. 26° 38′·9 S.; Long. 29° 37′·0 E.

Declination. D.

Date	G.M.T.	D (observed)	D	Observer	Instrument
1903 Aug. 7	12 54 p.m.	22° 12′·3 W.	22° 13′·1 W.	Beattie	73
	1 4 p.m.	22 11 ·5			

Dip. θ.

Date	G.M.T.	Needle	θ (observed)	θ	Observer	Instrument
1903 Aug. 7	11 20 a.m.	1	58° 46′·2 S.	58° 44′·2 S.	Beattie	142
	11 21 a.m.	4	58 43 ·5			

Horizontal Intensity. H.

Date	G.M.T.	H (observed)	H	Observer	Instrumen
1903 Aug. 7	12 30 p.m. V.	·19132	·19138	Beattie	73

113. Gordon's Bay. Lat. 34° 8′·0 S.; Long. 18° 55′·0 E.

Dip. θ.

Date	G.M.T.	Needle	θ (observed)	θ	Observer	Instrument
1899 Sept. 12	9 30 a.m.	1	58° 54′·0 S.	59° 23′·7 S.	Beattie	9
	9 30 a.m.	2	58 51 ·4			

Horizontal Intensity. H.

Date	G.M.T.	H (observed)	H	Observer	Instrument
1899 Aug. 11	12 7 p.m. V.	H_{30} ·18471	·18100	Morrison	31
	2 55 p.m. D.	H_{40} ·18479			

114. Graaff Reinet. Lat. 32° 16′·9 S.; Long. 24° 36′·0 E. In field adjoining camp on right-hand side of road going from town by road leading past the railway station.

Declination. D.

Date	G.M.T.	D (observed)	D	Observer	Instrument
1900 July 7	10 22 a.m.	27° 28′·8 W.	26° 57′·5 W.	Beattie Morrison	31
	10 35 a.m.	27 28 ·8			
	10 45 a.m.	27 28 ·2			
	10 54 a.m.	27 28 ·0			

Dip. θ.

Date	G.M.T.	Needle	θ (observed)	θ	Observer	Instrument
1900 July 7	10 18 a.m.	1	60° 10′·7 S.	60° 34′·6 S.	Beattie	9
	10 18 a.m.	2	60 10 ·4			

Horizontal Intensity. H.

Date	G.M.T.	H (observed)	H	Observer	Instrument
1900 July 7	1 11 p.m. V.	H_{30} ·18181	·17931	Morrison	31
	1 48 p.m. D.	H_{40} ·18184			
	2 22 p.m. V.				

115. Grahamstown. Lat. 33° 19′·7 S.; Long. 26° 32′·0 E. Between Drostdy and quarry, 100 yards on Drostdy side of quarry.

Declination. D.

Date	G.M.T.	D (observed)	D	Observer	Instrument
1902 July 8	8 14 a.m.	27° 15′·9 W.	27° 1′·1 W.	Beattie Morrison	31
	8 23 a.m.	27 16 ·0			
	8 32 a.m.	27 15 ·7			
	2 4 p.m.	27 12 ·9			
	2 15 p.m.	27 11 ·7			
	2 24 p.m.	27 12 ·4			
1902 July 9	7 13 a.m.	27 4 ·8			
	7 22 a.m.	27 5 ·3			
	8 48 a.m.	27 6 ·6			
	8 57 a.m.	27 7 ·4			

Dip. θ.

Date	G.M.T.	Needle	θ (observed)	θ	Observer	Instrument
1902 July 8	12 48 p.m.	1	61° 39′·9 S.	61° 43′·7 S.	Beattie	142
	12 48 p.m.	2	61 36 ·4			
	12 48 p.m.	3	61 32 ·8			
	12 48 p.m.	4	61 33 ·5			

Horizontal Intensity. H.

Date	G.M.T.	H (observed)	H	Observer	Instrument
1902 July 8	9 58 a.m. V.	H_{30} ·17675	·17589	Morrison	31
	10 13 a.m. D.	H_{40} ·17673			
	10 47 a.m. V.				
1902 July 9	6 42 a.m. V.	H_{30} ·17686			
	8 0 a.m. D.	H_{40} ·17682			
	9 24 a.m. V.				

116. Grange. Lat. 29° 7′·9 S.; Long. 30° 23′·0 E. Right-hand side of railway, Ladysmith to Pietermaritzburg. On edge of hill on Ladysmith side of railway station, about 50 yards from railway.

Dip. θ.

Date	G.M.T.	Needle	θ (observed)	θ	Observer	Instrument
1903 Oct. 9	11 13 a.m. 11 15 a.m.	1 4_9	61° 9′·4 S. 61 6 ·8	61° 6′·2 S.	Beattie	142

Horizontal Intensity. H.

Date	G.M.T.	H (observed)	H	Observer	Instrument
1903 Oct. 9	7 52 a.m. V. 8 26 a.m. D. 10 2 a.m. V.	H_{30} ·18079 H_{40} ·18083	·18097	Beattie	73

117. Gras Kop. Lat. 27° 15′·0 S.; Long. 29° 53′·0 E.

Declination. D.

Date	G.M.T.	D (observed)	D	Observer	Instrument
1903 Aug. 9	2 24 p.m. 2 33 p.m.	22° 33′·2 W. 22 33 ·6	22° 34′·6 W.	Beattie	73

Dip. θ.

Date	G.M.T.	Needle	θ (observed)	θ	Observer	Instrument
1903 Aug. 9	11 1 a.m. 11 1 a.m.	1 4	59° 19′·0 S. 59 17 ·7	59° 18′·4 S.	Beattie	142

Horizontal Intensity. H.

Date	G.M.T.	H (observed)	H	Observer	Instrument
1903 Aug. 9	11 47 a.m. V. 12 28 p.m. D. 1 3 p.m. V.	H_{30} ·18858 H_{40} ·18860	·18865	Beattie	73

118. GREYLINGSTAD. Lat. 26° 44′·6 S.; Long. 28° 45′·5 E. Left-hand side of railway, Germiston to Durban. Along railway 28 paces from dead end towards Germiston, then 200 paces at right angles to railway.

Declination. D.

Date	G.M.T.	D (observed)	D	Observer	Instrument
1903 Sept. 27	4 46 a.m.	22° 19′·4 W.			
	4 57 a.m.	22 20 ·2			
	6 50 a.m.	22 22 ·8	22° 23′·5 W.	Beattie	73
	7 0 a.m.	22 23 ·6			
	8 33 a.m.	22 24 ·5			

Dip. θ.

Date	G.M.T.	Needle	θ (observed)	θ	Observer	Instrument
1903 Sept. 27	9 47 a.m.	1	58° 58′·5 S.	58° 56′·7 S.	Beattie	142
	9 47 a.m.	4_9	58 58 ·2			

Horizontal Intensity. H.

Date	G.M.T.	H (observed)	H	Observer	Instrument
1903 Sept. 27	6 11 a.m. V.				
	7 49 a.m. D.	H_{30} ·18952	·18969	Beattie	73
	8 23 a.m. V.	H_{40} ·18957			

119. GREYTOWN. Lat. 29° 4′·9 S.; Long. 30° 38′·0 E. After leaving the station the railway to Pietermaritzburg forms a semicircle, the observations were taken at the centre of the semicircle on the opposite side of the sluit to the railway station and half-way between it and the police camp.

Declination. D.

Date	G.M.T.	D (observed)	D	Observer	Instrument
1903 Oct. 13	6 11 a.m.	23° 50′·5 W.			
	6 30 a.m.	23 50 ·4	23° 53′·6 W.	Beattie	73
	8 54 a.m.	23 53 ·6			
	9 1 a.m.	23 52 ·6			

Dip. θ.

Date	G.M.T.	Needle	θ (observed)	θ	Observer	Instrument
1903 Oct. 13	10 6 a.m.	1	60° 31′·4 S.	60° 28′·9 S.	Beattie	142
	10 9 a.m.	4	60 30 ·1			

Horizontal Intensity. H.

Date	G.M.T.	H (observed)	H	Observer	Instrument
1903 Oct. 13	6 58 a.m. V.				
	7 58 a.m. D.	H_{30} ·18672	·18690	Beattie	73
	8 32 a.m. D.	H_{40} ·18671			
	9 13 a.m. V.				

120. GROBLER'S BRIDGE. Lat. 25° 53'·5 S.; Long. 30° 13'·0 E. 100 paces from the bridge on the left bank of the Komati River. Right-hand side of road going to Machadodorp.

Declination. D.

Date	G.M.T.	D (observed)	D	Observer	Instrument
1903 Aug. 28	1 28 p.m.	20° 52'·3 W.			
	1 38 p.m.	20 53 ·7	20° 55'·2 W.	Beattie	73
	2 23 p.m.	20 54 ·6			

Dip. θ.

Date	G.M.T.	Needle	θ (observed)	θ	Observer	Instrument
1903 Aug. 28	10 18 a.m.	1	58° 27'·5 S.			
	10 20 a.m.	4	58 25 ·8	58° 25'·8 S.	Beattie	142
	10 29 a.m.	4_9	58 27 ·7			

Horizontal Intensity. H.

Date	G.M.T.	H (observed)	H	Observer	Instrument
1903 Aug. 28	11 26 a.m. V.				
	12 3 p.m. D.	H_{30} ·19304	·19310	Beattie	73
	12 35 p.m. D.	H_{40} ·19304			
	1 5 p.m. V.				

121. GROENKLOOF. Lat. 29° 28'·4 S.; Long. 27° 11'·4 E. Left-hand side of road, Wepener to Ladybrand. About 2½ miles on Ladybrand side of Constantia.

Declination. D

Date	G.M.T.	D (observed)	D	Observer	Instrument
1904 Jan. 19	8 9 a.m.	24° 28'·2 W.			
	8 20 a.m.	24 28 ·0	24° 31'·0 W.	Beattie	73
	8 35 a.m.	24 26 ·7			
	8 53 a.m.	24 26 ·8			

Dip. θ.

Date	G.M.T.	Needle	θ (observed)	θ	Observer	Instrument
1904 Jan. 19	1 8 p.m.	4_9	59° 58'·9 S.	59° 56'·6 S.	Beattie	142
	1 8 p.m.	6	60 1 ·9			

Horizontal Intensity. H.

Date	G.M.T.	H (observed)	H	Observer	Instrument
1904 Jan. 19	9 11 a.m. V.				
	10 0 a.m. D.	H_{30} ·18413	·18451	Beattie	73
	11 12 a.m. D.	H_{40} ·18408			
	11 57 a.m. V.				

122. Groenplaats. Lat. 27° 16′·0 S.; Long. 28° 3′·8 E. Right-hand side of road, Vrede to Frankfort, about six miles from Frankfort on Vrede side of spruit.

Declination. D.

Date	G.M.T.	D (observed)	D	Observer	Instrument
1904 Feb. 6	7 18 a.m. 7 26 a.m. 12 18 p.m. 12 26 p.m.	22° 24′·6 W. 22 24 ·4 22 13 ·3 22 13 ·6	22° 24′·9 W.	Beattie	73

Dip. θ.

Date	G.M.T.	Needle	θ (observed)	θ	Observer	Instrument
1904 Feb. 6	10 52 a.m. 10 53 a.m.	1 4_9	59° 4′·9 S. 59 6 ·1	59° 1′·3 S.	Beattie	142

Horizontal Intensity. H.

Date	G.M.T.	H (observed)	H	Observer	Instrument
1904 Feb. 6	8 36 a.m. V. 9 19 a.m. D. 9 50 a.m. V.	H_{30} ·19030 H_{40} ·19029	·19076	Beattie	73

123. Grootfontein. Lat. 33° 7′·6 S.; Long. 19° 15′·0 E. In a field adjoining the railway station, between station and hotel.

Declination. D.

Date	G.M.T.	D (observed)	D	Observer	Instrument
1900 Jan. 4	2 18 p.m. 2 34 p.m. 2 45 p.m.	28° 39′·0 W. 28 39 ·1 28 39 ·2	28° 16′·2 W.	Beattie Morrison	31

Dip. θ.

Date	G.M.T.	Needle	θ (observed)	θ	Observer	Instrument
1899 July 16	9 12 a.m. 9 17 a.m.	1 2	59° 31′·1 S. 59 27 ·9	60° 1′·5 S.	Beattie	9

Horizontal Intensity. H.

Date	G.M.T.	H (observed)	H	Observer	Instrument
1899 July 16	10 36 a.m. V. 11 4 p.m. D.	H_{30} ·18357 H_{40} ·18363	·18044	Morrison	31

124. Gwaai. Lat. 19° 17′·5 S.; Long. 27° 42′·2 E. Right-hand side of railway, Bulawayo to Victoria Falls. 34 paces on the Bulawayo side of the railway points and 118 paces from there at right angles to the railway line.

Declination. D.

Date	G.M.T.	D (observed)	D	Observer	Instrument
1904 July 3	12 55 p.m.	18° 6′·9 W.			
	2 3 p.m.	18 4·4			
	7 39 a.m.	18 9·7			
	8 46 a.m.	18 10·5	18° 17′·0 W.	Beattie	73
1904 July 4	6 34 a.m.	18 7·0			
	6 44 a.m.	18 6·7			
	12 46 p.m.	18 5·5			
	12 55 p.m.	18 5·0			

Dip. θ.

Date	G.M.T.	Needle	θ (observed)	θ	Observer	Instrument
1904 July 4	9 24 a.m.	1	52° 34′·0 S.	52° 27′·2 S.	Beattie	142
	9 23 a.m.	4	52 34·5			

Horizontal Intensity. H.

Date	G.M.T.	H (observed)	H	Observer	Instrument
1904 July 3	9 34 a.m. V.	H_{30} ·21691			
	10 28 a.m. D.	H_{40} ·21686	·21749	Beattie	73
1904 July 4	8 22 a.m. V.	H_{30} ·21695			
	10 22 a.m. V.	H_{40} ·21681			

125. Gwelo. Lat. 19° 28′·2 S.; Long. 29° 47′·0 E. On racecourse. 210 paces at right angles to railway, starting from a point 220 paces from Bulawayo end of platform, and going towards Bulawayo.

Declination. D.

Date	G.M.T.	D (observed)	D	Observer	Instrument
1903 April 7	3 3 p.m.	17° 58′·9 W.	17° 56′·9 W.	Beattie	73
	3 50 p.m.	17 59·6			

Dip. θ.

Date	G.M.T.	Needle	θ (observed)	θ	Observer	Instrument
1903 April 7	1 34 p.m.	3	54° 20′·0 S.			
	1 35 p.m.	4	54 20·3	54° 23′·1 S.	Beattie	142
	1 35 p.m.	1	54 24·3			

Horizontal Intensity. H.

Date	G.M.T.	H (observed)	H	Observer	Instrument
1903 April 7	9 3 a.m. V.	H_{30} ·22053			
	10 0 a.m. D.	H_{40} ·22062	·22034	Beattie	73
	11 3 a.m. V.	H_{25} ·22047			
		H_{35} ·22048			

126. Hamaan's Kraal. Lat. 25° 24′·3 S.; Long. 28° 17′·0 E. Right-hand side of railway, Pretoria to Pietersburg. 115 paces from dead end towards Pietersburg, then 237 paces at right angles to the railway.

Declination. *D.*

Date	G.M.T.	D (observed)	D	Observer	Instrument
1903 Sept. 24	5 11 a.m.	22° 20′·6 W.			
	5 23 a.m.	22 20 ·9			
	7 53 a.m.	22 26 ·0			
	8 2 a.m.	22 25 ·9	22° 5′·6 W.	Beattie	73
	12 15 p.m.	22 23 ·8			
	12 21 p.m.	22 22 ·6			
	12 34 p.m.	22 22 ·5			

Dip. *θ.*

Date	G.M.T.	Needle	θ (observed)	θ	Observer	Instrument
1903 Sept. 24	10 45 a.m.	1	57° 33′·7 S.	57° 29′·9 S.	Beattie	142
	10 45 a.m.	4_9	57 29 ·3			

Horizontal Intensity. *H.*

Date	G.M.T.	H (observed)	H	Observer	Instrument
1903 Sept. 24	8 33 a.m. V.	H_{30} ·19698	·19710	Beattie	73
	9 17 a.m. D.	H_{40} ·19692			
	9 43 a.m. V.				

127. Hankey. Lat. 33° 52′·0 S.; Long. 24° 53′·0 E. On the hill side south east of the post office.

Declination. *D.*

Date	G.M.T.	D (observed)	D	Observer	Instrument
1903 Feb. 7	9 56 a.m.	27° 27′·5 W.	27° 22′·6 W.	Beattie	31
	10 13 a.m.	27 28 ·0			

Dip. *θ.*

Date	G.M.T.	Needle	θ (observed)	θ	Observer	Instrument
1903 Feb. 7	1 10 p.m.	3	61° 9′·6 S.	61° 12′·9 S.	Beattie	142
	1 10 p.m.	4	61 9 ·6			

Horizontal Intensity. *H.*

Date	G.M.T.	H (observed)	H	Observer	Instrument
1903 Feb. 7	11 16 a.m. V.	H_{30} ·17810	·17773	Beattie	31
	12 15 p.m. D.	H_{40} ·17812			

128. Hartley. Lat. 18° 8′·3 S.; Long. 30° 8′·0 E. Right-hand side of railway, Gwelo to Salisbury. 210 paces at right angles to railway line starting from a point 100 paces from dead end of siding, and going along railway towards Salisbury.

Declination. D.

Date	G.M.T.	D (observed)	D	Observer	Instrument
1903 April 10	7 27 a.m.	16° 45′·2 W.	16° 40′·7 W.	Beattie	73
	7 40 a.m.	16 44 ·4			
	12 57 p.m.	16 40 ·8			
	1 10 p.m.	16 39 ·6			

Dip. θ.

Date	G.M.T.	Needle	θ (observed)	θ	Observer	Instrument
1903 April 10	11 20 a.m.	3	51° 53′·1 S.	51° 55′·7 S.	Beattie	142
	11 22 a.m.	4	51 54 ·3			
	11 21 a.m.	1	51 54 ·8			

Horizontal Intensity. H.

Date	G.M.T.	H (observed)	H	Observer	Instrument
1903 April 10	8 40 a.m. V.	H_{30} ·22295	·22273	Beattie	73
	9 20 a.m. D.	H_{40} ·22297			
	12 21 p.m. V.	H_{25} ·22284			
		H_{35} ·22287			

129. Hector Spruit. Lat. 25° 26′·2 S.; Long. 31° 40′·5 E. Left-hand side of railway, Waterval to Komati Poort. 247 paces along railway towards Komati Poort from blind end of dead end, then 241 paces at right angles to railway.

Declination. D.

Date	G.M.T.	D (observed)	D	Observer	Instrument
1903 Sept. 13	6 38 a.m.	20° 33′·2 W.	20° 34′·8 W.	Beattie	73
	7 15 a.m.	20 34 ·7			

Dip. θ.

Date	G.M.T.	Needle	θ (observed)	θ	Observer	Instrument
1903 Sept. 13	10 59 a.m.	1	58° 12′·3 S.	58° 10′·5 S.	Beattie	142
	10 59 a.m.	4	58 10 ·3			
	10 58 a.m.	4_9	58 13 ·5			

Horizontal Intensity. H.

Date	G.M.T.	H (observed)	H	Observer	Instrument
1903 Sept. 13	7 42 a.m. V.		·19673	Beattie	73
	8 27 a.m. D.	H_{30} ·19662			
	8 58 a.m. D.	H_{40} ·19660			
	9 35 a.m. V.				

130. HEIDELBERG, C. C. Lat. 34° 5′·3 S.; Long. 20° 58′·0 E. In field outside village nearly due west of cemetery on hill side. West of the Cape road and a quarter of a mile north of the Dutch Reformed Church.

Declination. D.

Date		G.M.T.	D (observed)	D	Observer	Instrument
1901	Jan. 30	6 50 a.m. 6 59 a.m. 7 7 a.m.	28° 39′·1 W. 28 39 ·3 28 39 ·8	28° 22′·3 W.	Beattie Morrison	31

Dip. θ.

Date		G.M.T.	Needle	θ (observed)	θ	Observer	Instrument
1901	Jan. 31	8 47 a.m. 8 47 a.m.	1 2	59° 43′·6 S. 59 45 ·1	60° 6′·8 S.	Beattie	142

Horizontal Intensity. H.

Date		G.M.T.	H (observed)	H	Observer	Instrument
1901	Jan. 31	10 12 a.m. V. 10 51 a.m. D. 11 20 a.m. V.	H_{30} ·18252 H_{40} ·18250	·18041	Morrison	31

131. HEILBRON. Lat. 27° 18′·2 S.; Long. 27° 58′·0 E. Left-hand side of railway, Wolvehoek to Heilbron. 190 paces at right angles to railway, starting from a point 19 paces from end of platform away from Wolvehoek, and going towards Wolvehoek.

Declination. D.

Date		G.M.T.	D (observed)	D	Observer	Instrument
1903	June 12	7 28 a.m. 7 36 a.m. 7 52 a.m. 12 16 p.m. 12 28 p.m. 12 38 p.m.	23° 17′·7 W. 23 17 ·7 23 18 ·0 23 17 ·6 23 18 ·6 23 18 ·4	23° 17′·5 W.	Beattie	73

Dip. θ.

Date		G.M.T.	Needle	θ (observed)	θ	Observer	Instrument
1903	June 12	8 55 a.m. 8 55 a.m. 8 54 a.m.	3 4 1	58° 52′·1 S. 58 50 ·4 58 49 ·0	58° 50′·9 S.	Beattie	142

Horizontal Intensity. H.

Date		G.M.T.	H (observed)	H	Observer	Instrument
1903	June 12	9 50 a.m. V. 10 19 a.m. D. 12 6 p.m. V.	H_{30} ·18998 H_{40} ·18998	·18998	Beattie	73

132. HELVETIA. Lat. 29° 52′·1 S.; Long. 26° 33′·0 E. In field adjoining store and on Smithfield side of store.

Declination. *D.*

Date	G.M.T.	D (observed)	D	Observer	Instrument
1904 Jan. 13	8 17 a.m.	25° 10′·1 W.			
	8 37 a.m.	25 12·4			
	8 51 a.m.	25 9·7	25° 15′·5 W.	Beattie	73
	1 16 p.m.	25 10·0			
	1 25 p.m.	25 10·0			

Dip. *θ.*

Date	G.M.T.	Needle	θ (observed)	θ	Observer	Instrument
1904 Jan. 13	9 51 a.m.	1	60° 1′·4 S.	59° 58′·9 S.	Beattie	142
	9 52 a.m.	4_9	60 3·9			

Horizontal Intensity. *H.*

Date	G.M.T.	H (observed)	H	Observer	Instrument
1904 Jan. 13	9 9 a.m. V.				
	11 50 a.m. D.	H_{30} ·18297	18336	Beattie	73
	12 17 p.m. D.	H_{40} ·18295			
	1 1 p.m. V.				

133. HERMANUS. Lat. 34° 25′·3 S.; Long. 19° 16′·0 E. On a flat open space lying near to and north of the road from Hawston to Hermanus, about 400 yards due north of the Dutch Reformed Church.

Declination. *D.*

Date	G.M.T.	D (observed)	D	Observer	Instrument
1901 Jan. 20	5 40 a.m.	28° 58′·0 W.			
	5 53 a.m.	28 59·1			
	6 4 a.m.	28 59·6			
	2 57 p.m.	28 54·6	28° 46′·3 W.	Beattie Morrison	31
	3 9 p.m.	28 54·4			
	3 20 p.m.	28 54·4			
	3 29 p.m.	28 54·6			

Dip. *θ.*

Date	G.M.T.	Needle	θ (observed)	θ	Observer	Instrument
1901 Jan. 20	1 0 p.m.	1	59° 14′·1 S.	59° 38′·2 S.	Beattie	142
	1 0 p.m.	2	59 18·1			

Horizontal Intensity. *H.*

Date	G.M.T.	H (observed)	H	Observer	Instrument
1901 Jan. 20	8 55 a.m. V.				
	9 35 a.m. D.	H_{30} ·18378	·18138	Morrison	31
	10 15 a.m. V.	H_{40} ·18379			

134. HERMON. Lat. 33° 26′·7 S.; Long. 18° 58′·0 E.

Declination. D.

Date	G.M.T.	D (observed)	D	Observer	Instrument
1902 Jan. 31	4 9 p.m.	28° 35′·0 W.	28° 29′·6 W.	Beattie	31
1902 Feb. 2	5 35 p.m.	28 36 ·5			

Dip. θ.

Date	G.M.T.	Needle	θ (observed)	θ	Observer	Instrument
1902 Feb. 1	3 25 p.m.	2	58° 56′·3 S.	59° 5′·4 S.	Beattie	142
	3 24 p.m.	1	58 52 ·0			

Horizontal Intensity. H.

Date	G.M.T.	H (observed)	H	Observer	Instrument
1902 Feb. 1	5 28 a.m. V.	·18432	·18309	Beattie	31
1902 Feb. 2	8 4 a.m. V.	·18425			
	9 55 a.m. V.	·18427			

135. HIGHLANDS. Lat. 27° 16′·0 S.; Long. 31° 23′·0 E. Right-hand side of road coming from Piet Retief. Just opposite big tree on left-hand side of road, and on Piet Retief side of Spekboom river.

Declination. D.

Date	G.M.T.	D (observed)	D	Observer	Instrument
1903 Aug. 21	12 32 p.m.	21° 51′·4 W.	21° 52′·8 W.	Beattie	73
	12 48 p.m.	21 50 ·9			

Dip. θ.

Date	G.M.T.	Needle	θ (observed)	θ	Observer	Instrument
1903 Aug. 19	11 16 a.m.	1	59° 56′·9 S.	59° 55′·8 S.	Beattie	142
	11 16 a.m.	4	59 56 ·8			
	11 11 a.m.	4_9	59 56 ·8			

Horizontal Intensity. H.

Date	G.M.T.	H (observed)	H	Observer	Instrument
1903 Aug. 19	12 12 p.m. V.	H_{30} ·18735	·18742	Beattie	73
	1 15 p.m. D.	H_{40} ·18737			
	1 54 p.m. D.				

136. HLABISA. Lat. 28° 18′·5 S.; Long. 32° 6′·0 E. Left-hand side of railway coming from Durban. 200 paces at right angles to railway, starting from a point opposite middle of goods shed.

Declination. D.

Date	G.M.T.	D (observed)	D	Observer	Instrument
1903 Oct. 23	6 57 a.m.	22° 9′·0 W.			
	7 7 a.m.	22 9 ·9			
	10 39 a.m.	22 5 ·7	22° 9′·1 W.	Beattie	73
	10 49 a.m.	22 5 ·4			
	1 55 p.m.	22 3 ·4			

Dip. θ.

Date	G.M.T.	Needle	θ (observed)	θ	Observer	Instrument
1903 Oct. 23	11 58 a.m.	1	60° 56′·1 S.			
	11 58 a.m.	4_9	60 54 ·0	60° 53′·8 S.	Beattie	142
	11 58 a.m.	4	60 57 ·4			

Horizontal Intensity. H.

Date	G.M.T.	H (observed)	H	Observer	Instrument
1903 Oct. 23	7 34 a.m. V.				
	8 39 a.m. D.	H_{30} ·18240	·18262	Beattie	73
	9 14 a.m. D.	H_{40} ·18242			
	9 58 a.m. V.				

137. HLUTI. Lat. 27° 11′·6 S.; Long. 31° 35′·0 E. In front of police camp and on opposite side of road.

Declination. D.

Date	G.M.T.	D (observed)	D	Observer	Instrument
1903 Aug. 20	1 58 p.m.	21° 44′·4 W.			
	2 13 p.m.	21 42 ·9	21° 45′·3 W.	Beattie	73
	2 25 p.m.	21 44 ·5			

Dip. θ.

Date	G.M.T.	Needle	θ (observed)	θ	Observer	Instrument
1903 Aug. 20	11 8 a.m.	1	59° 45′·4 S.			
	11 9 a.m.	4	59 44 ·4	59° 44′·1 S.	Beattie	142
	11 10 a.m.	4_9	59 45 ·4			

Horizontal Intensity. H.

Date	G.M.T.	H (observed)	H	Observer	Instrument
1903 Aug. 20	2 50 p.m. V.	·18754	·18760	Beattie	73

138. Hoetjes Bay. Lat. 33° 1′·0 S.; Long. 17° 57′·0 E. In field about 600 yards to north west of English Church.

Declination. D.

Date	G.M.T.	D (observed)	D	Observer	Instrument
1901 July 14	8 27 a.m.	28° 36′·0 W.	28° 26′·8 W.	Beattie Morrison	31
	8 38 a.m.	28 36 ·1			
	8 50 a.m.	28 36 ·0			
	2 27 p.m.	28 33 ·8			
	2 39 p.m.	28 34 ·7			
	2 51 p.m.	28 34 ·2			

Dip. θ.

Date	G.M.T.	Needle	θ (observed)	θ	Observer	Instrument
1901 July 14	1 14 p.m.	2	58° 18′·3 S.	58° 35′·0 S.	Beattie	142
	1 18 p.m.	1	58 16 ·0			

Horizontal Intensity. H.

Date	G.M.T.	H (observed)	H	Observer	Instrument
1901 July 14	9 9 a.m. V.	H_{30} ·18596	·18407	Morrison	31
	9 52 a.m. D.	H_{40} ·18593			
	10 26 a.m. V.				

139. Holfontein. Lat. 29° 14′·9 S.; Long. 27° 22′·5 E. One hour by cart from Ladybrand. On left-hand side of road going to Ladybrand, and just over the spruit.

Declination. D.

Date	G.M.T.	D (observed)	D	Observer	Instrument
1904 Jan. 20	7 54 a.m.	24° 35′·5 W.	24° 37′·6 W.	Beattie	73
	8 7 a.m.	24 33 ·8			

Dip. θ.

Date	G.M.T.	Needle	θ (observed)	θ	Observer	Instrument
1904 Jan. 20	12 44 p.m.	1	59° 52′·5 S.	59° 50′·1 S.	Beattie	142
	12 45 p.m.	4_9	59 55 ·5			
	12 45 p.m.	6	59 53 ·6			

Horizontal Intensity. H.

Date	G.M.T.	H (observed)	H	Observer	Instrument
1904 Jan. 20	8 28 a.m. V.	H_{30} ·18458	·18502	Beattie	73
	9 18 a.m. D.	H_{40} ·18461			
	10 37 a.m. V.				

140. Honey Nest Kloof. Lat. 29° 12′·2 S.; Long. 24° 33′·0 E.

Declination. D.

Date	G.M.T.	D (observed)	D	Observer	Instrument
1903 March 16	7 48 a.m.	25° 46′·8 W.			
	8 28 a.m.	25 46 ·5			
	8 45 a.m.	25 46 ·1	25° 36′·3 W.	Beattie	73
	1 26 p.m.	25 33 ·1			
	1 40 p.m.	25 32 ·3			

Dip. θ.

Date	G.M.T.	Needle	θ (observed)	θ	Observer	Instrument
1903 March 16	12 14 p.m.	3	58° 59′·8 S.	59° 1′·9 S.	Beattie	142
	12 13 p.m.	4	58 59′·4			

Horizontal Intensity. H.

Date	G.M.T.	H (observed)	H	Observer	Instrument
1903 March 16	9 7 a.m. V.				
	9 50 a.m. D.	H_{30} ·18644 H_{40} ·18652	·18623	Beattie	73
	10 53 a.m. V.				

141. Honing Spruit. Lat. 27° 27′·0 S.; Long. 27° 25′·0 E. Right-hand side of railway, Kroonstad to Pretoria. 210 paces at right angles to railway, starting from a point 79 paces from dead end, and walking towards Pretoria.

Declination. D.

Date	G.M.T.	D (observed)	D	Observer	Instrument
1903 June 11	7 46 a.m.	23° 15′·6 W.			
	8 0 a.m.	23 16 ·4			
	12 26 p.m.	23 15 ·0	23° 15′·0 W.	Beattie	73
	12 37 p.m.	23 15 ·2			
	12 50 p.m.	23 14 ·4			

Dip. θ.

Date	G.M.T.	Needle	θ (observed)	θ	Observer	Instrument
1903 June 11	9 9 a.m.	3	58° 42′·7 S.			
	9 9 a.m.	4	58 40 ·3	58° 42′·1 S.	Beattie	142
	9 9 a.m.	1	58 42 ·2			

Horizontal Intensity. H.

Date	G.M.T.	H (observed)	H	Observer	Instrumen
1903 June 11	10 30 a.m. V.				
	11 8 a.m. D.	H_{30} ·19007 H_{40} ·19004	·19005	Beattie	73
	11 47 a.m. V.				

142. HOPEFIELD. Lat. 33° 14′·4 S.; Long. 18° 21′·0 E. In field between church and Spes Bona Hotel.

Declination. D.

Date	G.M.T.	D (observed)	D	Observer	Instrument
1901 July 11	2 13 p.m.	28° 29′·1 W.			
	2 25 p.m.	28 29 ·6	28° 21′·5 W.	Beattie Morrison	31
	2 36 p.m.	28 29 ·2			

Dip. θ.

Date	G.M.T.	Needle	θ (observed)	θ	Observer	Instrument
1901 July 12	9 9 a.m.	2	58° 29′·9 S.	58° 46′·5 S.	Beattie	142
	9 9 a.m.	1	58 27 ·4			

Horizontal Intensity. H.

Date	G.M.T.	H (observed)	H	Observer	Instrument
1901 July 12	6 37 a.m. V.	H_{30} ·18540	·18348	Morrison	31
	7 17 a.m. D.	H_{40} ·18540			
	7 55 a.m. V.				

143. HOWHOEK. Lat. 34° 12′·7 S.; Long. 19° 10′·0 E. In vineyard belonging to hotel.

Declination. D.

Date	G.M.T.	D (observed)	D	Observer	Instrument
1901 Jan. 13	7 8 a.m.	28° 53′·4 W.			
	7 17 a.m.	28 53 ·1			
1901 Jan. 14	3 55 p.m.	28 44 ·5	28° 38′·3 W.	Beattie Morrison	31
	4 7 p.m.	28 45 ·4			
	4 19 p.m.	28 46 ·3			

Dip. θ.

Date	G.M.T.	Needle	θ (observed)	θ	Observer	Instrument
1901 Jan. 14	9 35 a.m.	1	59° 7′·7 S.	59° 31′·8 S.	Beattie	142
	9 35 a.m.	2	59 11 ·5			

Horizontal Intensity. H.

Date	G.M.T.	H (observed)	H	Observer	Instrument
1901 Jan. 14	1 46 p.m. V.	H_{30} ·18369	·18128	Beattie	31
	2 29 p.m. D.	H_{40} ·18367			
	3 7 p.m. V.				

144. Huguenot. Lat. 33° 45′·3 S.; Long. 19° 0′·0 E. On hill to the east of the railway station, and 20 minutes walk from the Commercial Hotel.

Declination. D.

Date	G.M.T.	D (observed)	D	Observer	Instrument
1902 Jan. 30	4 17 p.m.	28° 34′·4 W.	28° 29′·8 W.	Beattie	31
1902 Jan. 31	5 43 a.m.	28 37 ·5			

Dip. θ.

Date	G.M.T.	Needle	θ (observed)	θ	Observer	Instrument
1902 Jan. 30	2 26 p.m.	2	59° 5′·0 S.	59° 13′·6 S.	Beattie	142
	2 21 p.m.	1	58 59 ·7			

Horizontal Intensity. H.

Date	G.M.T.	H (observed)	H	Observer	Instrument
1902 Jan. 30	8 19 a.m. V.	H_{30} ·18422	·18286	Beattie	31
	9 18 a.m. D.	H_{40} ·18412			
	10 24 a.m. V.				

145. Humansdorp. Lat. 34° 2′·0 S.; Long. 24° 38′·5 E.

Declination. D.

Date	G.M.T.	D (observed)	D	Observer	Instrument
1903 Feb. 6	6 18 a.m.	27° 26′·8 W.	26° 58′·4 W.	Beattie	31
	6 34 a.m.	27° 26 ·7			

Dip. θ.

Date	G.M.T.	Needle	θ (observed)	θ	Observer	Instrument
1903 Feb. 6	3 38 p.m.	3	61° 13′·1 S.	61° 16′·0 S.	Beattie	142
	3 38 p.m.	4	61 12 ·5			

Horizontal Intensity. H.

Date	G.M.T.	H (observed)	H	Observer	Instrument
1903 Feb. 6	8 31 a.m. V.	H_{30} ·17738	·17702	Beattie	31
	9 30 a.m. D.	H_{40} ·17741			
	11 7 a.m. V.				

146. HUTCHINSON. Lat. 31° 29′·6 S.; Long. 23° 15′·0 E. Alongside sluit in field on opposite side of railway to hotel. Half a mile from station in a direction at right angles to platform and starting from the De Aar end of it.

Declination. D.

Date	G.M.T.	D (observed)	D	Observer	Instrument
1900 Jan. 11	10 59 a.m.	27° 17′·7 W.			
	11 15 a.m.	27 16 ·3			
	4 55 p.m.	27 14 ·4	26° 45′·5 W.	Beattie Morrison	31
	5 9 p.m.	27 15 ·2			
	5 21 p.m.	27 14 ·9			
1904 June 29	8 0 a.m.	26 42 ·9			
	8 14 a.m.	26 41 ·6			
	8 24 a.m.	26 42 ·0	26 48 ·6	Morrison	31
	1 52 p.m.	26 37 ·6			
	2 2 p.m.	26 37 ·3			
	2 11 p.m.	26 37 ·6			
			26 47 ·0 (mean adopted)		

Dip. θ.

Date	G.M.T.	Needle	(θ observed)	θ	Observer	Instrument
1898 April 16	9 28 a.m.	2	58° 51′·1 S.			
	2 27 p.m.	1	58 55 ·6	59° 34′·2 S.	Beattie Morrison	9
1898 April 17	5 20 p.m.	2	58 51 ·1			
1899 July 10	2 5 p.m.	1	59 5 ·2			
	2 48 p.m.	2	59 2 ·0	59 35 ·7	Beattie Morrison	9
1899 July 11	2 16 p.m.	1	59 4 ·7			
	3 20 p.m.	2	59 3 ·8			
1900 Jan. 11	2 40 p.m.	1	59 6 ·1	59 34 ·5	Beattie	9
	2 40 p.m.	2	59 6 ·5			
1904 June 28	3 14 p.m.	1	59 37 ·9			
	3 12 p.m.	2	59 40 ·9	59 32 ·0	Morrison	9
1904 June 29	10 0 a.m.	1	59 40 ·4			
	10 0 a.m.	2	59 40 ·6			
				59 34 ·1 (mean adopted)		

Horizontal Intensity. H.

Date	G.M.T.	H (observed)	H	Observer	Instrument
1899 July 11	9 49 a.m. V.	H_{30} ·18625	·18304	Morrison	31
	10 30 a.m. D.	H_{40} ·18622			
1900 Jan. 11	11 13 a.m. V.				
	12 15 p.m. D.	H_{30} ·18617 H_{40} ·18612	·18334	Morrison	31
	12 52 p.m. V.				
1904 June 28	8 38 a.m. V.				
	10 0 a.m. D.	H_{30} ·18229 H_{40} ·18233	·18311	Morrison	31
	10 26 a.m. V.				
			·18316 (mean adopted)		

147. Ibisi Bridge. Lat. 30° 24′·4 S.; Long. 29° 54′·5 E. Right-hand side of road, Harding to Ixopo. Harding side of river, 100 paces from road, and 100 paces from garden fence.

Declination. D.

Date	G.M.T.	D (observed)	D	Observer	Instrument
1903 Nov. 22	7 37 a.m.	24° 25′·7 W.			
	8 10 a.m.	24 26 ·9			
	8 19 a.m.	24 26 ·8	24° 26′·7 W.	Beattie	73
	11 19 a.m.	24 20 ·2			
	11 28 a.m.	24 19 ·9			

Dip. θ.

Date	G.M.T.	Needle	θ (observed)	θ	Observer	Instrument
1903 Nov. 22	12 55 p.m.	1	61° 4′·6 S.	61° 2′·0 S.	Beattie	142
	12 55 p.m.	4_9	61 4 ·7			

Horizontal Intensity. H.

Date	G.M.T.	H (observed)	H	Observer	Instrument
1903 Nov. 22	10 38 a.m. V.	H_{30} ·18144	·18170	Beattie	73
	11 10 a.m. D.	H_{40} ·18142			
	12 0 noon V.				

148. Idutywa. Lat. 32° 0′·8 S.; Long. 28° 20′·4 W. North west of main road, six miles east of Idutywa.

Declination. D.

Date	G.M.T.	D (observed)	D	Observer	Instrument
1906 Jan. 17	7 55 a.m.	25° 33′·1 W.			
	8 4 a.m.	25 33 ·2	25° 55′·8 W.	Brown Morrison	31
	8 15 a.m.	25 33 ·0			

Dip. θ.

Date	G.M.T.	Needle	θ (observed)	θ	Observer	Instrument
1906 Jan. 17	2 11 p.m.	1	62° 5′·1 S.	61° 44′·1 S.	Morrison	9
	2 9 p.m.	2	62 3 ·9			

Horizontal Intensity. H.

Date	G.M.T.	H (observed)	H	Observer	Instrument
1906 Jan. 17	9 59 a.m. V.	H_{30} ·17311	·17559	Morrison	31
	10 43 a.m. D.	H_{40} ·17314			
	11 26 a.m. V.				

149. IGUSI. Lat. 19° 40′·8 S.; Long. 28° 6′·0 E. Observations taken at a place on right-hand side of railway line, Victoria Falls to Bulawayo, 200 paces at right angles to railway starting from the Bulawayo end of ganger's cottage.

Declination. D.

Date	G.M.T.	D (observed)	D	Observer	Instrument
1904 July 13	7 42 a.m.	17° 9′·1 W.	17° 19′·3 W.	Beattie	73

Dip. θ.

Date	G.M.T.	Needle	θ (observed)	θ	Observer	Instrument
1904 July 13	10 57 a.m.	1	53° 7′·5 S.	53° 1′·0 S.	Beattie	142
	10 53 a.m.	4	53 9 ·1			

Horizontal Intensity. H.

Date	G.M.T.	H (observed)	H	Observer	Instrument
1904 July 13	8 35 a.m. V.	H_{30} ·21573	·21633	Beattie	73
	9 14 a.m. D.	H_{40} ·21569			

150. ILLOVO RIVER. Lat. 30° 6′·1 S.; Long. 30° 51′·0 E. Right-hand side of railway coming from Durban. 330 paces along road from point where it intersects railway line. Place of observation almost opposite railway station.

Declination. D.

Date	G.M.T.	D (observed)	D	Observer	Instrument
1903 Nov. 5	7 20 a.m.	23° 49′·1 W.	23° 49′·1 W.	Beattie	73
	7 31 a.m.	23 48 ·3			
	8 56 a.m.	23 46 ·0			
	9 8 a.m.	23 46 ·2			

Dip. θ.

Date	G.M.T.	Needle	θ (observed)	θ	Observer	Instrument
1903 Nov. 5	10 32 a.m.	1	61° 22′·4 S.	61° 19′·1 S.	Beattie	142
	10 33 a.m.	4_9	61 20 ·5			
	10 33 a.m.	4	61 21 ·8			

Horizontal Intensity. H.

Date	G.M.T.	H (observed)	H	Observer	Instrument
1903 Nov. 5	6 59 a.m. V.		·18001	Beattie	73
	8 28 a.m. D.	H_{30} ·17976			
	7 51 a.m. V.	H_{40} ·17977			
	9 20 a.m. V.				

151. Imvani. Lat. 32° 2′·0 S.; Long. 27° 5′·0 E. Observations taken in a field 60° E. of magnetic south from railway station, and at a distance of about 400 yards from it.

Declination. *D.*

Date	G.M.T.	D (observed)	D	Observer	Instrument
1906 Jan. 9	6 52 a.m.	25° 41′·1 W.	26° 3′·7 W.	Brown	31
	6 56 a.m.	25 41 ·0		Morrison	

Dip. θ.

Date	G.M.T.	Needle	θ (observed)	θ	Observer	Instrument
1906 Jan. 9	2 45 p.m.	1	61° 46′·9 S.	61° 28′·0 S.	Morrison	9
	2 48 p.m.	2	61 49 ·2			

Horizontal Intensity. *H.*

Date	G.M.T.	H (observed)	H	Observer	Instrument
1906 Jan. 9	9 46 a.m. D.	H_{30} ·17544	·17781	Morrison	31
		H_{40} ·17534			

152. Indowane. Lat. 29° 57′·5 S.; Long. 29° 26′·7 E.

Declination. *D.*

Date	G.M.T.	D (observed)	D	Observer	Instrument
1903 Nov. 17	4 39 a.m.	24° 13′·4 W.	24° 13′·7 W.	Beattie	73
	4 49 a.m.	24 11 ·1			
	5 0 a.m.	24 10 ·5			
	6 36 a.m.	24 13 ·2			
	6 46 a.m.	24 12 ·8			
	8 38 a.m.	24 12 ·7			
1903 Nov. 16	2 17 p.m.	24 6 ·2			
	2 24 p.m.	24 6 ·6			

Dip. θ.

Date	G.M.T.	Needle	θ (observed)	θ	Observer	Instrument
1903 Nov. 17	11 21 a.m.	1	61° 3′·0 S.	61° 0′·5 S.	Beattie	142

Horizontal Intensity. *H.*

Date	G.M.T.	H (observed)	H	Observer	Instrument
1903 Nov. 17	9 19 a.m. V.	H_{30} ·18048	·18074	Beattie	73
	9 55 a.m. D.	H_{40} ·18045			
	10 23 a.m. V.				

153. INDWE. Lat. 31° 27′·8 S.; Long. 27° 21′·0 E.

Declination. D.

Date	G.M.T.	D (observed)	D	Observer	Instrument
1902 Jan. 3	6 20 a.m.	25° 50′·3 W.	25° 32′·4 W.	Beattie	31
	8 32 a.m.	25 48 ·1			
	3 2 p.m.	25 46 ·7			
	4 8 p.m.	25 47 ·5			

Dip. θ.

Date	G.M.T.	Needle	θ (observed)	θ	Observer	Instrument
1902 Jan. 3	1 39 p.m.	2	60° 42′·9 S.	60° 56′·1 S.	Beattie	142
	1 24 p.m.	1	60 45 ·3			

Horizontal Intensity. H.

Date	G.M.T.	H (observed)	H	Observer	Instrument
1902 Jan. 3	8 40 a.m. V.	H_{30} ·18329	·18186	Beattie	31
	9 50 a.m. D.	H_{40} ·18313			

154. INOCULATION. Lat. 20° 49′·7 S.; Long. 27° 38′·0 E. Left-hand side of railway, Mafeking to Bulawayo. 180 paces at right angles to railway, starting from a point 120 paces from blind end of siding and going towards Bulawayo.

Declination. D.

Date	G.M.T.	D (observed)	D	Observer	Instrument
1903 March 31	2 57 p.m.	19° 14′·3 W.	19° 11′·6 W.	Beattie	73
	3 13 p.m.	19 13 ·6			

Dip. θ.

Date	G.M.T.	Needle	θ (observed)	θ	Observer	Instrument
1903 April 1	5 12 a.m.	3	53° 57′·7 S.	54° 0′·2 S.	Beattie	142
	5 14 a.m.	4	53 58 ·8			
	5 15 a.m.	1	53 59 ·0			

Horizontal Intensity. H.

Date	G.M.T.	H (observed)	H	Observer	Instrument
1903 March 31	11 36 a.m. V.	H_{30} ·21141	·21121	Beattie	73
	12 28 p.m. D.	H_{40} ·21141			
	1 20 p.m. V.	H_{25} ·21139			
		H_{35} ·21135			

155. INYANTUÉ. Lat. 18° 32'·5 S.; Long. 26° 41'·8 E. On right-hand side of railway, Bulawayo to Victoria Falls. 65 paces at right angles to siding reckoned from part of siding opposite telegraph pole which was 52 paces on Bulawayo side of water tank.

Declination. D.

Date	G.M.T.	D (observed)	D	Observer	Instrument
1904 July 6	1 58 p.m. 2 6 p.m.	18° 17'·6 W. 18 16 ·7	18° 27'·4 W.	Beattie	73

Dip. θ.

Date	G.M.T.	Needle	θ (observed)	θ	Observer	Instrument
1904 July 6	8 44 a.m. 8 44 a.m.	1 4	51° 17'·0 S. 51 14 ·7	51° 8'·8 S.	Beattie	142

Horizontal Intensity. H.

Date	G.M.T.	H (observed)	H	Observer	Instrument
1904 July 6	9 28 a.m. V. 10 0 a.m. D. 10 38 a.m. D. 12 6 p.m. V. 2 21 p.m. V.	H_{30} ·22286 H_{40} ·22276	·22341	Beattie	73

156. KAALFONTEIN. Lat. 26° 0'·5 S.; Long. 28° 16'·5 E. Right-hand side of railway, Germiston to Pretoria. 98 paces along railway line from dead end towards Pretoria, then 250 paces at right angles to railway line.

Dip. θ.

Date	G.M.T.	Needle	θ (observed)	θ	Observer	Instrument
1903 June 24	2 40 p.m. 2 41 p.m. 2 42 p.m. 2 42 p.m.	3 4 1 4_9	58° 15'·8 S. 58 12 ·4 58 12 ·8 58 11 ·2	58° 13'·1 S.	Beattie	142

Horizontal Intensity. H.

Date	G.M.T.	H (observed)	H	Observer	Instrument
1903 June 24	12 10 p.m. V. 12 55 p.m. D. 1 27 p.m. V.	H_{30} ·19456 H_{40} ·19463	·19459	Beattie	73

157. Kaalkop Farm. Lat. 27° 47′·3 S.; Long. 28° 58′·3 E. Right-hand side of main road, Harrismith to Vrede. First spruit after passing store at Cornelius River Bridge.

Declination. D.

Date	G.M.T.	D (observed)	D	Observer	Instrument
1904 Feb. 2	8 56 a.m.	22° 49′·4 W.	22° 54′·6 W.	Beattie	73
	9 30 a.m.	22 46 ·4			
	12 28 p.m.	22 48 ·4			
	12 35 p.m.	22 49 ·6			

Dip. θ.

Date	G.M.T.	Needle	θ (observed)	θ	Observer	Instrument
1904 Feb. 2	12 2 p.m.	1	59° 29′·7 S.	59° 26′·8 S.	Beattie	142
	12 2 p.m.	4_9	59 32 ·0			

Horizontal Intensity. H.

Date	G.M.T.	H (observed)	H	Observer	Instrument
1904 Feb. 2	9 22 a.m. V.	H_{30} ·18723	·18770	Beattie	73
	9 53 a.m. D.	H_{40} ·18726			
	11 52 a.m. V.				

158. Kaapmuiden. Lat. 25° 31′·7 S.; Long. 31° 19′·0 E. Left-hand side of railway, Waterval to Komati Poort. From end of platform towards Komati Poort along the railway 75 paces, then 316 paces at right angles to the railway.

Declination. D.

Date	G.M.T.	D (observed)	D	Observer	Instrument
1903 Sept. 11	6 44 a.m.	20° 56′·0 W.	20° 57′·9 W.	Beattie	73
	7 18 a.m.	20 57 ·7			
	7 29 a.m.	20 58 ·3			

Dip. θ.

Date	G.M.T.	Needle	θ (observed)	θ	Observer	Instrument
1903 Sept. 11	10 37 a.m.	1	58° 32′·2 S.	58° 29′·4 S.	Beattie	142
	10 37 a.m.	4_9	58 29 ·2			

Horizontal Intensity. H.

Date	G.M.T.	H (observed)	H	Observer	Instrument
1903 Sept. 11	8 30 a.m. V.	H_{30} ·19298	·19310	Beattie	73
	9 6 a.m. D.	H_{40} ·19299			
	9 33 a.m. V.				

159. KALKBANK (WATT'S STORE). Lat. 23° 31′·5 S.; Long. 29° 20′·0 E. Left-hand side of road, Pietersburg to Tuli. 50 paces back from house along road towards Pietersburg, then 300 yards from road.

Declination. D.

Date	G.M.T.	D (observed)	D	Observer	Instrument
1903 June 30	2 19 p.m.	20° 9′·4 W.	20° 8′·7 W.	Beattie Löwinger	73
	2 37 p.m.	20 9 ·3			
1903 July 1	6 8 a.m.	20 7 ·3			
	6 25 a.m.	20 8 ·6			

Dip. θ.

Date	G.M.T.	Needle	θ (observed)	θ	Observer	Instrument
1903 July 1	9 38 a.m.	1	56° 21′·6 S.	56° 22′·4 S.	Beattie	142
	9 39 a.m.	4	56 23 ·2			

Horizontal Intensity. H.

Date	G.M.T.	H (observed)	H	Observer	Instrument
1903 June 30	3 3 p.m. V.	·20281	·20283	Beattie	73
1903 July 1	7 44 a.m. D.	H_{30} ·20284			
	8 28 a.m. V.	H_{40} ·20283			

160. KALOOMBIES (South of). Lat. 22° 39′·3 S.; Long. 29° 14′·0 E. About 10 miles south of Kaloombies. Right-hand side of road, Warmbad to Pont drift. Opposite side of road to outspan.

Declination. D.

Date	G.M.T.	D (observed)	D	Observer	Instrument
1903 July 10	11 41 a.m.	20° 2′·3 W.	20° 2′·1 W.	Beattie Löwinger	73
	12 2 p.m.	20 1 ·1			

Dip. θ.

Date	G.M.T.	Needle	θ (observed)	θ	Observer	Instrument
1903 July 10	10 51 a.m.	1	56° 34′·6 S.	56° 35′·4 S.	Beattie	142
	10 51 a.m.	4	56 36 ·6			

Horizontal Intensity. H.

Date	G.M.T.	H (observed)	H	Observer	Instrument
1903 July 10	11 30 a.m. V.	·20150	·20150	Beattie	73

161. KARREE. Lat. 28° 52′·5 S.; Long. 26° 21′·0 E. Right-hand side of railway going to Kroonstad. 164 paces from railway, starting from a point 187 paces from dead end and going towards Kroonstad.

Declination. D.

Date	G.M.T.	D (observed)	D	Observer	Instrument
1903 June 6	7 5 a.m.	23° 27′·9 W.	23° 27′·4 W.	Beattie	73
	7 20 a.m.	23 28 ·1			
	12 24 p.m.	23 28 ·3			
	12 40 p.m.	23 27 ·5			

Dip. θ.

Date	G.M.T.	Needle	θ (observed)	θ	Observer	Instrument
1903 June 6	8 45 a.m.	3	59° 19′·4 S.	59° 19′·5 S.	Beattie	142
	8 45 a.m.	4	59 18 ·3			
	8 45 a.m.	1	59 19 ·3			

Horizontal Intensity. H.

Date	G.M.T.	H (observed)	H	Observer	Instrument
1903 June 6	9 37 a.m. V.	H_{30} ·18545	·18548	Beattie	73
	10 46 a.m. D.	H_{25} ·18548			
	12 0 noon V.	H_{35} ·18550			

162. KATHOEK. Lat. 34° 23′·3 S.; Long. 20° 20′·0 E. In open space round the farmhouse. About 100 yards north of the house.

Declination. D.

Date	G.M.T.	D (observed)	D	Observer	Instrument
1901 Jan. 28	3 20 p.m.	28° 48′·4 W.	28° 37′·1 W.	Beattie Morrison	31
	3 32 p.m.	28 48 ·9			
	3 49 p.m.	28 48 ·9			
	4 2 p.m.	28 48 ·9			
	4 42 p.m.	28 49 ·6			
	5 2 p.m.	28 49 ·8			

Dip. θ.

Date	G.M.T.	Needle	θ (observed)	θ	Observer	Instrument
1901 Jan. 28	11 0 a.m.	1	59° 39′·8 S.	60° 2′·7 S.	Beattie	142
	11 0 a.m.	2	59 42 ·0			

Horizontal Intensity. H.

Date	G.M.T.	H (observed)	H	Observer	Instrument
1901 Jan. 28	1 1 p.m. V.	H_{30} ·18243	·18012	Morrison	31
	1 42 p.m. D.	H_{40} ·18241			
	2 17 p.m. V.				

163. KENHARDT. Lat. 29° 18′·0 S.; Long. 21° 9′·0 E. In bed of river, west from Kenhardt Hotel.

Declination. D.

Date		G.M.T.	D (observed)	D	Observer	Instrument
1904	Dec. 17	5 43 a.m.	26° 9′·2 W.	26° 15′·4 W.	Beattie	73

Dip. θ.

Date		G.M.T.	Needle	θ (observed)	θ	Observer	Instrument
1904	Dec. 17	8 37 a.m.	3_9	58° 13′·7 S.	58° 1′·4 S.	Beattie	142
		8 37 a.m.	4_9	58 12 ·6			

Horizontal Intensity. H.

Date		G.M.T.	H (observed)	H	Observer	Instrument
1904	Dec. 17	6 54 a.m. V.	H_{30} ·18812	·18925	Beattie	73
		7 40 a.m. D.	H_{40} ·18804			

163A. KENILWORTH (KIMBERLEY). Lat. 28° 42′·0 S.; Long. 24° 27′·0 E. (By Sutton in Cape Meteorological Commission Report.) In old grave-yard at de Beers. 2 miles S. from Kimberley.

Dip. θ.

Date		G.M.T.	Needle	θ (observed)	θ	Observer	Instrument
1898	Jan. 25	1 45 p.m.	2	57° 57′·9 S.	57° 39′·2 S.	Beattie	9

164. KIMBERLEY. Lat. 28° 43′·0 S.; Long. 24° 46′·0 E. (From Cape Meteorological Commission Report.) In Kimberley public gardens.

Dip. θ.

Date		G.M.T.	Needle	θ (observed)	θ	Observer	Instrument
1898	Jan. 3	4 41 p.m.	1	58° 9′·7 S.	58° 48′·5 S.	Beattie, Morrison	9
1898	Jan. 5	10 0 a.m.	2	58 6 ·1			
		1 52 p.m.	1	58 10 ·5			
1898	Jan. 6	8 0 a.m.	2	58 7 ·1			
1898	Jan. 24	9 20 a.m.	1	58 5 ·6			
		9 20 a.m.	2	58 4 ·2			

Horizontal Intensity. H.

Date		G.M.T.	H (observed)	H	Observer	Instrument
1898	Jan. 4	7 9 a.m. V.	H_{30} ·19140	·18727	Morrison	31
		10 0 a.m. D.	H_{40} ·19128			
1898	Jan. 5	4 9 p.m. V.	H_{30} ·19108		Beattie	31
		4 50 p.m. D.	H_{40} ·19115			

165. King William's Town. Lat. 32° 52′·5 S.; Long. 27° 25′·0 E.

Declination. D.

Date	G.M.T.	D (observed)	D	Observer	Instrument
1902 Jan. 6	6 8 a.m.	26° 37′·6 W.	26° 19′·1 W.	Beattie Morrison	31
	6 20 a.m.	26 36 ·6			
	6 30 a.m.	26 36 ·1			
	4 0 p.m.	26 33 ·0			
	4 17 p.m.	26 32 ·8			
	4 24 p.m.	26 32 ·4			

Dip. θ.

Date	G.M.T.	Needle	θ (observed)	θ	Observer	Instrument
1902 Jan. 6	5 24 a.m.	2	61° 26′·0 S.	61° 39′·0 S.	Beattie	142
	5 27 a.m.	1	61 28 ·1			

Horizontal Intensity. H.

Date	G.M.T.	H (observed)	H	Observer	Instrument
1902 Jan. 6	1 4 p.m. V.	H_{30} ·17862	·17726	Morrison	31
	1 50 p.m. D.	H_{40} ·17859			
	2 28 p.m. V.				

166. Klaarstroom. Lat. 33° 20′·0 S.; Long. 22° 32′·5 E. On left-hand side of road to Meiring's Poort on Meiring's Poort side of the hotel, 100 yards beyond the native location.

Declination. D.

Date	G.M.T.	D (observed)	D	Observer	Instrument
1903 Jan. 5	2 10 p.m.	27° 56′·9 W.	27° 56′·0 W.	Beattie	31
	2 22 p.m.	27 57 ·2			
	3 56 p.m.	27 59 ·6			
	4 8 p.m.	27 58 ·8			

Dip. θ.

Date	G.M.T.	Needle	θ (observed)	θ	Observer	Instrument
1903 Jan. 6	5 45 a.m.	3	60° 24′·3 S.	60° 28′·3 S.	Beattie	142
	5 45 a.m.	4	60 24 ·2			

Horizontal Intensity. H.

Date	G.M.T.	H (observed)	H	Observer	Instrument
1903 Jan. 6	8 30 a.m. V.	H_{30} ·17951	·17906	Beattie	31
	9 33 a.m. D.	H_{40} ·17951			

167. KLERKSDORP. Lat. 26° 52′·3 S.; Long. 26° 38′·0 E. Right-hand side of railway, Klerksdorp to Potchefstroom. 320 paces from railway line starting from dead end.

Declination. D.

Date	G.M.T.	D (observed)	D	Observer	Instrument
1903 June 21	7 44 a.m. 7 55 a.m.	23° 42′·1 W. 23 41 ·8	23° 41′·9 W.	Beattie Löwinger	73

Dip. θ.

Date	G.M.T.	Needle	θ (observed)	θ	Observer	Instrument
1903 June 21	10 24 a.m. 10 22 a.m. 10 24 a.m.	4 1 4_9	58° 13′·0 S. 58 12 ·2 58 10 ·8	58° 12′·2 S.	Beattie	142

Horizontal Intensity. H.

Date	G.M.T.	H (observed)	H	Observer	Instrument
1903 June 21	11 34 a.m. V. 12 3 p.m. D. 12 34 p.m. V.	H_{30} ·19128 H_{40} ·19119	·19123	Beattie	73

168. KLIPFONTEIN, C. C. Lat. 30° 42′·1 S.; Long. 22° 23′·5 E. Alongside garden wall between garden and house at house end of garden wall.

Declination. D.

Date	G.M.T.	D (observed)	D	Observer	Instrument
1904 Dec. 8	9 32 a.m.	22° 55′·2 W.	23° 6′·3 W.	Beattie Hough	73

Dip. θ.

Date	G.M.T.	Needle	θ (observed)	θ	Observer	Instrument
1904 Dec. 8	9 42 a.m. 9 42 a.m.	3_9 4_9	59° 26′·8 S. 59 28 ·8	59° 16′·1 S.	Beattie	142

Horizontal Intensity. H.

Date	G.M.T.	H (observed)	H	Observer	Instrument
1904 Dec. 8	7 58 a.m. V. 8 41 a.m. D.	H_{30} ·18225 H_{40} ·18226	·18340	Beattie	73

169. Klipfontein (Spelonken). Lat. 23° 5′·7 S.; Long. 30° 10′·0 E. At a point on road from post office to Birthday. A little beyond intersection of road from Fort Edward with the first road, on opposite side of Fort Edward road to post office.

Declination. D.

Date	G.M.T.	D (observed)	D	Observer	Instrument
1903 July 15	6 13 a.m.	19° 30′·5 W.	19° 31′·3 W.	Beattie Löwinger	73
	6 28 a.m.	19 30 ·7			
	1 27 p.m.	19 31 ·3			
	1 40 p.m.	19 31 ·3			

Dip. θ.

Date	G.M.T.	Needle	θ (observed)	θ	Observer	Instrument
1903 July 15	5 25 a.m.	1	56° 20′·4 S.	56° 21′·1 S.	Beattie	·142
	5 26 a.m.	4	56 22 ·4			

Horizontal Intensity. H.

Date	G.M.T.	H (observed)	H	Observer	Instrument
1903 July 15	6 53 a.m. V.	H_{30} ·20347	·20344	Beattie	73
	3 19 p.m. D.	H_{40} ·20341			
	2 21 p.m. V.				

170. Klipplaat. Lat. 33° 2′·0 S.; Long. 24° 26′·0 E. Opposite Grand Junction Railway station on other side of Graaff Reinet railway.

Declination. D.

Date	G.M.T.	D (observed)	D	Observer	Instrument
1900 July 18	2 52 p.m.	27° 54′·9 W.	27° 27′·0 W.	Beattie Morrison	31
	3 4 p.m.	27 54 ·9			
	3 15 p.m.	27 55 ·6			
1900 July 19	8 3 a.m.	27 57 ·6			
	8 14 a.m.	27 58 ·3			
	8 25 a.m.	27 58 ·9			

Dip. θ.

Date	G.M.T.	Needle	θ (observed)	θ	Observer	Instrument
1900 July 18	10 0 a.m.	1	60° 48′·4 S.	61° 11′·1 S.	Beattie	9
	10 0 a.m.	2	60 46 ·9			
	11 30 a.m.	1	60 48 ·4			
	11 30 a.m.	2	60 46 ·1			

Horizontal Intensity. H.

Date	G.M.T.	H (observed)	H	Observer	Instrument
1900 July 19	12 49 p.m. V.	H_{30} ·17994	·17745	Morrison	31
	1 32 p.m. D.	H_{40} ·17994			
	2 7 p.m. V.				

171. Knysna. Lat. 34° 1′·7 S.; Long. 23° 3′·0 E. On erven on hill north east of the Royal Hotel. Place close to the bush, and 150 yards from well towards Knysna Heads.

Declination. D.

Date	G.M.T.	D (observed)	D	Observer	Instrument
1903 Jan. 28	6 11 a.m.	28° 0′·5 W.			
1903 Jan. 29	5 45 a.m.	28 1 ·3	27° 52′·8 W.	Beattie	31
	8 18 a.m.	27 56 ·3			

Dip. θ.

Date	G.M.T.	Needle	θ (observed)	θ	Observer	Instrument
1903 Jan. 29	9 25 a.m.	3	60° 47′·9 S.	60° 50′·7 S.	Beattie	142
	9 24 a.m.	4	60 46 ·0			

Horizontal Intensity. H.

Date	G.M.T.	H (observed)	H	Observer	Instrument
1903 Jan. 28	8 24 a.m. V. 10 8 a.m. V.	·17885	·17840	Beattie	31

172. Kokstad. Lat. 30° 32′·8 S.; Long. 29° 28′·0 E. On right-hand side of prolongation of street in which the Royal Hotel is. 100 paces along road from spruit and away from village, then 85 paces from road.

Declination. D.

Date	G.M.T.	D (observed)	D	Observer	Instrument
1903 Nov. 19	7 21 a.m.	24° 37′·6 W.			
	7 33 a.m.	24 38 ·6	24° 40′·1 W.	Beattie	73
	7 44 a.m.	24 38 ·9			

Dip. θ.

Date	G.M.T.	Needle	θ (observed)	θ	Observer	Instrument
1903 Nov. 19	12 8 p.m.	1	61° 15′·1 S.	61° 11′·9 S.	Beattie	142
	12 8 p.m.	2	61 13 ·8			

Horizontal Intensity. H.

Date	G.M.T.	H (observed)	H	Observer	Instrument
1903 Nov. 19	9 59 a.m. V. 10 30 a.m. D. 10 55 a.m. V.	H_{30} ·17988 H_{40} ·17989	·18016	Beattie	73

173. Komati Poort. Lat. 25° 26′·0 S.; Long. 31° 54′·0 E. On right-hand side of railway going to Lorenço Marques. 336 paces from railway from a point 326 paces from Lorenço Marques end of platform along railway line towards Lorenço Marques.

Dip. θ.

Date	G.M.T.	Needle	θ (observed)	θ	Observer	Instrument
1903 Sept. 14	9 34 a.m. 9 35 a.m.	1 4_9	59° 46′·6 S. 59 47 ·0	59° 45′·3 S.	Beattie	142

Horizontal Intensity. H.

Date	G.M.T.	H (observed)	H	Observer	Instrument
1903 Sept. 14	10 25 a.m. V. 11 7 a.m. D. 11 41 a.m. D. 12 21 p.m. V.	H_{30} ·18860 H_{40} ·18854	·18869	Beattie	73

174. Komgha. Lat. 32° 35′·6 S.; Lat. 27° 54′·5 E.

Declination. D.

Date	G.M.T.	D (observed)	D	Observer	Instrument
1906 Jan. 13	3 28 p.m. 3 38 p.m. 3 46 p.m.	25° 46′·0 W. 25 45 ·6 25 45 ·8	26° 12′·0 W.	Brown Morrison	31

Dip. θ.

Date	G.M.T.	Needle	θ (observed)	θ	Observer	Instrument
1906 Jan. 13	10 32 a.m 10 31 a.m.	1 2	61° 54′·2 S. 61 56 ·3	61° 35′·0 S.	Morrison	9

Horizontal Intensity. H.

Date	G.M.T.	H (observed)	H	Observer	Instrument
1906 Jan. 13	7 53 a.m. V. 8 36 a.m. D. 9 12 a.m. V.	H_{30} ·17537 H_{40} ·17541	·17782	Morrison	31

175. Kraal. Lat. 26° 25′·1 S.; Long. 28° 26′·0 E.

Declination. D.

Date	G.M.T.	D (observed)	D	Observer	Instrument
1903 Sept. 26	4 54 a.m.	21° 58′·9 W.			
	5 12 a.m.	21 58·6			
	6 59 a.m.	22 4·6	22° 5′·7 W.	Beattie	73
	7 9 a.m.	22 4·8			
	8 49 a.m.	22 9·0			
	8 57 a.m.	22 8·8			

Dip. θ.

Date	G.M.T.	Needle	θ (observed)	θ	Observer	Instrument
1903 Sept. 26	10 1 a.m.	1	58° 40′·3 S.	58° 38′·6 S.	Beattie	142
	10 1 a.m.	4_9	58 40·2			

Horizontal Intensity. H.

Date	G.M.T.	H (observed)	H	Observer	Instrument
1903 Sept. 26	7 29 a.m. V.				
	8 10 a.m. D.	H_{30} ·19051 H_{40} ·19056	·19068	Beattie	73
	8 38 a.m. V.				

176. Krantz Kloof. Lat. 29° 48′·0 S.; Long. 30° 54′·0 E. Right-hand side of railway, Pietermaritzburg to Durban. 250 paces at right angles to railway, starting from a point 176 paces along the railway from Durban end of platform, and going towards Durban.

Declination. D.

Date	G.M.T.	D (observed)	D	Observer	Instrument
1903 Oct. 17	6 56 a.m.	23° 31′·0 W.			
	7 3 a.m.	23 32·1			
	7 18 a.m.	23 31·9	23° 31′·8 W.	Beattie	73
	11 36 a.m.	23 26·5			
	11 48 a.m.	23 25·9			

Dip. θ.

Date	G.M.T.	Needle	θ (observed)	θ	Observer	Instrument
1903 Oct. 17	9 49 a.m.	1	61° 24′·2 S.			
	9 48 a.m.	4_9	61 21·0	61° 20′·8 S.	Beattie	142
	9 48 a.m.	4	61 22·7			

Horizontal Intensity. H.

Date	G.M.T.	H (observed)	H	Observer	Instrument
1903 Oct. 17	7 42 a.m. V.				
	8 20 a.m. D.	H_{30} ·17978 H_{40} ·17983	·18001	Beattie	73
	8 52 a.m. V.				

177. Krantz Kop. Lat. 30° 48′·8 S.; Long. 20° 45′·4 E. On left-hand side of road, Rietpoort to Krantzkop, where road bifurcates.

Declination. D.

Date	G.M.T.	D (observed)	D	Observer	Instrument
1905 Jan. 23	5 50 a.m. 6 7 a.m.	26° 44′·6 W. 26 45·4	26° 54′·7 W.	Beattie Brown	73

Dip. θ.

Date	G.M.T.	Needle	θ (observed)	θ	Observer	Instrument
1905 Jan. 23	9 47 a.m. 9 47 a.m.	3_9 4_9	58° 39′·8 S. 58 38·9	58° 26′·9 S.	Beattie	142

Horizontal Intensity. H.

Date	G.M.T.	H (observed)	H	Observer	Instrument
1905 Jan. 23	8 18 a.m. V. 8 56 a.m. D.	H_{30} ·18552 H_{40} ·18552	·18677	Beattie	73

178. Kromm River. Lat. 27° 19′·0 S.; Long. 28° 18′·8 E. Left-hand side of road, Frankfort to Heilbron, just over Kromm River drift.

Declination. D.

Date	G.M.T.	D (observed)	D	Observer	Instrument
1904 Feb. 7	7 44 a.m. 7 51 a.m.	24° 1′·7 W. 24 1·8	24° 5′·8 W.	Beattie	73

Dip. θ.

Date	G.M.T.	Needle	θ (observed)	θ	Observer	Instrument
1904 Feb. 7	11 11 a.m. 11 11 a.m.	1 4_9	58° 59′·9 S. 59 0·3	58° 55′·9 S.	Beattie	142

Horizontal Intensity. H.

Date	G.M.T.	H (observed)	H	Observer	Instrument
1904 Feb. 7	8 43 a.m. V. 9 20 a.m. D. 9 57 a.m. V.	H_{30} ·18893 H_{40} ·18896	·18940	Beattie	73

179. KRUGERS. Lat. 29° 57′·1 S.; Long. 25° 50′·0 E. 220 paces from railway, on right-hand side going from Springfontein to Bloemfontein, starting from a point 147 paces from dead end and going towards Bloemfontein.

Declination. D.

Date	G.M.T.	D (observed)	D	Observer	Instrument
1903 May 30	7 28 a.m.	25° 2′·9 W.	25° 0′·2 W.	Beattie	73
	7 42 a.m.	25 3 ·1			
	2 27 p.m.	24 58 ·7			
	2 41 p.m.	24 58 ·9			

Dip. θ.

Date	G.M.T.	Needle	θ (observed)	θ	Observer	Instrument
1903 May 30	9 13 a.m.	3	59° 52′·9 S.	59° 52′·5 S.	Beattie	142
	9 14 a.m.	4	59 51 ·1			
	9 13 a.m.	1	59 51 ·6			

Horizontal Intensity. H.

Date	G.M.T.	H (observed)	H	Observer	Instrument
1903 May 30	2 59 p.m. V.	H_{30} ·18491	·18490	Beattie	73
	1 1 p.m. D.	H_{40} ·18489			
		H_{25} ·18487			
		H_{35} ·18494			

180. KRUISPAD. Lat. 32° 56′·8 S.; Long. 20° 33′·3 E.

Declination. D.

Date	G.M.T.	D (observed)	D	Observer	Instrument
1905 Feb. 2	7 50 a.m.	28° 2′·2 W.	28° 10′·4 W.	Beattie	73
	7 58 a.m.	28 2 ·4		Brown	

Dip. θ.

Date	G.M.T.	Needle	θ (observed)	θ	Observer	Instrument
1905 Feb. 2	5 38 a.m.	4_9	59° 57′·4 S.	59° 45′·7 S.	Beattie	142
	5 38 a.m.	3_9	59 57 ·0			
	5 38 a.m.	5	60 0 ·7			

Horizontal Intensity. H.

Date	G.M.T.	H (observed)	H	Observer	Instrument
1905 Feb. 2	8 18 a.m. V.	H_{30} ·18023	·18146	Beattie	73
	8 42 a.m. D.	H_{40} ·18015			

181. Kwambonambi. Lat. 28° 36′·2 S.; Long. 32° 5′·0 E. Right-hand side of railway, Durban to Hlabisa. 230 paces at right angles to railway starting from a point opposite middle of railway platform.

Declination. D.

Date	G.M.T.	D (observed)	D	Observer	Instrument
1903 Oct. 24	7 22 a.m.	22° 54′·8 W.			
	7 34 a.m.	22 55 ·0			
	1 36 p.m.	22 46 ·7			
	1 44 p.m.	22 46 ·5			
1903 Oct. 25	7 34 a.m.	23 1 ·6	22° 57′·7 W.	Beattie	73
	7 44 a.m.	23 0 ·3			
	8 34 a.m.	22 59 ·5			
	8 49 a.m.	23 0 ·0			
	10 59 a.m.	22 53 ·0			

Dip. θ.

Date	G.M.T.	Needle	θ (observed)	θ	Observer	Instrument
1903 Oct. 25	5 50 a.m.	1	60° 38′·7 S.			
	5 50 a.m.	4_9	60 38 ·7	60° 37′·5 S.	Beattie	142
	5 50 a.m.	4	60 41 ·0			

Horizontal Intensity. H.

Date	G.M.T.	H (observed)	H	Observer	Instrument
1903 Oct. 24	8 26 a.m. V.				
	9 13 a.m. D.	H_{30} ·18420			
	12 16 p.m. D.	H_{40} ·18417	·18454	Beattie	73
	12 45 p.m. V.				
1903 Oct. 25	8 4 a.m. V.	·18463			

182. Laat Rivier. Lat. 29° 38′·2 S.; Long. 21° 19′·3 E.

Declination. D.

Date	G.M.T.	D (observed)	D	Observer	Instrument
1904 Dec. 14	7 40 a.m.	26° 14′·7 W.	26° 25′·7 W.	Beattie Hough	73

Dip. θ.

Date	G.M.T.	Needle	θ (observed)	θ	Observer	Instrument
1904 Dec. 14	11 13 a.m.	3_9	58° 24′·9 S.	58° 13′·3 S.	Beattie	142
	11 13 a.m.	4_9	58 25 ·0			

Horizontal Intensity. H.

Date	G.M.T.	H (observed)	H	Observer	Instrument
1904 Dec. 14	8 30 a.m. V.	H_{30} ·18677	·18793	Beattie	73
	9 14 a.m. D.	H_{40} ·18678			

183. LADISMITH, C. C. Lat. 33° 29′·0 S.; Long. 21° 17′·0 E. Station at reservoir end of continuation of street containing the post office and the residency, and 60 yards across the sluit.

Declination. D.

Date	G.M.T.	D (observed)	D	Observer	Instrument
1903 Jan. 14	5 48 a.m.	28° 22′·6 W.	28° 14′·7 W.	Beattie Morrison	31
	6 9 a.m.	28 22 ·7			
	6 30 a.m.	28 22.0			
	2 46 p.m.	28 15 ·0			
	3 6 p.m.	28 15 ·5			

Dip. θ.

Date	G.M.T.	Needle	θ (observed)	θ	Observer	Instrument
1903 Jan. 14	9 41 a.m.	1	60° 4′·9 S.	60° 6′·0 S.	Beattie	142
	9 41 a.m.	3	60 1 ·5			
	9 37 a.m.	4	60 0 ·9			
	1 18 p.m.	1	60 4 ·1			
	1 20 p.m.	3	60 0 ·0			
	1 19 p.m.	4	60 2 ·2			

Horizontal Intensity. H.

Date	G.M.T.	H (observed)	H	Observer	Instrument
1903 Jan. 13	7 34 a.m. V.	H_{30} ·18017	·17974	Morrison	31
	8 12 a.m. D.	H_{40} ·18020			

184. L'AGULHAS. Lat. 34° 50′·0 S.; Long. 20° 0′·0 E. 200 yards in front of Lighthouse.

Declination. D.

Date	G.M.T.	D (observed)	D	Observer	Instrument
1901 Jan. 25	6 33 a.m.	28° 55′·7 W.	28° 41′·4 W.	Beattie Morrison	31
	6 46 a.m.	28 56 ·0			
	6 59 a.m.	28 55 ·9			
	7 8 a.m.	28 55 ·2			
	3 34 p.m.	28 52 ·3			
	3 47 p.m.	28 52 ·8			
	4 3 p.m.	28 53 ·0			
	4 19 p.m.	28 53 ·3			

Dip. θ.

Date	G.M.T.	Needle	θ (observed)	θ	Observer	Instrument
1901 Jan. 25	11 56 a.m.	1	59° 41′·5 S.	60° 4′·7 S.	Beattie	142
	12 5 p.m.	2	59 44 ·1			

Horizontal Intensity. H.

Date	G.M.T.	H (observed)	H	Observer	Instrument
1901 Jan. 25	1 51 p.m. V.	H_{30} ·18201	·17970	Morrison	31
	2 30 p.m. D.	H_{40} ·18199			
	3 2 p.m. V.				

185. Laingsburg. Lat. 33° 12′·0 S.; Long. 20° 52′·0 E.

Declination. D.

Date	G.M.T.	D (observed)	D	Observer	Instrument
1902 Feb. 4	6 51 a.m.	28° 29′·4 W.	28° 16′·9 W.	Beattie	31

Dip. θ.

Date	G.M.T.	Needle	θ (observed)	θ	Observer	Instrument
1902 Feb. 4	8 39 a.m.	2	59° 45′·4 S.	59° 55′·8 S.	Beattie	142
	8 42 a.m.	1	59 43 ·6			

Horizontal Intensity. H.

Date	G.M.T.	H (observed)	H	Observer	Instrument
1902 Feb. 3	1 49 p.m. V.	H_{30} ·18142	·18022	Beattie	31
	2 57 p.m. D.	H_{40} ·18140			
	3 28 p.m. V.				

186. Lake Banagher. Lat. 26° 22′·0 S.; Long. 30° 19′·0 E. On left-hand side of vlei going from farmhouse towards north.

Declination. D.

Date	G.M.T.	D (observed)	D	Observer	Instrument
1903 Aug. 26	1 45 p.m.	21° 36′·9 W.	21° 38′·8 W.	Beattie	73
	1 57 p.m.	21 37 ·5			

Dip. θ.

Date	G.M.T.	Needle	θ (observed)	θ	Observer	Instrument
1903 Aug. 26	11 7 a.m.	1	58° 57′·4 S.	58° 54′·6 S.	Beattie	142
	11 8 a.m.	4	58 53 ·5			
	11 8 a.m.	4_9	58 56 ·2			

Horizontal Intensity. H.

Date	G.M.T.	H (observed)	H	Observer	Instrument
1903 Aug. 26	12 12 p.m. V.	H_{30} ·19048	·19056	Beattie	73
	12 45 p.m. D.	H_{40} ·19051			

187. LANGLAAGTE. Lat. 26° 11′·8 S.; Long. 28° 1′·0 E. Left-hand side of railway, Randfontein to Johannesburg. 276 paces at right angles to railway, starting from Randfontein end of platform.

Declination. D.

Date	G.M.T.	D (observed)	D	Observer	Instrument
1903 June 23	6 17 a.m.	23° 4′·0 W.			
	6 35 a.m.	23 4 ·2	23° 4′·6 W.	Beattie	73
	6 50 a.m.	23 5 ·6			

Dip. θ.

Date	G.M.T.	Needle	θ (observed)	θ	Observer	Instrument
1903 June 23	8 20 a.m.	3	58° 57′·1 S.			
	8 20 a.m.	4	58 55 ·8	58° 56′·2 S.	Beattie	142
	8 21 a.m.	1	58 56 ·6			
	8 21 a.m.	4_9	58 54 ·9			

Horizontal Intensity. H.

Date	G.M.T.	H (observed)	H	Observer	Instrument
1903 June 23	9 20 a.m. V.	H_{30} ·18914	·18907	Beattie	73
	9 51 a.m. D.	H_{40} ·18900			

188. LETJESBOSCH. Lat. 32° 34′·0 S.; Long. 22° 18′·0 E.

Dip. θ.

Date	G.M.T.	Needle	θ (observed)	θ	Observer	Instrument
1902 Feb. 7	1 9 p.m.	2	59° 29′·9 S.	59° 39′·4 S.	Beattie	142
	1 8 p.m.	1	59 26 ·3			

Horizontal Intensity. H.

Date	G.M.T.	H (observed)	H	Observer	Instrument
1902 Feb. 7	7 30 a.m. V.	·18460	·18346	Beattie	31
	9 32 a.m. V.				

189. LIBODE. Lat. 31° 32′·1 S.; Long. 29° 1′·5 E. About 24 miles from Umtata on road to Port St John.

Declination. D.

Date	G.M.T.	D (observed)	D	Observer	Instrument
1906 Jan. 21	5 47 a.m.	24° 49′·1 W.	25° 13′·0 W.	Brown Morrison	31
	5 55 a.m.	24 48 ·6			
	6 4 a.m.	24 51 ·0			

Dip. θ.

Date	G.M.T.	Needle	θ (observed)	θ	Observer	Instrument
1906 Jan. 21	10 16 a.m.	1	61° 50′·4 S.	61° 31′·1 S.	Morrison	9
	10 36 a.m.	2	61 52 ·6			

Horizontal Intensity. H.

Date	G.M.T.	H (observed)	H	Observer	Instrument
1906 Jan. 21	7 45 a.m. V.	H_{30} ·17769	·18006	Morrison	31
	8 24 a.m. D.	H_{40} ·17753			
	8 54 a.m. V.				

190. LOBATSI. Lat. 25° 13′·8 S.; Long. 25° 40′·0 E. At a point 330 paces from railway line starting from the middle of the platform.

Declination. D.

Date	G.M.T.	D (observed)	D	Observer	Instrument
1903 March 21	2 55 p.m.	22° 26′·9 W.	22° 27′·8 W.	Beattie	73
	3 9 p.m.	22 27 ·6			

Dip. θ.

Date	G.M.T.	Needle	θ (observed)	θ	Observer	Instrument
1903 March 21	12 20 p.m.	3	56° 56′·9 S.	57° 0′·4 S.	Beattie	142
	12 20 p.m.	4	56 57 ·8			
	12 26 p.m.	1	56 57 ·5			
	12 26 p.m.	2	57 1 ·3			

Horizontal Intensity. H.

Date	G.M.T.	H (observed)	H	Observer	Instrument
1903 March 21	8 43 a.m. V.	H_{30} ·19564	·19542	Beattie	73
	9 24 a.m. D.	H_{40} ·19560			
	9 58 a.m. V.				

191. LOCHARD. Lat. 19° 55′·3 S.; Long. 29° 3′·0 E. Left-hand side of railway, Bulawayo to Gwelo. 247 paces at right angles to railway, starting from a point 130 paces from Bulawayo end of siding, and going towards Gwelo.

Declination. D.

Date	G.M.T.	D (observed)	D	Observer	Instrument
1903 April 6	8 16 a.m.	18° 20′·4 W.			
	8 40 a.m.	18 19 ·4	18° 17′·0 W.	Beattie	73
	2 13 p.m.	18 19 ·1			
	2 30 p.m.	18 18 ·1			

Dip. θ.

Date	G.M.T.	Needle	θ (observed)	θ	Observer	Instrument
1903 April 6	12 22 p.m.	4	53° 30′·9 S.			
	12 23 p.m.	3	53 33 ·1	53° 33′·6 S.	Beattie	142
	12 24 p.m.	1	53 31 ·7			

Horizontal Intensity. H.

Date	G.M.T.	H (observed)	H	Observer	Instrument
1903 April 6	9 9 a.m. V.	H_{30} ·21524			
	10 20 a.m. D.	H_{40} ·21523	21506	Beattie	73
	11 4 a.m. V.	H_{25} ·21519			

192. LYDENBURG. Lat. 25° 5′·8 S.; Long. 30° 26′·0 E. South of post office on low ground in prolongation of cross street.

Declination. D.

Date	G.M.T.	D (observed)	D	Observer	Instrument
1903 Sept. 1	5 34 a.m.	20° 51′·0 W.			
	5 44 a.m.	20 51 ·0	20° 53′·5 W.	Beattie	73
	6 44 a.m.	20 54 ·1			
	6 55 a.m.	20 54 ·9			

Dip. θ.

Date	G.M.T.	Needle	θ (observed)	θ	Observer	Instrument
1903 Sept. 1	10 6 a.m.	1	58° 6′·1 S.	58° 5′·3 S.	Beattie	142
	10 3 a.m.	4_9	58 6 ·9			

Horizontal Intensity. H.

Date	G.M.T.	H (observed)	H	Observer	Instrument
1903 Sept. 1	7 53 a.m. V.				
	8 29 a.m. D.	H_{30} ·19465	·19483	Beattie	73
	8 57 a.m. V.	H_{40} ·19476			

193. MACHADODORP. Lat. 25° 39′·9 S.; Long. 30° 15′·0 E. On level patch north of village in corner surrounded by spruit.

Declination. D.

Date	G.M.T.	D (observed)	D	Observer	Instrument
1903 Aug. 29	1 30 p.m. 1 37 p.m.	20° 47′·0 W. 20 48 ·1	20° 50′·1 W.	Beattie	73

Dip. θ.

Date	G.M.T.	Needle	θ (observed)	θ	Observer	Instrument
1903 Aug. 29	10 48 a.m. 10 50 a.m.	1 4_9	58° 44′·7 S. 58 45 ·4	58° 43′·9 S.	Beattie	142

Horizontal Intensity. H.

Date	G.M.T.	H (observed)	H	Observer	Instrument
1903 Aug. 29	11 39 a.m. V. 12 40 p.m. D. 1 10 p.m. V.	H_{30} ·19061 H_{40} ·19063	·19073	Beattie	73

194. MACHEKE. Lat. 18° 8′·3 S.; Long. 31° 51′·0 E. Left-hand side of railway, Salisbury to Umtali. 200 yards from railway, starting at milestone 308.

Declination. D.

Date	G.M.T.	D (observed)	D	Observer	Instrument
1903 April 26	8 20 a.m. 8 33 a.m.	16° 17′·4 W. 16 15 ·8	16° 14′·8 W.	Beattie	73

Dip. θ.

Date	G.M.T.	Needle	θ (observed)	θ	Observer	Instrument
1903 April 26	9 44 a.m. 9 45 a.m. 9 46 a.m.	3 4 1	52° 6′·6 S. 52 7 ·9 52 7 ·9	52° 8′·9 S.	Beattie	142

Horizontal Intensity. H.

Date	G.M.T.	H (observed)	H	Observer	Instrument
1903 April 26	5 48 a.m. V. 7 40 a.m. D.	H_{30} ·22437 H_{40} ·22430 H_{25} ·22433 H_{35} ·22437	·22424	Beattie	73

195. MAFEKING. Lat. 25° 52′·0 S.; Long. 25° 39′·0 E. On right-hand side of main street coming from railway station to commonage. 45 paces at right angles to the road, and 540 paces along road from school.

Declination. D.

Date	G.M.T.	D (observed)	D	Observer	Instrument
1903 March 19	2 4 p.m.	23° 4′·9 W.			
	2 16 p.m.	23 4 ·9	23° 5′·6 W.	Beattie	73
1903 March 20	5 4 a.m.	23 11 ·5			
	6 5 a.m.	23 12 ·0			
1906 Jan. 24	7 40 a.m.	22 44 ·0			
	7 56 a.m.	22 43 ·6			
	10 16 a.m.	22 36 ·1			
	10 27 a.m.	22 35 ·7	23 4 ·4	Beattie	73
	12 22 p.m.	22 33 ·8			
	12 33 p.m.	22 34 ·3			
	2 37 p.m.	22 36 ·0			
1906 Jan. 25	7 57 a.m.	22 44 ·3			
	8 13 a.m.	22 44 ·6			
	10 29 a.m.	22 40 ·1			
	10 36 a.m.	22 40 ·0	23 4 ·4	Beattie	73
	12 50 p.m.	22 35 ·8			
	12 58 p.m.	22 35 ·7			
			23 4 ·8 (mean adopted)		

Dip. θ.

Date	G.M.T.	Needle	θ (observed)	θ	Observer	Instrument
1898 Jan. 7	9 30 a.m.	1	56° 30′·6 S.		Beattie	
	10 55 a.m.	2	56 29 ·8	57° 9′·1 S.		9
1898 Jan. 8	9 57 a.m.	1	56 31 ·4		Morrison	
	11 0 a.m.	2	56 32 ·5			
1903 March 20	9 27 a.m.	3	57 11 ·8			
	9 28 a.m.	4	57 10 ·8	57 12 ·5	Beattie	142
	9 29 a.m.	1	57 9 ·6			
1906 Jan. 26	8 48 a.m.	3_9	57 33 ·8	57 14 ·1	Beattie	142
				57 11 ·9 (mean adopted)		

Horizontal Intensity. H.

Date	G.M.T.	H (observed)	H	Observer	Instrument
1898 Jan. 7	3 20 p.m. V.	H_{40} ·19891		Morrison	31
	5 0 p.m. D.	H_{30} ·19865	·19494		
1898 Jan. 8	2 53 p.m. V.	H_{40} ·19858		Beattie	31
	3 58 p.m. D.				
1903 March 19	11 59 a.m. V.				
	12 32 p.m. D.	H_{30} ·19502	·19480	Beattie	73
	1 5 p.m. D.	H_{40} ·19500			
	1 41 p.m. V.				
1906 Jan. 24	8 26 a.m. V.	H_{30} ·19325			
	9 20 a.m. D.	H_{40} ·19320			
	9 57 a.m. V.				
1906 Jan. 24	1 55 p.m. V.	H_{30} ·19290	·19487	Beattie	73
	1 37 p.m. D.	H_{40} ·19288			
	2 17 p.m. V.				
1906 Jan. 25	8 33 a.m. V.	·19301			
1906 Jan. 26	9 35 a.m. V.	·19322			
			·19487 (mean adopted)		

196. MAGALAPYE. Lat. 23° 6′·8 S.; Long. 26° 50′·0 E. On left-hand side of railway, Mafeking to Bulawayo. 325 yards at right angles to railway, starting from a point on railway 90 yards from refreshment room, and towards Bulawayo.

Declination. D.

Date	G.M.T.	D (observed)	D	Observer	Instrument
1903 March 27	5 24 a.m.	20° 26′·3 W.			
	5 51 a.m.	20 27 ·4			
	6 4 a.m.	20 28 ·2	20° 23′·5 W	Beattie	73
	2 44 p.m.	20 25 ·1			
	2 59 p.m.	20 24 ·9			

Dip. θ.

Date	G.M.T.	Needle	θ (observed)	θ	Observer	Instrument
1903 March 27	12 55 p.m.	3	55° 32′·4 S.			
	12 55 p.m.	4	55 32 ·6	55° 34′·0 S.	Beattie	142
	12 54 p.m.	1	55 31 ·8			

Horizontal Intensity. H.

Date	G.M.T.	H (observed)	H	Observer	Instrument
1903 March 27	7 54 a.m. V.	H_{30} ·20517			
	9 42 a.m. D.	H_{40} ·20527	·20496	Beattie	31
	10 58 a.m. V.	H_{25} ·20518			

197. MAGNET HEIGHTS (Store below). Lat. 24° 44′·8 S.; Long. 29° 58′·0 E. About five miles on the Pietersburg side of store on Magnet Heights.

Declination. D.

Date	G.M.T.	D (observed)	D	Observer	Instrument
1903 July 27	7 26 a.m.	19° 44′·4 W.	19° 44′·9 W.	Beattie	73
	7 40 a.m.	19 44 ·1		Löwinger	

Dip. θ.

Date	G.M.T.	Needle	θ (observed)	θ	Observer	Instrument
1903 July 27	9 20 a.m.	1	57° 59′·2 S.	57° 58′·6 S.	Beattie	142
	9 21 a.m.	4	57 59 ·1			

Horizontal Intensity. H.

Date	G.M.T.	H (observed)	H	Observer	Instrument
1903 July 27	7 57 a.m. V.	·19271	·19272	Beattie	73

198. MAKWIRO SIDING. Lat. 17° 57′·3 S.; Long. 30° 25′·0 E. Left-hand side of railway, Gwelo to Salisbury. 165 paces at right angles to railway, starting from a point 192 paces along railway from Gwelo end of siding.

Declination. D.

Date	G.M.T.	D (observed)	D	Observer	Instrument
1903 May 3	2 4 p.m.	16° 31′·5 W.			
	2 21 p.m.	16 30 ·7	16° 34′·4 W.	Beattie	73
1903 May 4	7 29 a.m.	16 41 ·2			
	7 45 a.m.	16 41 ·1			

Dip. θ.

Date	G.M.T.	Needle	θ (observed)	θ	Observer	Instrument
1903 May 4	8 53 a.m.	3	51° 34′·2 S.			
	8 54 a.m.	4	51 34 ·9	51° 36′·3 S.	Beattie	142
	8 54 a.m.	1	51 36 ·3			

Horizontal Intensity. H.

Date	G.M.T.	H (observed)	H	Observer	Instrument
1903 May 3	10 42 a.m. V.	H_{30} ·22457			
	11 33 a.m. D.	H_{40} ·22455	·22447	Beattie	73
	12 49 p.m. V.	H_{25} ·22461			
		H_{35} ·22453			

199. MALAGAS. Lat. 34° 18′·5 S.; Long. 20° 36′·0 E. In open space between the houses, about 70 yards east of hotel, and 30 yards from the river.

Declination. D.

Date	G.M.T.	D (observed)	D	Observer	Instrument
1901 Jan. 29	3 30 p.m.	28° 53′·6 W.			
	3 42 p.m.	28 54 ·4	28° 40′·9 W.	Beattie Morrison	31
	3 53 p.m.	28 53 ·4			
	4 26 p.m.	28 53 ·2			

Dip. θ.

Date	G.M.T.	Needle	θ (observed)	θ	Observer	Instrument
1901 Jan. 29	1 41 p.m.	1	59° 42′·4 S.	60° 5′·0 S.	Beattie	142
	1 41 p.m.	2	59 44 ·1			

Horizontal Intensity. H.

Date	G.M.T.	H (observed)	H	Observer	Instrument
1901 Jan. 29	10 34 a.m. V.				
	11 13 a.m. D.	H_{30} ·18275	·18066	Morrison	31
	11 47 a.m. V.				

200. Malenje Siding. Lat. 18° 55′·2 S.; Long. 32° 15′·0 E. Left-hand side of railway, Salisbury to Umtali. 292 paces at right angles to railway, starting from a point on railway at middle of siding.

Declination. D.

Date	G.M.T.	D (observed)	D	Observer	Instrument
1903 April 17	7 10 a.m.	16° 12′·0 W.			
	8 28 a.m.	16 12·4	16° 7′·0 W.	Beattie	73
	1 4 p.m.	16 5·1			
	1 18 p.m.	16 5·0			

Dip. θ.

Date	G.M.T.	Needle	θ (observed)	θ	Observer	Instrument
1903 April 17	9 53 a.m.	3	53° 1′·6 S.			
	9 58 a.m.	4	53 2·3	53° 3′·4 S.	Beattie	142
	9 57 a.m.	1	53 1·7			

Horizontal Intensity. H.

Date	G.M.T.	H (observed)	H	Observer	Instrument
1903 April 17	10 4 a.m. V.	H_{30} ·22049			
	11 57 a.m. D.	H_{40} ·22052	·22040	Beattie	73
	12 41 p.m. V.	H_{25} ·22059			
		H_{35} ·22052			

201. Malinde. Lat. 18° 45′·0 S.; Long. 27° 1′·3 E. Right-hand side of railway, Bulawayo to Victoria Falls. 140 paces at right angles to railway from a point 48 paces along railway starting from Falls end of sign-board and going towards Falls.

Declination. D.

Date	G.M.T.	D (observed)	D	Observer	Instrument
1904 July 5	6 47 a.m.	17° 52′·1 W.			
	6 55 a.m.	17 52·2			
	12 57 p.m.	17 51·0	18° 1′·6 W.	Beattie	73
	1 8 p.m.	17 51·4			
	2 10 p.m.	17 50·2			

Dip. θ.

Date	G.M.T.	Needle	θ (observed)	θ	Observer	Instrument
1904 July 5	10 37 a.m.	1	51° 58′·2 S.	51° 50′·3 S.	Beattie	142
	10 37 a.m.	4	51 56·4			

Horizontal Intensity. H.

Date	G.M.T.	H (observed)	H	Observer	Instrument
1904 July 5	8 31 a.m. V.	H_{30} ·21800			
	8 50 a.m. D.	H_{40} ·21798	·21859	Beattie	73
	2 29 p.m. V.				

202. MALMESBURY. Lat. 33° 28′·0 S.; Long. 18° 43′·0 E. In field to north of Mineral Baths Hotel.

Declination. D.

Date	G.M.T.	D (observed)	D	Observer	Instrument
1901 Aug. 17	3 25 p.m.	28° 36′·3 W.			
	3 37 p.m.	28 36 ·8			
	3 43 p.m.	28 36 ·4	28° 28′·9 W.	Beattie Morrison	31
1901 Aug. 18	7 45 a.m.	28 36 ·1			
	8 2 a.m.	28 35 ·6			
	8 12 a.m.	28 36 ·8			

Dip. θ.

Date	G.M.T.	Needle	θ (observed)	θ	Observer	Instrument
1898 March 19	11 0 a.m.	1	58° 16′·5 S.			
	9 35 a.m.	2	58 18 ·1	59° 5′·2 S.	Beattie Morrison	9
1898 March 20	6 30 a.m.	2	58 16 ·9			
1901 Aug. 18	10 28 a.m.	2	58 50 ·8	59 6 ·1	Beattie	142
	10 28 a.m.	1	58 47 ·5			
				59 5 ·7	(mean adopted)	

Horizontal Intensity. H.

Date	G.M.T.	H (observed)	H	Observer	Instrument
1898 March 20	9 47 a.m. V.	·18696	·18188	Beattie	31
1901 Aug. 17	1 48 p.m. V.	H_{30} ·18430			
	2 34 p.m. D.		·18243	Morrison	31
	3 46 p.m. V.	H_{40} ·18432			
1901 Aug. 18	1 44 p.m. V.	H_{30} ·18415			
	2 28 p.m. D.		·18229	Morrison	31
	2 58 p.m. V.	H_{40} ·18419			
			·18220	(mean adopted)	

203. MANDEGOS. Lat. 19° 7′·0 S.; Long. 33° 28′·0 E. Under tree at end of path on right-hand side of railway, Umtali to Beira. About 100 yards from railway.

Declination. D.

Date	G.M.T.	D (observed)	D	Observer	Instrument
1903 April 20	2 57 p.m.	16° 0′·2 W.	15° 58′·1 W.	Beattie	73
	3 10 p.m.	16 0 ·0			

Dip. θ.

Date	G.M.T.	Needle	θ (observed)	θ	Observer	Instrument
1903 April 20	10 32 a.m.	3	53° 23′·6 S.			
	10 33 a.m.	4	53 25 ·1	53° 25′·6 S.	Beattie	142
	10 32 a.m.	1	53 23 ·6			

Horizontal Intensity. H.

Date	G.M.T.	H (observed)	H	Observer	Instrument
1903 April 20	11 48 a.m. V.	H_{30} ·21941			
	12 36 p.m. D.	H_{40} ·21945	·21930	Beattie	73
	1 18 p.m. V.	H_{25} ·21941			
		H_{35} ·21941			

204. Mapani Loep. Lat. 22° 17′·5 S.; Long. 29° 3′·0 E. Right-hand side of road to Pont drift. 60 paces on Brak River side of stables.

Declination. D.

Date	G.M.T.	D (observed)	D	Observer	Instrument
1903 July 7	7 47 a.m. 7 56 a.m.	19° 4′·3 W. 19 3·9	19° 4′·2 W.	Beattie Löwinger	73

Dip. θ.

Date	G.M.T.	Needle	θ (observed)	θ	Observer	Instrument
1903 July 7	10 38 a.m. 10 37 a.m.	1 4	55° 47′·5 S. 55 48·0	55° 47′·8 S.	Beattie	142

Horizontal Intensity. H.

Date	G.M.T.	H (observed)	H	Observer	Instrument
1903 July 7	9 17 a.m. V. 9 46 a.m. D.	H_{30} ·20528 H_{40} ·20535	·20532	Beattie	73

205. Mara. Lat. 23° 8′·0 S.; Long. 29° 21′·0 E.

Dip. θ.

Date	G.M.T.	Needle	θ (observed)	θ	Observer	Instrument
1903 July 12	6 25 a.m. 6 25 a.m.	1 4	57° 16′·0 S. 57 16·5	57° 16′·1 S.	Beattie	142

Horizontal Intensity. H.

Date	G.M.T.	H (observed)	H	Observer	Instrument
1903 July 12	8 32 a.m. V. 9 16 a.m. D.	H_{30} ·20179 H_{40} ·20183	·20181	Beattie	73

206. MARANDELLAS. Lat. 18° 11′·3 S.; Long. 31° 32′·9 E. Right-hand side of railway, Salisbury to Umtali. 209 paces from railway, starting at Salisbury end of siding.

Declination. D.

Date	G.M.T.	D (observed)	D	Observer	Instrument
1903 April 14	3 10 p.m.	15° 54′·4 W.			
	3 29 p.m.	15 55 ·4	15° 55′·2 W.	Beattie	73
1903 April 15	8 21 a.m.	16 0 ·3			
	8 36 a.m.	15 59 ·2			

Dip. θ.

Date	G.M.T.	Needle	θ (observed)	θ	Observer	Instrument
1903 April 14	1 38 p.m.	3	52° 52′·0 S.			
	1 39 p.m.	4	52 53 ·9	52° 54′·6 S.	Beattie	142
	1 37 p.m.	1	52 53 ·4			

Horizontal Intensity. H.

Date	G.M.T.	H (observed)	H	Observer	Instrument
1903 April 15	5 11 a.m. V.	H_{30} ·22230			
	7 35 a.m. D.	H_{40} ·22219	·22211	Beattie	73
	8 53 a.m. V.	H_{25} ·22232			
		H_{35} ·22221			

207. MARIBOGO. Lat. 26° 25′·1 S.; Long. 25° 15′·0 E. On opposite side of railway to railway station, about 300 paces from it.

Declination. D.

Date	G.M.T.	D (observed)	D	Observer	Instrument
1906 Jan. 27	2 36 p.m.	23° 6′·1 W.			
1906 Jan. 28	7 47 a.m.	23 16 ·6	23° 35′·4 W.	Beattie	73
	7 57 a.m.	23 14 ·8			
	10 22 a.m.	23 7 ·8			

Dip. θ.

Date	G.M.T.	Needle	θ (observed)	θ	Observer	Instrument
1906 Jan. 28	9 20 a.m.	3_9	57° 38′·1 S.			
	9 21 a.m.	4_9	57 38 ·6	57° 17′·7 S.	Beattie	142
	9 21 a.m.	5	57 35 ·6			

Horizontal Intensity. H.

Date	G.M.T.	H (observed)	H	Observer	Instrument
1906 Jan. 27	9 9 a.m. V.				
	12 23 p.m. D.	H_{30} ·19189			
	12 57 p.m. D.	H_{40} ·19189	·19366	Beattie	73
	1 26 p.m. V.				
1906 Jan. 28	8 15 a.m. V.	·19189			
	10 6 a.m. V.				

208. Matetsi. Lat. 18° 12′·5 S.; Long. 26° 1′·5 E. Right-hand side of railway, Bulawayo to Victoria Falls. 102 paces at right angles to railway, starting from a point 91 paces from end of tank distant from Bulawayo, and going along line towards the falls.

Declination. D.

Date	G.M.T.	D (observed)	D	Observer	Instrument
1904 July 10	2 26 p.m.	17° 48′·8 W.			
	2 34 p.m.	17 48 ·7	17° 57′·9 W.	Beattie	73
1904 July 11	7 21 a.m.	17 46 ·7			

Dip. θ.

Date	G.M.T.	Needle	θ (observed)	θ	Observer	Instrument
1904 July 11	6 21 a.m.	1	51° 48′·1 S.	51° 41′·1 S.	Beattie	142
	6 21 a.m.	4	51 48 ·3			

Horizontal Intensity. H.

Date	G.M.T.	H (observed)	H	Observer	Instrument
1904 July 10	11 17 a.m. V.				
	12 38 p.m. D.	H_{30} ·21634 H_{40} ·21639	·21696	Beattie	73
	1 32 p.m. V.				

209. Matjesfontein. Lat. 33° 14′·2 S.; Long. 20°. 36′·0 E. Place of observation reached from village by going along the road towards Sutherland till spruit reached, then taking path to left passing large tree on right to top of rise on golf course. On bare patch on right-hand side of path.

Declination. D.

Date	G.M.T.	D (observed)	D	Observer	Instrument
1900 April 8	9 48 a.m.	28° 49′·1 W.			
	10 5 a.m.	28 48 ·3			
	10 18 a.m.	28 46 ·9			
	10 33 a.m.	28 46 ·7	28° 23′·4 W.	Beattie	31
	4 18 p.m.	28 45 ·2			
	4 28 p.m.	28 45 ·0			
1900 April 9	7 47 a.m.	28 49 ·1			
	8 42 a.m.	28 50 ·0			
	9 37 a.m.	28 50 ·4			
	10 8 a.m.	28 48 ·9			
	10 31 a.m.	28 47 ·0			
	11 0 a.m.	28 45 ·1			
	11 25 a.m.	28 44 ·4			
	12 2 p.m.	28 43 ·3	28 23 ·4	Beattie	31
	12 28 p.m.	28 42 ·4			
	1 12 p.m.	28 42 ·2			
	1 47 p.m.	28 42 ·0			
	2 1 p.m.	28 41 ·8			
	2 47 p.m.	28 41 ·4			
	3 20 p.m.	28 42 ·7			
	4 0 p.m.	28 43 ·6			

MATJESFONTEIN (*continued*).

Date	G.M.T.	D (observed)	D	Observer	Instrument
1900 April 10	7 47 a.m.	28° 47′·7 W.			
	8 30 a.m.	28 47 ·8			
	9 0 a.m.	28 47 ·6			
	9 30 a.m.	28 46 ·5			
	10 0 a.m.	28 45 ·6			
	10 30 a.m.	28 45 ·0	28° 21′·9 W.	Beattie	31
	11 0 a.m.	28 44 ·1			
	11 30 a.m.	28 44 ·3			
	12 20 p.m.	28 43 ·5			
	1 50 p.m.	28 44 ·0			
	4 0 p.m.	28 45 ·4			
1900 April 11	7 45 a.m.	28 47 ·3			
	8 50 a.m.	28 48 ·4			
	9 30 a.m.	28 47 ·9			
	10 0 a.m.	28 47 ·0			
	10 30 a.m.	28 46 ·7			
	11 0 a.m.	28 45 ·3			
	11 30 a.m.	28 44 ·9	28 22 ·5	Beattie	31
	12 0 noon	28 44 ·3			
	12 45 p.m.	28 43 ·6			
	1 30 p.m.	28 44 ·0			
	2 0 p.m.	28 44 ·0			
	3 30 p.m.	28 44 ·4			
	3 42 p.m.	28 44 ·4			
	4 0 p.m.	28 44 ·8			
1900 April 12	5 45 a.m.	28 47 ·3			
	7 45 a.m.	28 50 ·0			
	8 45 a.m.	28 48 ·9			
	9 4 a.m.	28 48 ·4			
	9 39 a.m.	28 47 ·8			
	9 50 a.m.	28 46 ·6			
	10 34 a.m.	28 45 ·2	28 22 ·7	Beattie	31
	11 3 a.m.	28 44 ·1			
	11 45 a.m.	28 43 ·7			
	12 37 p.m.	28 43 ·3			
	1 24 p.m.	28 43 ·3			
	2 6 p.m.	28 43 ·3			
	3 6 p.m.	28 43 ·1			
1900 April 13	6 40 a.m.	28 49 ·8			
	8 0 a.m.	28 53 ·6			
	8 30 a.m.	28 54 ·3			
	9 5 a.m.	28 53 ·9			
	9 30 a.m.	28 53 ·5			
	10 5 a.m.	28 51 ·7			
	10 33 a.m.	28 50 ·8			
	11 0 a.m.	28 49 ·8			
	11 30 a.m.	28 48 ·4	28 26 ·0	Beattie	31
	12 0 noon	28 47 ·4			
	12 43 p.m.	28 45 ·7			
	1 10 p.m.	28 45 ·3			
	1 30 p.m.	28 45 ·1			
	2 15 p.m.	28 45 ·3			
	2 45 p.m.	28 45 ·9			
	3 15 p.m.	28 47 ·0			
	4 0 p.m.	28 48 ·1			
1900 April 14	5 30 a.m.	28 48 ·8			
	6 10 a.m.	28 48 ·6	28 23 ·6	Beattie	31
	6 55 a.m.	28 49 ·5			
	7 50 a.m.	28 51 ·7			

MATJESFONTEIN (*continued*).

Date	G.M.T.	D (observed)	D	Observer	Instrument
1900 April 14	8 15 a.m.	28° 52′·3 W.			
	8 35 a.m.	28 53 ·0			
	8 57 a.m.	28 52 ·4			
	9 32 a.m.	28 50 ·1			
	10 9 a.m.	28 47 ·2			
	10 32 a.m.	28 45 ·7	28° 23′·6 W.	Beattie	31
	11 0 a.m.	28 44 ·3			
	11 18 a.m.	28 43 ·5			
	11 50 a.m.	28 42 ·3			
	12 47 p.m.	28 41 ·6			
	1 50 p.m.	28 42 ·5			
	2 8 p.m.	28 43 ·0			
1900 May 13	9 6 a.m.	28 53 ·1			
	9 18 a.m.	28 53 ·1	28 29 ·7	Beattie Morrison	31
	9 25 a.m.	28 53 ·6			
1900 July 28	8 37 a.m.	28 47 ·7			
	8 47 a.m.	28 47 ·5	28 26 ·0	Beattie Morrison	31
	8 58 a.m.	28 47 ·7			
1902 Dec. 26	3 42 p.m.	28 20 ·6	28 17 ·9	Beattie	73
1902 Dec. 27	6 36 a.m.	28 31 ·2			
	6 45 a.m.	28 29 ·7			
	8 13 a.m.	28 24 ·3			
	8 30 a.m.	28 25 ·9			
	8 41 a.m.	28 26 ·8	28 23 ·5	Beattie	73
	2 2 p.m.	28 25 ·3			
	2 11 p.m.	28 26 ·5			
	2 19 p.m.	28 25 ·3			
	10 14 a.m.	28 27 ·4			
	10 27 a.m.	28 27 ·2			
	1 17 p.m.	28 24 ·5	28 22 ·6	Beattie	31
	1 28 p.m.	28 25 ·1			
	2 46 p.m.	28 23 ·5			
	2 58 p.m.	28 24 ·5			
1902 Dec. 29	4 42 a.m.	28 28 ·4			
	4 53 a.m.	28 29 ·4			
	6 33 a.m.	28 29 ·0			
	6 43 a.m.	28 28 ·3			
	6 58 a.m.	28 27 ·9	28 24 ·3	Beattie	31
	12 57 p.m.	28 25 ·4			
	1 9 p.m.	28 25 ·2			
	2 52 p.m.	28 26 ·9			
	3 3 p.m.	28 27 ·7			
	2 12 p.m.	28 27 ·3			
	2 20 p.m.	28 27 ·3	28 25 ·4	Beattie	73
	2 28 p.m.	28 27 ·4			
1903 Dec. 25	5 12 a.m.	28 25 ·8			
	5 23 a.m.	28 26 ·1			
	6 19 a.m.	28 18 ·4			
	6 29 a.m.	28 19 ·9	28 24 ·0	Beattie	73
	7 21 a.m.	28 19 ·4			
1903 Dec. 26	8 2 a.m.	28 24 ·1			
	8 15 a.m.	28 24 ·0			
	12 43 p.m.	28 20 ·5			
1903 Dec. 30	7 32 a.m.	28 26 ·6			
	8 2 a.m.	28 25 ·2			
	9 47 a.m.	28 24 ·2			
	2 18 p.m.	28 22 ·2	28 26 ·4	Beattie	73
	3 4 p.m.	28 20 ·8			
	3 13 p.m.	28 20 ·6			

MATJESFONTEIN (*continued*).

Date	G.M.T.	D (observed)	D	Observer	Instrument
1904 Jan. 1	7 51 a.m.	28° 26′·2 W.	28° 26′·4 W.	Beattie	73
	8 0 a.m.	28 27 ·4			
	2 35 p.m.	28 20 ·0			
	3 15 p.m.	28 21 ·3			
1904 Sept. 18	7 41 a.m.	28 15 ·5	28 20 ·8	Beattie	73
	8 20 a.m.	28 15 ·0			
1904 Sept. 20	6 24 a.m.	28 12 ·4			
	6 37 a.m.	28 10 ·8			
1904 Sept. 19	7 47 a.m.	28 16 ·9	28 22 ·7	Beattie	73
	7 56 a.m.	28 15 ·9			
	10 19 a.m.	28 16 ·9			
	10 30 a.m.	28 18 ·2			
	11 41 a.m.	28 12 ·5			
	11 56 a.m.	28 13 ·5			
	2 16 p.m.	28 11 ·4			
	2 25 p.m.	28 11 ·0			
1904 Oct. 29	6 9 a.m.	28 12 ·0	28 20 ·6	Beattie	73
	5 58 a.m.	28 13 ·5	28 21 ·6	Morrison	31
	6 20 a.m.	28 12 ·5			
1904 Oct. 30	6 36 a.m.	28 12 ·2	28 20 ·9	Beattie	73
	6 45 a.m.	28 11 ·8			
	7 22 a.m.	28 11 ·2			
	7 30 a.m.	28 11 ·8			
	8 27 a.m.	28 14 ·6			
	9 12 a.m.	28 14 ·5			
	9 24 a.m.	28 13 ·4			
	9 37 a.m.	28 12 ·8			
1904 Oct. 30	6 58 a.m.	28 8 ·2	28 19 ·3	Morrison	31
	7 7 a.m.	28 9 ·2			
	8 12 a.m.	28 11 ·8			
	8 38 a.m.	28 13 ·3			
	9 11 a.m.	28 12 ·6			
	9 21 a.m.	28 12 ·6			
	9 36 a.m.	28 11 ·0			
1905 Feb. 3	6 19 a.m.	28 12 ·3	28 18 ·7	Beattie Brown	73
	8 14 a.m.	28 9 ·0			
			28 22 ·9 (mean adopted)		

Dip. θ.

Date	G.M.T.	Needle	θ (observed)	θ	Observer	Instrument
1899 July 18	10 0 a.m.	1	59° 18′·4 S.	59° 50′·1 S.	Beattie	9
	10 1 a.m.	2	59 15 ·8			
	1 22 p.m.	1	59 20 ·5			
	1 22 p.m.	2	59 17 ·5			
1900 April 7	3 22 p.m.	1	59 23 ·2	59 49 ·7	Beattie	9
	3 22 p.m.	2	59 24 ·2			
1900 May 12	9 55 a.m.	1	59 25 ·6	59 50 ·0	Beattie	9
	9 55 a.m.	2	59 24 ·3			
1900 July 28	10 55 a.m.	1	59 26 ·0	59 49 ·0	Beattie	9
	10 49 a.m.	2	59 25 ·1			

Matjesfontein (*continued*).

Date		G.M.T.	Needle	θ (observed)	θ	Observer	Instrument
1902	Dec. 25	9 52 a.m.	1	59° 45′·4 S.			
		9 52 a.m.	2	59 46 ·2			
		9 52 a.m.	3	59 47 ·9			
		9 51 a.m.	4	59 48 ·8		Beattie	9
1902	Dec. 26	5 39 a.m.	1	59 47 ·2			
		5 41 a.m.	2	59 44 ·6			
		5 37 a.m.	3	59 45 ·5			
		5 38 a.m.	4	59 47 ·7	59° 51′·3 S.		
1902	Dec. 25	6 7 a.m.	1	59 47 ·3			
		6 9 a.m.	2	59 51 ·2			
		6 13 a.m.	3	59 46 ·1			
		6 11 a.m.	4	59 47 ·2		Beattie	142
1902	Dec. 26	8 58 a.m.	1	59 49 ·9			
		8 58 a.m.	2	59 49 ·6			
		8 58 a.m.	3	59 45 ·9			
		9 0 a.m.	4	59 45 ·6			
1903	Dec. 30	11 18 a.m.	1	59 52 ·1			
		11 18 a.m.	4_9	59 54 ·2			
1904	Jan. 1	11 19 a.m.	1	59 55 ·7	59 52 ·1	Beattie	142
		11 19 a.m.	4_9	59 58 ·3			
		11 19 a.m.	6	59 56 ·9			
		11 19 a.m.	5	59 59 ·2			
1904	Sept. 21	7 43 a.m.	1_{142}	59 58 ·4			
		7 43 a.m.	4	59 59 ·0			
		9 56 a.m.	1_{142}	59 56 ·8			
		9 56 a.m.	4	59 58 ·8			
1904	Sept. 22	6 37 a.m.	1_{142}	59 58 ·1			
		6 36 a.m.	4	59 59 ·8			
		6 36 a.m.	3	60 0 ·9		Beattie	9
		10 16 a.m.	1_{142}	59 57 ·7			
		10 16 a.m.	4	60 0 ·7			
		10 16 a.m.	3	59 59 ·7			
1904	Sept. 23	8 24 a.m.	4	60 2 ·0			
		8 25 a.m.	3	60 0 ·7			
		9 0 a.m.	4	60 0 ·0			
		9 0 a.m.	3	59 59 ·6	59 49 ·5		
1904	Sept. 21	6 33 a.m.	1	59 59 ·0			
		6 33 a.m.	4_9	59 58 ·5			
		8 58 a.m.	1	60 0 ·0			
		8 57 a.m.	4_9	59 58 ·9			
1904	Sept. 22	8 3 a.m.	1	59 59 ·3			
		8 3 a.m.	4_9	59 59 ·6			
		8 4 a.m.	3_9	60 0 ·6		Beattie	142
		9 4 a.m.	1	59 58 ·8			
		9 4 a.m.	4_9	59 59 ·6			
		9 3 a.m.	3_9	59 58 ·2			
1904	Sept. 23	7 28 a.m.	4_9	60 1 ·6			
		7 30 a.m.	3_9	59 59 ·5			
		10 6 a.m.	4_9	59 58 ·3			
		10 6 a.m.	3_9	60 1 ·0			
1902	Feb. 3	5 22 a.m.	4_9	60 4 ·7			
		5 23 a.m.	3_9	60 7 ·7	59 53 ·5	Beattie	142
		5 24 a.m.	5	60 6 ·3			
1906	June 28	9 22 a.m.	4	60 14 ·5			
		9 23 a.m.	3	60 13 ·0			
		10 24 a.m.	4	60 15 ·4		Beattie	142
		10 25 a.m.	3	60 11 ·7			
1906	June 30	1 13 p.m.	4	60 12 ·8			
		1 13 p.m.	3	60 10 ·6	59 50 ·1		
1906	June 28	9 26 a.m.	2	60 10 ·7			
		9 27 a.m.	1	60 12 ·1			
		10 30 a.m.	2	60 11 ·1		Morrison	9
		10 30 a.m.	1	60 12 ·2			
1906	June 30	1 8 p.m.	2	60 10.1			
		1 8 p.m.	1	60 11 ·3			
					59 50 ·5 (mean adopted)		

MATJESFONTEIN (*continued*).

Horizontal Intensity. II.

Date	G.M.T.	H (observed)	H	Observer	Instrument
1899 July 17	8 32 a.m. V.				
	11 0 a.m. D.	H_{30} ·18358	·18044	Morrison	31
	12 50 p.m. D.	H_{40} ·18362			
	12 39 p.m. V.				
1900 May 12	1 11 p.m. V.	H_{30} ·18293			
	2 50 p.m. D.	H_{40} ·18290	·18027	Morrison	31
	3 24 p.m. V.				
1900 July 28	1 26 p.m. V.	H_{30} ·18270			
	2 30 p.m. D.	H_{40} ·18274	·18039	Morrison	31
	3 7 p.m. V.				
1903 Dec. 29	11 8 a.m. V.	H_{30} ·17993			
	12 7 p.m. D.	H_{40} ·17995	·18028	Beattie	73
	5 47 p.m. V.	H_{30} ·17975			
	4 53 p.m. D.	H_{40} ·17975			
1903 Dec. 30	10 40 a.m. V.	·18010	·18021		
	5 34 p.m. V.	·17947			
1904 Jan. 1	10 21 a.m. V.	·17975		Beattie	73
	5 36 p.m. V.	·17934	·17998		
1904 Sept. 22	12 20 p.m. V.	H_{30} ·17906			
	12 46 p.m. D.	H_{40} ·17910		Beattie	73
	1 12 p.m. V.		·18007		
	1 40 p.m. V.	H_{30} ·17905			
	2 0 p.m. D.	H_{40} ·17902		Beattie	31
	2 43 p.m. V.				
1904 Sept. 19	12 14 p.m. V.	H_{30} ·17912			
	12 50 p.m. D.	H_{40} ·17918			
	1 26 p.m. V.				
1904 Sept. 20	8 15 a.m. V.		·18024	Beattie	73
	9 37 a.m. D.	H_{30} ·17930			
	10 10 a.m. V.	H_{40} ·17935			
	7 56 a.m. V.	H_{30} ·17965		Beattie	73
	8 31 a.m. D.	H_{40} ·17955			
	9 22 a.m. V.	H_{30} ·17965			
	10 29 a.m. D.	H_{40} ·17965	·18029	Beattie	31
	1 10 p.m. V.	H_{30} ·17898			
	12 31 p.m. D.	H_{40} ·17899			
	2 17 p.m. V.	H_{30} ·17890		Beattie	73
	1 46 p.m. D.	H_{40} ·17894			
1904 Sept. 21	12 40 p.m. V.	H_{30} ·17910			
	1 19 p.m. D.	H_{40} ·17914		Beattie	31
	1 50 p.m. V.		·18005		
	2 19 p.m. V.	H_{30} ·17896			
	2 37 p.m. D.	H_{40} ·17895		Beattie	73
	3 16 p.m. V.				
1904 Sept. 23	12 41 p.m. V.	H_{30} ·17906			
	1 13 p.m. D.	H_{40} ·17910		Beattie	31
	1 44 p.m. V.		·18008		
	2 14 p.m. V.	H_{30} ·17906			
	2 43 p.m. D.	H_{40} ·17908		Beattie	73
	3 10 p.m. V.				
1904 Sept 24	8 4 a.m. V.	H_{30} ·17951			
	8 35 a.m. D.	H_{40} ·17950		Beattie	73
	9 27 a.m. V.		·18040		
	10 0 a.m. V.	H_{30} ·17929			
	10 35 a.m. D.	H_{40} ·17927		Beattie	31
	11 7 a.m. V.				

MATJESFONTEIN (*continued*).

Date	G.M.T.	H (observed)	H	Observer	Instrument
1904 Oct. 29	9 35 a.m. V. 10 20 a.m. D. 10 54 a.m. V.	H_{30} ·17904 H_{40} ·17902		Morrison	31
	9 24 a.m. V. 10 7 a.m. D. 10 50 a.m. V.	H_{30} ·17908 H_{40} ·17902	·18011	Beattie	73
	2 2 p.m. V. 2 35 p.m. D. 3 26 p.m. V.	H_{30} ·17895 H_{40} ·17895		Morrison	31
	1 33 p.m. V. 2 47 p.m. D. 3 11 p.m. V.	H_{30} ·17894 H_{40} ·17896			
1904 Oct. 30	10 6 a.m. V. 10 36 a.m. D. 11 7 a.m. V.	H_{30} ·17892 H_{40} ·17889	·18009	Morrison	31
	9 57 a.m. V. 10 29 a.m. D. 11 2 a.m. V.	H_{30} ·17903 H_{40} ·17903		Beattie	73
1905 Feb. 3	9 22 a.m. V.	H ·17928	·18081	Beattie	73
1906 June 29	8 12 a.m. V. 8 54 a.m. D. 9 24 a.m. V.	H_{30} ·17819 H_{40} ·17817	·18059	Morrison	31
1906 June 30	8 40 a.m. V. 9 36 a.m. D. 10 8 a.m. V.	H_{30} ·17797 H_{40} ·17797		Beattie	73
1906 July 2	8 28 a.m. V. 9 6 a.m. D. 10 3 a.m. V.	H_{30} ·17805 H_{40} ·17805		Morrison	31
	8 17 a.m. V. 9 8 a.m. D. 9 58 a.m. V.	H_{30} ·17793 H_{40} ·17789		Beattie	73
	10 3 a.m. V. 10 34 a.m. D. 11 2 a.m. V.	H_{30} ·17796 H_{40} ·17801		Morrison	31
	9 58 a.m. V. 10 30 a.m. D. 11 0 a.m. V.	H_{30} ·17793 H_{40} ·17790	·18040	Beattie	73
	12 48 p.m. V. 1 22 p.m. D. 1 52 p.m. V.	H_{30} ·17788 H_{40} ·17788		Morrison	31
	12 45 p.m. V. 1 20 p.m. D. 1 52 p.m. V.	H_{30} ·17767 H_{40} ·17767		Beattie	73
	1 52 p.m. V. 2 21 p.m. D. 3 5 p.m. V.	H_{30} ·17794 H_{40} ·17794		Morrison	31
	1 52 p.m. V. 2 32 p.m. D. 3 2 p.m. V.	H_{30} ·17772 H_{40} ·17775		Beattie	73
			·18028 (mean adopted)		

210. Meyerton. Lat. 26° 33′·2 S.; Long. 28° 1′·0 E. Right-hand side of railway, Kroonstad to Pretoria. 234 paces at right angles to railway, starting from a point 45 paces from Pretoria end of platform and going towards Pretoria.

Declination. D.

Date	G.M.T.	D (observed)	D	Observer	Instrument
1903 June 14	6 18 a.m.	22° 42′·1 W.	22° 42′·4 W.	Beattie Löwinger	73
	6 31 a.m.	22 42·8			
	6 40 a.m.	22 43·0			

Dip. θ.

Date	G.M.T.	Needle	θ (observed)	θ	Observer	Instrument
1903 June 14	8 28 a.m.	3	58° 31′·8 S.	58° 30′·3 S.	Beattie	142
	8 29 a.m.	4	58 28·7			
	8 31 a.m.	1	58 29·4			

Horizontal Intensity. H.

Date	G.M.T.	H (observed)	H	Observer	Instrument
1903 June 14	9 10 a.m. V.		·19148	Beattie	73
	9 45 a.m. D.	H_{30} ·19145			
	10 9 a.m. D.	H_{40} ·19150			
	10 36 a.m. V.				

211. Middelberg (Tzitzikama). Lat. 34° 0′·0 S.; Long. 24° 9′·0 E. Near Shepherd's store.

Declination. D.

Date	G.M.T.	D (observed)	D	Observer	Instrument
1903 Feb. 4	5 16 a.m.	27° 47′·7 W.	27° 43′·8 W.	Beattie	31
	7 50 a.m.	27 48·1			

Dip. θ.

Date	G.M.T.	Needle	θ (observed)	θ	Observer	Instrument
1903 Feb. 3	3 9 p.m.	3	60° 57′·0 S.	61° 2′·3 S.	Beattie	142
	3 8 p.m.	4	61 0·9			

Horizontal Intensity. H.

Date	G.M.T.	H (observed)	H	Observer	Instrument
1903 Feb. 4	8 18 a.m. V.	H_{30} ·17815	·17779	Beattie	31
	7 8 a.m. D.	H_{40} ·17818			
	10 12 a.m. V.				

212. MIDDLEPOST. Lat. 31° 54′·2 S.; Long. 20° 14′·0 E. Left-hand side of road coming from De Drift. About 800 paces from road, opposite house on hill, on De Drift side of store.

Declination. D.

Date	G.M.T.	D (observed)	D	Observer	Instrument
1905 Jan. 28	6 35 a.m. 6 47 a.m.	27° 20′·6 W. 27 20 ·6	27° 29′·0 W.	Beattie Brown	73

Dip. θ.

Date	G.M.T.	Needle	θ (observed)	θ	Observer	Instrument
1905 Jan. 28	9 49 a.m. 12 0 noon	4_9 3_9	59° 1′·4 S. 59 3 ·1	58° 49′·5 S.	Beattie	142

Horizontal Intensity. H.

Date	G.M.T.	H (observed)	H	Observer	Instrument
1905 Jan. 28	7 58 a.m. V. 8 46 a.m. D.	H_{30} ·18324 H_{40} ·18332	·18455	Beattie	73

213. MIDDLETON. Lat. 32° 57′·8 S.; Long. 25° 51′·0 E. On flat piece of ground on north west of the railway. On left of railway coming from Port Elizabeth, and on side of hotel away from Port Elizabeth, about 80 yards from railway line.

Declination. D.

Date	G.M.T.	D (observed)	D	Observer	Instrument
1902 July 12	1 52 p.m. 2 0 p.m. 6 30 a.m. 6 42 a.m. 6 50 a.m.	27° 4′·9 W. 27 5 ·1 27 4 ·9 27 5 ·9 27 5 ·9	26° 56′·1 W.	Beattie Morrison	31

Dip. θ.

Date	G.M.T.	Needle	θ (observed)	θ	Observer	Instrument
1902 July 13	8 54 a.m. 8 53 a.m. 8 54 a.m. 8 53 a.m.	1 2 3 4	61° 14′·9 S. 61 16 ·6 61 12 ·3 61 10 ·9	61° 22′·0 S.	Beattie	142

Horizontal Intensity. H.

Date	G.M.T.	H (observed)	H	Observer	Instrument
1902 July 12	1 28 p.m. V. 2 37 p.m. D. 3 8 p.m. V.	H_{30} ·17827 H_{40} ·17836	·17742	Morrison	31

214. Mill River. Lat. 33° 36′·0 S.; Long. 22° 55′·0 E. In field next store, in corner of field away from the store.

Dip. θ.

Date	G.M.T.	Needle	θ (observed)	θ	Observer	Instrument
1903 Feb. 15	7 29 a.m.	3	60° 36′·6 S.	60° 38′·4 S.	Beattie	142
	7 29 a.m.	4	60 34 ·2			

Horizontal Intensity. H.

Date	G.M.T.	H (observed)	H	Observer	Instrument
1903 Feb. 15	9 29 a.m. V.		·17839	Beattie	31
	10 14 a.m. D.	H_{30} ·17876			
	10 53 a.m. D.	H_{40} ·17878			
	11 58 a.m. V.				

215. Miller Siding. Lat. 33° 5′·4 S.; Long. 24° 8′·0 E. In field in front of Miller's store, about 200 yards from store.

Declination. D.

Date	G.M.T.	D (observed)	D	Observer	Instrument
1900 July 17	8 48 a.m.	28° 4′·9 W.	27° 33′·5 W.	Beattie Morrison	31
	8 59 a.m.	28 4 ·5			
	9 12 a.m.	28 4 ·3			
1904 July 4	1 8 p.m.	27 23 ·5	27 29 ·1	Morrison	31
	1 16 p.m.	27 23 ·5			
	1 24 p.m.	27 23 ·9			
1904 July 5	5 25 a.m.	27 13 ·8			
	5 35 a.m.	27 13 ·9			
	5 44 a.m.	27 12 ·2			
			27 31 ·3 (mean adopted)		

Dip. θ.

Date	G.M.T.	Needle	θ (observed)	θ	Observer	Instrument
1900 July 17	1 45 p.m.	1	60° 37′·4 S.	61° 0′·4 S.	Beattie	9
	1 45 p.m.	2	60 36 ·1			
1904 July 4	8 12 a.m.	1	61 6 ·6	61 0 ·0	Morrison	9
	8 9 a.m.	2	61 7 ·2			
1904 July 5	8 49 a.m.	1	61 9 ·2			
	8 46 a.m.	2	61 8 ·1			
				61 0 ·2 (mean adopted)		

Horizontal Intensity. H.

Date	G.M.T.	H (observed)	H	Observer	Instrument
1900 July 17	9 45 a.m. V.	H_{30} ·18058	·17792	Morrison	31
	10 41 a.m. D.	H_{40} ·18059			
	11 34 a.m. V.				
1904 July 5	12 38 p.m. V.	H_{30} ·17706	·17796	Morrison	31
	1 21 p.m. D.	H_{40} ·17705			
	1 53 p.m. V.				
			·17794 (mean adopted)		

216. MILLER'S POINT. Lat. 34° 14′·2 S.; Long. 18° 26′·0 E. On grass plot beyond Molteno's house, round tower at Simonstown due south, and about two miles away.

Declination. D.

Date	G.M.T.	D (observed)	D	Observer	Instrument
1901 Jan. 8	4 0 p.m. 4 15 p.m.	28° 49′·9 W. 28 50 ·4	28° 41′·1 W.	Beattie Morrison	31

Dip. θ.

Date	G.M.T.	Needle	θ (observed)	θ	Observer	Instrument
1901 Jan. 9	2 12 p.m. 2 6 p.m.	1 2	58° 53′·2 S. 58 56 ·2	59° 17′·2 S.	Beattie	9

Horizontal Intensity. H.

Date	G.M.T.	H (observed)	H	Observer	Instrument
1901 Jan. 9	8 20 a.m. V. 9 58 a.m. D. 10 29 a.m. V.	H_{30} ·18499 H_{40} ·18500	·18259	Morrison	31

217. MISGUND. Lat. 33° 45′·5 S.; Long. 23° 32′·0 E. Behind the hotel, and about 100 yards from the stable.

Declination. D.

Date	G.M.T.	D (observed)	D	Observer	Instrument
1903 Feb. 12	1 57 p.m. 3 6 p.m.	27° 49′·7 W. 27 49 ·1	27° 53′·4 W.	Beattie	31
1903 Feb. 13	5 0 a.m.	28 4 ·9			

Dip. θ.

Date	G.M.T.	Needle	θ (observed)	θ	Observer	Instrument
1903 Feb. 12	1 4 p.m. 1 5 p.m.	3 4	60° 45′·2 S. 60 46 ·0	60° 48′·6 S.	Beattie	142

Horizontal Intensity. H.

Date	G.M.T.	H (observed)	H	Observer	Instrument
1903 Feb. 12	3 23 p.m. V. 4 22 p.m. D. 4 41 p.m. V.	H_{30} ·17862 H_{40} ·17864	·17825	Beattie	31

218. Mission Station. Lat. 23° 12′·7 S.; Long. 30° 27′·0 E. On same side of road as large fig tree. On opposite side of path—from house to water—to the house, a little nearer the water than the fig tree.

Declination. D.

Date	G.M.T.	D (observed)	D	Observer	Instrument
1903 July 16	2 2 p.m. 2 13 p.m.	19° 31′·3 W. 19 31 ·5	19° 31′·8 W.	Beattie Löwinger	73

Dip. θ.

Date	G.M.T.	Needle	θ (observed)	θ	Observer	Instrument
1903 July 16	3 5 a.m. 3 6 a.m.	1 4	57° 6′·2 S. 57 6 ·2	57° 5′·9 S.	Beattie	142

Horizontal Intensity. H.

Date	G.M.T.	H (observed)	H	Observer	Instrument
1903 July 16	1 39 p.m. V.	·19910	·19910	Beattie	73

219. Modder Spruit. Lat. 28° 28′·9 S.; Long. 29° 53′·0 E. Left-hand side of railway, Newcastle to Ladysmith. 247 paces along railway from old station towards Newcastle, and then 126 paces at right angles to railway.

Declination. D.

Date	G.M.T.	D (observed)	D	Observer	Instrument
1903 Oct. 6	4 44 a.m. 4 53 a.m. 8 10 a.m. 8 18 a.m.	23° 5′·4 W. 23 5 ·4 23 10 ·3 23 10 ·0	23° 9′·8 W.	Beattie	73

Dip. θ.

Date	G.M.T.	Needle	θ (observed)	θ	Observer	Instrument
1903 Oct. 6	9 17 a.m.	1	60° 2′·2 S.	60° 0′·3 S.	Beattie	142

Horizontal Intensity. H.

Date	G.M.T.	H (observed)	H	Observer	Instrument
1903 Oct. 6	6 47 a.m. V. 7 47 a.m. D. 8 36 a.m. V.	H_{30} ·18542 H_{40} ·18542	·18560	Beattie	73

220. MOLTENO. Lat. 31° 24′·0 S.; Long. 26° 21′·0 E.

Declination. D.

Date	G.M.T.	D (observed)	D	Observer	Instrument
1901 Dec. 28	6 26 a.m.	26° 16′·2 W.	25° 58′·2 W.	Beattie Morrison	31
	6 36 a.m.	26 14 ·8			
	6 46 a.m.	26 15 ·3			

Dip. θ.

Date	G.M.T.	Needle	θ (observed)	θ	Observer	Instrument
1901 Dec. 28	9 5 a.m.	2	60° 24′·3 S.	60° 35′·9 S.	Beattie	142
	9 5 a.m.	1	60 23 ·5			

221. MOSSEL BAY. Lat. 34° 10′·8 S.; Long. 22° 9′·5 E. In field on Cape Town side of town. Between road going along the coast and the shore.

Declination. D.

Date	G.M.T.	D (observed)	D	Observer	Instrument
1903 Jan. 20	4 24 p.m.	28° 15′·4 W.	28° 12′·1 W.	Beattie	31
	4 30 p.m.	28 15 ·7			
1903 Jan. 21	8 24 a.m.	28 18 ·6			
	8 33 a.m.	28 18 ·1			
	8 43 a.m.	28 18 ·5			

Dip. θ.

Date	G.M.T.	Needle	θ (observed)	θ	Observer	Instrument
1903 Jan. 21	9 55 a.m.	3	60° 28′·3 S.	60° 33′·3 S.	Beattie	142
	9 56 a.m.	4	60 31 ·3			

Horizontal Intensity. H.

Date	G.M.T.	H (observed)	H	Observer	Instrument
1903 Jan. 20	9 24 a.m. V.	H_{30} ·17996	·17930	Beattie	31
	10 15 a.m. D.	H_{40} ·18000			
1903 Jan. 21	3 10 p.m. V.	H_{30} ·17950			
	4 3 p.m. D.	H_{40} ·17952			

222. MOUNT AYLIFF. Lat. 30° 48′·2 S.; Long. 29° 31′·5 E. 10 miles from Mount Ayliff, on road from St Johns to Mount Ayliff.

Declination. D.

Date	G.M.T.	D (observed)	D	Observer	Instrument
1906 Jan. 28	8 20 a.m. 8 28 a.m. 8 34 a.m.	24° 9′·2 W. 24 10 ·0 24 9 ·2	24° 33′·1 W.	Brown Morrison	31

Dip. θ.

Date	G.M.T.	Needle	θ (observed)	θ	Observer	Instrument
1906 Jan. 28	12 2 p.m. 12 2 p.m.	1 2	61° 36′·1 S. 61 38 ·1	61° 16′·4 S.	Morrison	9

Horizontal Intensity. H.

Date	G.M.T.	H (observed)	H	Observer	Instrument
1906 Jan. 28	9 56 a.m. V. 10 26 a.m. D. 10 58 a.m. V.	H_{30} ·17645 H_{40} ·17640	·17890	Morrison	31

223. MOUNT FRERE. Lat. 30° 53′·5 S.; Long. 28° 59′·0 E.

Declination. D.

Date	G.M.T.	D (observed)	D	Observer	Instrument
1906 Jan. 30	9 21 a.m. 9 30 a.m.	24° 8′·0 W. 24 9 ·5	24° 33′·6 W.	Brown Morrison	31

Dip. θ.

Date	G.M.T.	Needle	θ (observed)	θ	Observer	Instrument
1906 Jan. 30	12 55 p.m. 12 55 p.m.	1 2	61° 21′·7 S. 61 22 ·8	61° 1′·6 S.	Morrison	9

Horizontal Intensity. H.

Date	G.M.T.	H (observed)	H	Observer	Instrument
1906 Jan. 30	9 53 a.m. V. 10 39 a.m. D. 11 12 a.m. V.	H_{30} ·17821 H_{40} ·17816	·18067	Morrison	31

223A. MOVENE. Lat. 25° 34′·0 S.; Long. 32° 7′·0 E. Right-hand side of railway, Lorenço Marques to Komati Poort. 49 paces from Waterval end of platform towards Delagoa, and 220 paces at right angles to the railway.

Declination. D.

Date	G.M.T.	D (observed)	D	Observer	Instrument
1903 Sept. 15	1 0 p.m. 1 7 p.m.	17° 39′·6 W. 17 39 ·0	17° 42′·5 W.	Beattie	73

Dip. θ.

Date	G.M.T.	Needle	θ (observed)	θ	Observer	Instrument
1903 Sept. 15	8 26 a.m. 8 26 a.m.	1 4_9	59° 7′·1 S. 59 9 ·0	59° 6′·5 S.	Beattie	142

Horizontal Intensity. H.

Date	G.M.T.	H (observed)	H	Observer	Instrument
1903 Sept. 15	9 19 a.m. V. 10 38 a.m. D. 11 33 a.m. D. 1 42 p.m. V.	H_{30} ·19562 H_{40} ·19559	·19573	Beattie	73

224. MOUNT MORELAND. Lat. 29° 38′·4 S.; Long. 31° 11′·0 E. Right-hand side of railway, Durban to Hlabisa. 140 paces at right angles to railway, starting from Durban end of platform.

Declination. D.

Date	G.M.T.	D (observed)	D	Observer	Instrument
1903 Oct. 19	6 39 a.m. 6 55 a.m.	23° 10′·0 W. 23 10 ·3	23° 12′·1 W.	Beattie	73

Dip. θ.

Date	G.M.T.	Needle	θ (observed)	θ	Observer	Instrument
1903 Oct. 19	9 51 a.m. 9 52 a.m. 9 51 a.m.	1 4_9 4	61° 20′·6 S. 61 21 ·9 61 22 ·0	61° 19′·6 S.	Beattie	142

Horizontal Intensity. H.

Date	G.M.T.	H (observed)	H	Observer	Instrument
1903 Oct. 19	7 33 a.m. V. 8 14 a.m. D. 8 42 a.m. V.	H_{30} ·18047 H_{40} ·18039	·18064	Beattie	73

225. M'Phateles Location. Lat. 24° 19′·8 S.; Long. 29° 41′·0 E. About two hours by cart from Chunie's Poort.

Declination. D.

Date	G.M.T.	D (observed)	D	Observer	Instrument
1903 July 25	12 23 p.m.	20° 57′·6 W.	20° 58′·8 W.	Beattie	73
	12 41 p.m.	20 58 ·3		Löwinger	

Dip. θ.

Date	G.M.T.	Needle	θ (observed)	θ	Observer	Instrument
1903 July 25	10 50 a.m.	1	59° 11′·8 S.	59° 11′·7 S.	Beattie	142
	10 50 a.m.	4	59 11 ·6			

Horizontal Intensity. H.

Date	G.M.T.	H (observed)	H	Observer	Instrument
1903 July 25	11 45 a.m. V.	·19535	·19535	Beattie	73

226. Naauwpoort. Lat. 31° 14′·0 S.; Long. 24° 55′·0 E. On veld across the sluit running alongside Elsworth's hotel, 400 yards on other side of sluit from hotel. 600 yards west of railway station.

Declination. D.

Date	G.M.T.	D (observed)	D	Observer	Instrument
1901 Dec. 20	4 5 p.m.	26° 12′·9 W.	26° 1′·6 W.	Beattie	31
	4 15 p.m.	26 12 ·6		Morrison	
	4 22 p.m.	26 13 ·0			
1901 Dec. 21	5 57 a.m.	26 22 ·0			
	6 8 a.m.	26 22 ·7			
	6 15 a.m.	26 22 ·2			

Dip. θ.

Date	G.M.T.	Needle	θ (observed)	θ	Observer	Instrument
1898 April 9	2 0 p.m.	1	59° 24′·6 S.	60° 1′·3 S.	Beattie	9
	3 30 p.m.	2	59 24 ·0			
1901 Dec. 21	9 20 a.m.	1	59 48 ·2	60 2 ·2	Beattie	142
	9 10 a.m.	2	59 52 ·1			
				60 1 ·8 (mean adopted)		

Horizontal Intensity. H.

Date	G.M.T.	H (observed)	H	Observer	Instrument
1898 April 10	7 30 a.m. V.	·18599	·18127	Beattie	31
	3 40 p.m. V.				
1901 Dec. 21	6 11 a.m. V.	H_{30} ·18338	·18205	Morrison	31
	6 41 a.m. D.	H_{40} ·18341			
			·18166 (mean adopted)		

227. NABOOMSPRUIT. Lat. 24° 31′·3 S.; Long. 28° 43′·0 E. Right-hand side of railway, Pretoria to Pietersburg. 19 paces along railway from Pietersburg end of platform towards Pietersburg, then 225 paces at right angles to railway.

Declination. D.

Date	G.M.T.	D (observed)	D	Observer	Instrument
1903 Sept. 23	4 47 a.m.	20° 56′·1 W.			
	5 0 a.m.	20 57 ·0			
	6 13 a.m.	21 0 ·3	21° 3′·4 W.	Beattie	73
	8 21 a.m.	21 5 ·6			
	8 33 a.m.	21 6 ·4			

Dip. θ.

Date	G.M.T.	Needle	θ (observed)	θ	Observer	Instrument
1903 Sept. 23	9 22 a.m.	1	57° 6′·8 S.	57° 4′·9 S.	Beattie	142
	9 23 a.m.	4_9	57 6 ·2			

Horizontal Intensity. H.

Date	G.M.T.	H (observed)	H	Observer	Instrument
1903 Sept. 23	7 12 a.m. V.				
	7 42 a.m. D.	H_{30} ·19877 H_{40} ·19871	·19889	Beattie	73
	8 10 a.m. V.				

228. NELSPOORT. Lat. 32° 7′·7 S.; Long. 23° 1′·0 E. On right-hand side of railway, Cape Town to De Aar. 300 paces from railway starting from Cape Town end of platform.

Declination. D.

Date	G.M.T.	D (observed)	D	Observer	Instrument
1902 Feb. 9	2 48 p.m.	27° 44′·5 W.	27° 35′·2 W.	Beattie	31
1902 Feb. 10	5 55 a.m.	27 49 ·3			

Dip. θ.

Date	G.M.T.	Needle	θ (observed)	θ	Observer	Instrument
1902 Feb. 9	9 33 a.m.	2	60° 7′·7 S.	60° 18′·1 S.	Beattie	142
	9 35 a.m.	1	60 6 ·2			

Horizontal Intensity. H.

Date	G.M.T.	H (observed)	H	Observer	Instrument
1902 Feb. 9	1 44 p.m. V	·17914	·17795	Beattie	31

229. NELSPRUIT. Lat. 25° 28′·1 S.; Long. 30° 58′·5 E. Right-hand side of railway, Waterval Onder to Komati Poort. On Waterval side of railway station, 70 yards at right angles to railway, starting from blind end of siding.

Declination. D.

Date	G.M.T.	D (observed)	D	Observer	Instrument
1903 Sept. 9	7 25 a.m.	20° 59′·7 W.			
	7 41 a.m.	21 0·1			
	12 53 p.m.	20 55·0	20° 59′·5 W.	Beattie	73
	1 3 p.m.	20 55·3			
	2 31 p.m.	20 55·2			

Dip. θ.

Date	G.M.T.	Needle	θ (observed)	θ	Observer	Instrument
1903 Sept. 9	8 54 a.m.	1	58° 29′·8 S.	58° 28′·4 S.	Beattie	142
	8 54 a.m.	4_9	58 29·8			

Horizontal Intensity. H.

Date	G.M.T.	H (observed)	H	Observer	Instrument
1903 Sept. 9	9 47 a.m. V.				
	11 12 a.m. D.	H_{30} ·19279	·19290	Beattie	73
	11 45 a.m. D.	H_{40} ·19277			
	12 40 p.m. V.				

230. NEWCASTLE. Lat. 27° 45′·3 S.; Long. 29° 58′·0 E. Left-hand side of railway, Newcastle to Glencoe. Along railway from Glencoe end of platform, 85 paces towards Glencoe, then 215 paces at right angles to railway.

Declination. D.

Date	G.M.T.	D (observed)	D	Observer	Instrument
1903 Oct. 1	4 50 a.m.	22° 27′·9 W.			
	4 58 a.m.	22 27·8	22° 33′·1 W.	Beattie	73
	7 42 a.m.	22 34·0			
	7 50 a.m.	22 34·9			

Dip. θ.

Date	G.M.T.	Needle	θ (observed)	θ	Observer	Instrument
1903 Oct. 1	9 10 a.m.	1	59° 39′·7 S.			
	9 10 a.m.	4_9	59 42·1	59° 38′·6 S.	Beattie	142
	9 10 a.m.	4	59 41·3			

Horizontal Intensity. H.

Date	G.M.T.	H (observed)	H	Observer	Instrument
1903 Oct. 1	6 49 a.m. V.	H_{30} ·18828			
	7 17 a.m. D.		·18845	Beattie	73
	8 7 a.m. V.	H_{40} ·18826			

231. Newcastle (Transvaal). Lat. 26° 32′·1 S.; Long. 30° 27′·0 E. On south west side of farmhouse behind lower clump of trees.

Declination. D.

Date	G.M.T.	D (observed)	D	Observer	Instrument
1903 Aug. 25	1 42 p.m.	21° 56′·2 W.	21° 58′·1 W.	Beattie	73
	1 54 p.m.	21 56 ·9			

Dip. θ.

Date	G.M.T.	Needle	θ (observed)	θ	Observer	Instrument
1903 Aug. 25	11 22 a.m.	1	58° 58′·0 S.	58° 56′·6 S.	Beattie	142
	11 23 a.m.	4	58 56 ·1			
	11 24 a.m.	4_0	58 59 ·0			

Horizontal Intensity. H.

Date	G.M.T.	H (observed)	H	Observer	Instrument
1903 Aug. 25	12 47 p.m. V.	·18976	·18982	Beattie	73

232. Nooitgedacht. Lat. 25° 38′·1 S.; Long. 30° 31′·0 E. Left-hand side of railway, Waterval to Komati Poort. 67 paces along railway, starting from blind end of siding and going towards Komati Poort, then 257 paces at right angles to the railway.

Declination. D.

Date	G.M.T.	D (observed)	D	Observer	Instrument
1903 Sept. 10	2 38 p.m.	18° 58′·2 W.	19° 0′·7 W.	Beattie	73

Dip. θ.

Date	G.M.T.	Needle	θ (observed)	θ	Observer	Instrument
1903 Sept. 10	3 39 p.m.	1	57° 23′·5 S.	57° 22′·1 S.	Beattie	142

Horizontal Intensity. H.

Date	G.M.T.	H (observed)	H	Observer	Instrument
1903 Sept. 10	3 4 p.m. V.	·19192	·19204	Beattie	73

233. Norval's Pont. Lat. 30° 39′·0 S.; Long. 25° 27′·0 E. Right-hand side of railway going to Bloemfontein, opposite railway station and 500 paces from it.

Dip. θ.

Date	G.M.T.	Needle	θ (observed)	θ	Observer	Instrument
1903 May 25	11 4 a.m.	3	59° 51′·9 S.	59° 50′·1 S.	Beattie	142
	11 5 a.m.	4	59 47 ·7			
	11 2 a.m.	1	59 48 ·9			

Horizontal Intensity. H.

Date	G.M.T.	H (observed)	H	Observer	Instrument
1903 May 25	12 32 p.m. V.	·18323	·18320	Beattie	73

234. 'NQUTU ROAD. Lat. 28° 5'·0 S.; Long. 30° 26'·0 E. Left-hand side of railway, Glencoe to 'Nqutu. 105 paces at right angles to railway from a point 85 paces from Glencoe end of platform towards Glencoe.

Dip. θ.

Date	G.M.T.	Needle	θ (observed)	θ	Observer	Instrument
1903 Oct. 3	10 47 a.m.	1	60° 14'·8 S.	60° 13'·0 S.	Beattie	142

Horizontal Intensity. H.

Date	G.M.T.	H (observed)	H	Observer	Instrument
1903 Oct. 3	9 45 a.m. V.	·18568	·18584	Beattie	73

235. NYLSTROOM Lat. 24° 42'·4 S.; Long. 28° 26'·0 E. On prolongation of dead end of triangle, right-hand side of railway, Pretoria to Pietersburg, 186 paces from dead end.

Declination. D.

Date	G.M.T.	D (observed)	D	Observer	Instrument
1903 June 26	2 11 p.m.	21° 16'·9 W.			
	2 24 p.m.	21 16 ·8	21° 11'·8 W.	Beattie Löwinger	73
1903 June 27	6 8 a.m.	21 6 ·9			
	6 37 a.m.	21 7 ·0			

Dip. θ.

Date	G.M.T.	Needle	θ (observed)	θ	Observer	Instrument
1903 June 27	8 39 a.m.	1	57° 28'·6 S.			
	8 37 a.m.	3	57 30 ·5	57° 28'·9 S.	Beattie	142
	8 39 a.m.	4	57 27 ·7			

Horizontal Intensity. H.

Date	G.M.T.	H (observed)	H	Observer	Instrument
1903 June 26	3 15 p.m. V.	·19840			
1903 June 27	5 57 a.m. V.				
	10 17 a.m. D.	H_{30} ·19842			
	8 43 a.m. V.	H_{40} ·19840	·19843	Beattie	73
	11 16 a.m. V.	H_{30} ·19850			
	10 52 a.m. D.	H_{40} ·19847			

235A. ORANGE RIVER. Lat. 29° 38′·0 S.; Long. 24° 16′·0 E. Tent about 650 yards due west from Railway Station (1902). The place of observation in 1898 was different and could not be re-occupied on account of presence of corrugated iron buildings.

Declination. D.

Date	G.M.T.	D (observed)	D	Observer	Instrument
1902 July 19	1 38 p.m.	25° 31′·5 W.			
	1 56 p.m.	25 32 ·3	25° 24′·0 W.	Beattie Morrison	31
	2 52 p.m.	25 34 ·4			
	3 0 p.m.	25 35 ·1			

Dip. θ.

Date	G.M.T.	Needle	θ (observed)	θ	Observer	Instrument
1898 Jan. 1	9 30 a.m.	2	58° 23′·6 S.			
	10 50 a.m.	1	58 23 ·5	59° 6′·7 S.	Beattie Morrison	9
1898 Jan. 2	9 20 a.m.	1	58 21 ·8			
1902 July 19	11 2 a.m.	1	59 0 ·1			
	11 0 a.m.	3	59 1 ·4	59 8 ·0	Beattie	142
	11 0 a.m.	4	59 1 ·4			
				59 7 ·3 (mean adopted)		

Horizontal Intensity. H.

Date	G.M.T.	H (observed)	H	Observer	Instrument
1898 Jan. 1	1 45 p.m. V.	H_{30} ·18956	·18554	Beattie	31
	3 45 p.m. D.	H_{40} ·18943			
1902 July 19	1 14 p.m. V.				
	2 24 p.m. D.	H_{30} ·18634	·18563	Morrison	31
	3 18 p.m. V.	H_{40} ·18635			
			·18559 (mean adopted)		

236. ORJIDA. Lat. 33° 26′·0 S.; Long. 23° 19′·0 E. 400 yards from farmhouse in camp on left-hand side of road from Uniondale to Willowmore.

Declination. D.

Date	G.M.T.	D (observed)	D	Observer	Instrument
1903 Feb. 18	1 36 p.m.	27° 47′·1 W.	27° 48′·7 W.	Beattie	31
	1 54 p.m.	27 46 ·6			

Dip. θ.

Date	G.M.T.	Needle	θ (observed)	θ	Observer	Instrument
1903 Feb. 18	12 22 p.m.	3	60° 41′·0 S.	60° 45′·1 S.	Beattie	142
	12 22 p.m.	4	60 43 ·2			

Horizontal Intensity. H.

Date	G.M.T.	H (observed)	H	Observer	Instrument
1903 Feb. 18	1 6 p.m. V.	·17874	·17836	Beattie	31

237. OUDEMUUR. Lat. 31° 5′·8 S.; Long. 20° 19′·1 E. The road from Brandvlei to De Drift forms a triangle at Oudemuur farm. Observations taken in corner of triangle nearest De Drift.

Declination. D.

Date	G.M.T.	D (observed)	D	Observer	Instrument
1905 Jan. 26	6 5 a.m.	27° 10′·6 W.	28° 19′·4 W.	Beattie	73
	6 20 a.m.	27 11 ·1		Brown	

Dip. θ.

Date	G.M.T.	Needle	θ (observed)	θ	Observer	Instrument
1905 Jan. 26	9 32 a.m.	3_9	58° 53′·7 S.	58° 42′·2 S.	Beattie	142
	9 32 a.m.	4_9	58 56 ·0			

Horizontal Intensity. H.

Date	G.M.T.	H (observed)	H	Observer	Instrument
1905 Jan. 26	8 0 a.m. V.	H_{30} ·18422	·18549	Beattie	73
	8 44 a.m. D.	H_{40} ·18424			

238. OUDTSHOORN. Lat. 33° 35′·2 S.; Long. 22° 12′·5 E. On the recreation ground.

Declination. D.

Date	G.M.T.	D (observed)	D	Observer	Instrument
1903 Jan. 8	3 26 p.m.	28° 14′·0 W.	28° 11′·4 W.	Beattie	31
	3 52 p.m.	28 12 ·9		Morrison	
1903 Jan. 9	6 27 a.m.	28 18 ·6			
	6 45 a.m.	28 17 ·7			

Dip. θ.

Date	G.M.T.	Needle	θ (observed)	θ	Observer	Instrument
1903 Jan. 9	5 26 a.m.	1	60° 19′·6 S.	60° 24′·5 S.	Beattie	142
	5 24 a.m.	2	60 25 ·4			
	5 30 a.m.	3	60 18 ·0			
	5 30 a.m.	4	60 19 ·3			

Horizontal Intensity. H.

Date	G.M.T.	H (observed)	H	Observer	Instrument
1903 Jan. 9	8 59 a.m. V.	H_{40} ·17989	·17944	Morrison	31
	10 12 a.m. D.				
	11 51 a.m. V.				

239. Paardevlei. Lat. 30° 36′·1 S.; Long. 21° 54′·0 E. Right-hand side of road, Carnarvon to Van Wyk's Vlei. 250 paces from stable.

Declination. D.

Date	G.M.T.	D (observed)	D	Observer	Instrument
1904 Dec. 11	6 28 a.m.	26° 49′·8 W.	27° 0′·7 W.	Beattie	73
	6 37 a.m.	26 49 ·5		Hough	

Dip. θ.

Date	G.M.T.	Needle	θ (observed)	θ	Observer	Instrument
1904 Dec. 11	5 10 a.m.	3_9	58° 54′·2 S.	58° 43′·6 S.	Beattie	142
	5 10 a.m.	4_9	58 55 ·1			

Horizontal Intensity. H.

Date	G.M.T.	H (observed)	H	Observer	Instrument
1904 Dec. 11	7 8 a.m. V.	H_{30} ·18469	·18588	Beattie	73
	7 44 a.m. D.	H_{40} ·18476			

240. Paarl. Lat. 33° 45′·0 S.; Long. 18° 57′·0 E. (From Cape Meteorological Commission Report.) Up the hill from Le Roux's farm, under tree near reservoir.

Dip. θ.

Date	G.M.T.	Needle	θ (observed)	θ	Observer	Instrument
1899 Dec. 16	2 36 p.m.	1	58° 41′·5 S.	59° 13′·5 S.	Beattie	9
	2 36 p.m.	2	58 39 ·5			
1899 Dec. 17	9 0 a.m.	1	58 43 ·0			
	9 0 a.m.	2	58 41 ·9			

241. PALAPYE. Lat. 22° 33′·4 S.; Long. 27° 7′·0 E. Left-hand side of railway, Mafeking to Bulawayo. 430 paces at right angles to railway, starting from a point 79 paces from Bulawayo end of platform, and going towards Bulawayo.

Declination. D.

Date	G.M.T.	D (observed)	D	Observer	Instrument
1903 March 28	5 46 a.m.	20° 39′·0 W.			
	6 2 a.m.	20 37 ·8	20° 33′·4 W.	Beattie	73
	3 3 p.m.	20 34 ·0			
	3 17 p.m.	20 34 ·4			

Dip. θ.

Date	G.M.T.	Needle	θ (observed)	θ	Observer	Instrument
1898 Jan. 10	9 12 a.m.	1	54° 24′·7 S.		Beattie	
	2 0 p.m.	2	54 28 ·3	55° 5′·9 S.	Morrison	9
1898 Jan. 11	1 50 p.m.	2	54 30 ·7			
1903 March 29	1 9 p.m.	3	55 8 ·1			
	1 9 p.m.	4	55 7 ·6	55 9 ·5	Beattie	142
	1 9 p.m.	1	55 7 ·6			
				55 7 ·7	(mean adopted)	

Horizontal Intensity. H.

Date	G.M.T.	H (observed)	H	Observer	Instrument
1898 Jan. 11	9 11 a.m. V.	H_{30} ·20964	·20636	Beattie	31
	10 8 a.m. D.	H_{40} ·20968			
1903 March 28	8 6 a.m. V.	H_{30} ·20659			
	8 25 a.m. D.	H_{40} ·20658	·20641	Beattie	142
	9 23 a.m. V.	H_{25} ·20660			
		H_{35} ·20660			
			·20639	(mean adopted)	

242. PAMPOENPOORT. Lat. 31° 3′·5 S.; Long. 22° 39′·1 E. In bed of river. Right-hand side of road coming from Victoria West, about 200 paces from place where road crosses the river.

Declination. D.

Date	G.M.T.	D (observed)	D	Observer	Instrument
1904 Dec. 7	7 49 a.m.	26° 37′·1 W.	26° 47′·7 W.	Beattie Hough	73

Dip. θ.

Date	G.M.T.	Needle	θ (observed)	θ	Observer	Instrument
1904 Dec. 7	10 1 a.m.	3_9	59° 31′·0 S.	59° 19′·0 S.	Beattie	142
	10 1 a.m.	4_9	59 30 ·4			

Horizontal Intensity. H.

Date	G.M.T.	H (observed)	H	Observer	Instrument
1904 Dec. 7	8 16 a.m. V.	H_{30} ·18207	·18324	Beattie	73
	8 54 a.m. D.	H_{40} ·18213			

243. Payne's Farm, Ingela. (Between Ingela and Harding.) Lat. 30° 36′·9 S.; Long. 29° 47′·5 E. Right-hand side of road, Kokstad to Harding. Opposite clump of trees on road side. Farmhouse about 1½ miles off.

Declination. D.

Date	G.M.T.	D (observed)	D	Observer	Instrument
1903 Nov. 20	12 28 p.m. 12 36 p.m.	24° 32′·1 W. 24 31 ·0	24° 36′·7 W.	Beattie	73

Dip. θ.

Date	G.M.T.	Needle	θ (observed)	θ	Observer	Instrument
1903 Nov. 20	12 46 p.m. 12 48 p.m.	1 4_9	61° 37′·1 S. 61 36 ·8	61° 34′·4 S.	Beattie	142

Horizontal Intensity. H.

Date	G.M.T.	H (observed)	H	Observer	Instrument
1903 Nov. 20	1 31 p.m. V.	·17799	·17826	Beattie	73

244. Picene. Lat. 25° 40′·8 S.; Long. 32° 18′·5 E. Right-hand side of railway, Lorenço Marques to Komati Poort. Start from Lorenço Marques end of platform and go towards Lorenço Marques 136 paces, then 211 paces at right angles to railway.

Declination. D.

Date	G.M.T.	D (observed)	D	Observer	Instrument
1903 Sept. 16	7 20 a.m. 7 31 a.m.	20° 47′·4 W. 20 47 ·0	20° 48′·1 W.	Beattie	73

Dip. θ.

Date	G.M.T.	Needle	θ (observed)	θ	Observer	Instrument
1903 Sept. 16	10 32 a.m.	1	59° 1′·5 S.	59° 0′·0 S.	Beattie	142

Horizontal Intensity. H.

Date	G.M.T.	H (observed)	H	Observer	Instrument
1903 Sept. 16	8 10 a.m. V. 8 54 a.m. D. 9 31 a.m. V.	H_{30} ·19296 H_{40} ·19296	·19308	Beattie	73

245. PIENAAR'S RIVER. Lat. 25° 12′·7 S.; Long. 28° 19′·0 E. Left-hand side of railway, Pretoria to Pietersburg. 205 paces from railway, starting from Pietersburg end of loop.

Declination. D.

Date	G.M.T.	D (observed)	D	Observer	Instrument
1903 June 26	6 7 a.m.	23° 20′·2 W.	23° 21′·5 W.	Beattie Löwinger	73
	6 20 a.m.	23 21 ·6			
	6 36 a.m.	23 22 ·0			

Dip. θ.

Date	G.M.T.	Needle	θ (observed)	θ	Observer	Instrument
19?3 June 26	7 59 a.m.	4	57° 48′·0 S.	57° 48′·8 S.	Beattie	142
	7 56 a.m.	1	57 49 ·6			

Horizontal Intensity. H.

Date	G.M.T.	H (observed)	H	Observer	Instrument
1903 June 26	8 35 a.m. V.	H_{30} ·19729	·19729	Beattie	73
	9 4 a.m. D.	H_{40} ·19730			

246. PIETERSBURG. Lat. 23° 50′·3 S.; Long. 29° 27′·0 E. Right-hand side of railway, Pretoria to Pietersburg. Start from end of platform distant from Pretoria, go 80 paces in prolongation of platform, then 220 paces from railway.

Declination. D.

Date	G.M.T.	D (observed)	D	Observer	Instrument
1903 June 28	6 26 a.m.	20° 10′·9 W.	20° 12′·0 W.	Beattie Löwinger	73
	6 42 a.m.	20 11 ·1			
	1 34 p.m.	20 13 ·3			
	1 48 p.m.	20 12 ·8			

Dip. θ.

Date	G.M.T.	Needle	θ (observed)	θ	Observer	Instrument
1903 June 28	3 3 p.m.	1	56° 55′·7 S.	56° 57′·0 S.	Beattie	142
	3 3 p.m.	4	56 58 ·8			
	3 3 p.m.	4_9	56 56 ·5			

Horizontal Intensity. H.

Date	G.M.T.	H (observed)	H	Observer	Instrument
1903 June 28	8 44 a.m. V.		·19855	Beattie	73
	9 55 a.m. D.	H_{30} ·19854			
	10 44 a.m. D.	H_{40} ·19855			
	11 0 a.m. V.				

247. Piet Potgietersrust. Lat. 24° 11′·2 S.; Long. 29° 1′·0 E. Left-hand side of railway, Pretoria to Pietersburg. 299 paces at right angles to railway, starting from a point 224 paces along the railway towards Pretoria from Pretoria end of platform.

Declination. D.

Date	G.M.T.	D (observed)	D	Observer	Instrument
1903 June 29	6 36 a.m.	20° 43′·4 W.			
	6 51 a.m.	20 43 ·0			
	1 19 p.m.	20 35 ·9	20° 40′·2 W.	Beattie Löwinger	73
	1 35 p.m.	20 36 ·7			
	1 52 p.m.	20 38 ·3			

Dip. θ.

Date	G.M.T.	Needle	θ (observed)	θ	Observer	Instrument
1903 June 29	9 32 a.m.	1	56° 58′·2 S.	56° 58′·9 S.	Beattie	142
	9 33 a.m.	4_9	56 59 ·6			

Horizontal Intensity. H.

Date	G.M.T.	H (observed)	H	Observer	Instrument
1903 June 29	10 13 a.m. V.	H_{30} ·19831	·19831	Beattie	73
	10 33 a.m. D.	H_{40} ·19832			
	11 48 a.m. D.				

248. Piet Retief. Lat. 27° 0′·5 S.; Long. 30° 48′·5 E. On east side of town in hollow near slaughter sticks.

Declination. D.

Date	G.M.T.	D (observed)	D	Observer	Instrument
1903 Aug. 17	6 48 a.m.	22° 13′·2 W.	22° 14′·5 W.	Beattie	73

Dip. θ.

Date	G.M.T.	Needle	θ (observed)	θ	Observer	Instrument
1903 Aug. 17	10 58 a.m.	1	59° 33′·7 S.	59° 32′·6 S.	Beattie	142
	11 0 a.m.	4	59 33 ·3			

Horizontal Intensity. H.

Date	G.M.T.	H (observed)	H	Observer	Instrument
1903 Aug. 17	8 35 a.m. V.	H_{30} ·18826	·18831	Beattie	73
	9 10 a.m. D.	H_{40} ·18824			
	9 41 a.m. V.				

249. Pilgrim's Rest. Lat. 24° 56′·8 S.; Long. 30° 45′·0 E. About six miles from Pilgrim's Rest on road to Nelspruit. On a level patch of ground surrounded by hills, road passes across this from north to south, on bank of Blyde river.

Declination. D.

Date	G.M.T.	D (observed)	D	Observer	Instrument
1903 Sept. 6	1 43 p.m.	20° 56′·9 W.			
	1 55 p.m.	20 55 ·8	20° 59′·2 W.	Beattie	73
	2 8 p.m.	20 57 ·0			

Dip. θ.

Date	G.M.T.	Needle	θ (observed)	θ	Observer	Instrument
1903 Sept. 6	11 0 a.m.	1	57° 51′·1 S.			
	11 1 a.m.	4	57 50 ·3	57° 49′·8 S.	Beattie	142
	11 1 a.m.	4_9	57 51 ·8			

Horizontal Intensity. H.

Date	G.M.T.	H (observed)	H	Observer	Instrument
1903 Sept. 6	11 56 a.m. V.	·19600	·19612	Beattie	73

250. Piquetberg. Lat. 32° 55′·0 S.; Long. 18° 43′·0 E. On a level with and to the south east of the Dutch reformed manse, and about 400 yards from it.

Declination. D.

Date	G.M.T.	D (observed)	D	Observer	Instrument
1901 July 18	7 52 a.m.	28° 18′·7 W.			
	8 0 a.m.	28 19 ·4	28° 10′·7 W.	Beattie Morrison	31
	8 10 a.m.	28 19 ·8			

Dip. θ.

Date	G.M.T.	Needle	θ (observed)	θ	Observer	Instrument
1901 July 18	9 13 a.m.	2	58° 39′·8 S.	58° 56′·4 S.	Beattie	142
	9 13 a.m.	1	58 37 ·6			

Horizontal Intensity. H.

Date	G.M.T.	H (observed)	H	Observer	Instrument
1901 July 18	10 12 a.m. V.				
	10 43 a.m. D.	H_{30} ·18528 H_{40} ·18527	·18340	Morrison	31
	11 32 a.m. V.				

251. PIVAANS POORT. Lat. 27° 33′·8 S.; Long. 30° 28′·0 E. Just over the Pivaans river, left-hand side of road coming from Utrecht to Piet Retief.

Declination. D.

Date		G.M.T.	D (observed)	D	Observer	Instrument
1903	Aug. 13	7 58 a.m. 8 8 a.m.	22° 38′·8 W. 22 38 ·3	22° 39′·2 W.	Beattie	73

Dip. θ.

Date		G.M.T.	Needle	θ (observed)	θ	Observer	Instrument
1903	Aug. 13	10 48 a.m.	1	59° 43′·9 S.	59° 43′·0 S.	Beattie	142

Horizontal Intensity. H.

Date		G.M.T.	H (observed)	H	Observer	Instrument
1903	Aug. 13	8 35 a.m. V. 9 18 a.m. D.	H_{30} ·18789 H_{40} ·18791	·18796	Beattie	73

252. PLATRAND. Lat. 27° 6′·4 S.; Long. 29° 29′·0 E. Right-hand side of railway, Germiston to Charlestown. 245 paces along railway from dead end towards Germiston, then 220 paces at right angles to railway.

Declination. D.

Date		G.M.T.	D (observed)	D	Observer	Instrument
1903	Sept. 29	4 21 a.m. 4 35 a.m. 4 52 a.m.	22° 34′·8 W. 22 34 ·1 22 33 ·2	22° 36′·4 W.	Beattie	73

Dip. θ.

Date		G.M.T.	Needle	θ (observed)	θ	Observer	Instrument
1903	Sept. 29	9 3 a.m. 10 26 a.m. 10 27 a.m.	1 1 4	59° 14′·8 S. 59 13 ·7 59 17 ·4	59° 12′·2 S.	Beattie	142

Horizontal Intensity. H.

Date		G.M.T.	H (observed)	H	Observer	Instrument
1903	Sept. 29	5 17 a.m. V. 7 11 a.m. D. 7 43 a.m. D. 8 13 a.m. V.	H_{30} ·18892 H_{40} ·18892	·18907	Beattie	73

253. Plettenberg Bay. Lat. 34° 2′·2 S.; Long. 23° 21′·0 E. 200 yards west of 'Welcome.'

Declination. D.

Date	G.M.T.	D (observed)	D	Observer	Instrument
1903 Jan. 30	5 20 a.m.	27° 55′·2 W.			
	9 38 a.m.	27 50 ·8			
	9 52 a.m.	27 49 ·1	27° 48′·1 W.	Beattie	31
	3 7 p.m.	27 52 ·9			
	4 12 p.m.	27 54 ·0			

Dip. θ.

Date	G.M.T.	Needle	θ (observed)	θ	Observer	Instrument
1903 Jan. 30	12 34 p.m.	3	60° 49′·7 S.	60° 54′·9 S.	Beattie	142
	12 34 p.m.	4	60 53 ·6			

Horizontal Intensity. H.

Date	G.M.T.	H (observed)	H	Observer	Instrument
1903 Jan. 30	1 34 p.m. V.	H_{30} ·17859	·17812	Beattie	31
	2 17 p.m. D.	H_{40} ·17854			
	2 54 p.m. V.				

254. Plumtree. Lat. 20° 30′ S.; Long. 27° 50′ E. Right-hand side of railway, Mafeking to Bulawayo. 173 paces at right angles to railway, starting from a point 333 paces from blind end of siding, and going towards Bulawayo.

Dip. θ.

Date	G.M.T.	Needle	θ (observed)	θ	Observer	Instrument
1903 April 2	5 44 a.m.	3	53° 44′·1 S.			
	5 44 a.m.	4	53 45 ·6	53° 46′·9 S.	Beattie	142
	5 44 a.m.	1	53 45 ·8			

Horizontal Intensity. H.

Date	G.M.T.	H (observed)	H	Observer	Instrument
1903 April 1	11 48 a.m. V.	H_{30} ·21236			
	12 32 p.m. D.	H_{40} ·21234	·21219	Beattie	73
	1 19 p.m. V.	H_{25} ·21240			
		H_{35} ·21238			

255. Pokwani, Transvaal. Lat. 24° 52′·2 S.; Long. 29° 46′·0 E. Left-hand side of main road, Pietersburg to Middelburg. 96 paces from store along road to Middelburg, then 59 paces at right angles to road.

Declination. D.

Date	G.M.T.	D (observed)	D	Observer	Instrument
1903 July 28	12 58 p.m. 1 13 p.m.	21° 17′·2 W. 21 17 ·0	21° 17′·9 W.	Beattie Löwinger	73

Dip. θ.

Date	G.M.T.	Needle	θ (observed)	θ	Observer	Instrument
1903 July 28	10 42 a.m. 10 43 a.m.	1 4	57° 42′·1 S. 57 40 ·4	57° 41′·3 S.	Beattie	142

Horizontal Intensity. H.

Date	G.M.T.	H (observed)	H	Observer	Instrument
1903 July 28	11 33 a.m. V. 12 15 p.m. D. 12 49 p.m. V.	H_{30} ·19756 H_{40} ·19750	·19753	Beattie	73

256. Port Alfred. Lat. 33° 35′·8 S.; Long. 26° 54′·0 E. On level ground at foot of hill in front of Macdonald's Hotel, about 200 yards from the bathing house.

Declination. D.

Date	G.M.T.	D (observed)	D	Observer	Instrument
1902 July 10	9 21 a.m. 9 26 a.m. 9 31 a.m. 1 18 p.m. 1 24 p.m.	27° 8′·6 W. 27 7 ·7 27 6 ·8 27 2 ·7 27 2 ·7	26° 54′·8 W.	Beattie Morrison	31

Dip. θ.

Date	G.M.T.	Needle	θ (observed)	θ	Observer	Instrument
1902 July 10	1 20 p.m. 1 21 p.m. 1 20 p.m. 1 21 p.m.	1 2 3 4	61° 34′·1 S. 61 35 ·2 61 34 ·7 61 31 ·7	61° 42′·0 S.	Beattie	142

Horizontal Intensity. H.

Date	G.M.T.	H (observed)	H	Observer	Instrument
1902 July 10	9 58 a.m. V. 10 40 a.m. D. 11 12 a.m. V.	H_{30} ·17687 H_{40} ·17687	·17595	Morrison	31

257. PORT BEAUFORT. Lat. 34° 23′·8 S.; Long. 20° 49′·0 E. About 140 yards west of church.

Declination. D.

Date	G.M.T.	D (observed)	D	Observer	Instrument
1901 Jan. 29	8 36 a.m. 8 48 a.m. 9 1 a.m.	28° 44′·2 W. 28 44 ·4 28 44 ·2	28° 27′·3 W.	Beattie Morrison	31

Dip. θ.

Date	G.M.T.	Needle	θ (observed)	θ	Observer	Instrument
1901 Jan. 30	11 12 a.m. 11 21 a.m.	1 2	59° 49′·1 S. 59 52 ·1	60° 12′·4 S.	Beattie	142

Horizontal Intensity. H.

Date	G.M.T.	H (observed)	H	Observer	Instrument
1901 Jan. 30	9 26 a.m. V. 9 57 a.m. D. 10 25 a.m. V.	H_{30} ·18200 H_{40} ·18201	·17990	Morrison	31

258. PORT ELIZABETH. Lat. 33° 58′·0 S.; Long. 25° 37′·0 E. Place of observation in 1898 on right-hand side of road, near clump of trees round farm, on common land beyond end of tram lines.

Declination. D.

Date	G.M.T.	D (observed)	D	Observer	Instrument
1900 Jan. 23	8 0 a.m. 8 15 a.m. 8 30 a.m.	28° 15′·7 W. 28 17 ·0 28 16 ·4	27° 39′·4 W.	Beattie Morrison	31

Dip. θ.

Date	G.M.T.	Needle	θ (observed)	θ	Observer	Instrument
1898 April 12	10 30 a.m. 10 30 a.m.	1 2	60° 53′·6 S. 60 53 ·2	61° 35′·4 S.	Beattie	9
1900 Jan. 23	10 0 a.m. 10 0 a.m.	1 2	61 1 ·8 61 0 ·0	61 28 ·0	Beattie	9

The two stations of observation were not the same.

61° 35′·4 S. (mean adopted)

Horizontal Intensity. H.

Date	G.M.T.	H (observed)	H	Observer	Instrument
1898 April 13	2 19 p.m. V. 3 8 p.m. D.	H_{30} ·18105 H_{40} ·18111	·17638	Morrison	31
1900 Jan. 22	2 31 p.m. V. 3 10 p.m. D. 3 47 p.m. V.	H_{30} ·17977 H_{40} ·17979	·17642	Morrison	31

·17640 (mean adopted)

259. PORT SHEPSTONE. Lat. 30° 43′·7 S.; Long. 30° 27′·0 E. 140 paces at right angles to railway from a point 363 paces towards Durban, measured from Durban end of platform. Left-hand side of railway coming from Durban.

Declination. D.

Date		G.M.T.	D (observed)	D	Observer	Instrument
1903	Nov. 1	7 23 a.m.	24° 15′·5 W.			
		7 36 a.m.	24 14 ·9	24° 16′·2 W.	Beattie	73
		8 2 a.m.	24 14 ·6			

Dip. θ.

Date		G.M.T.	Needle	θ (observed)	θ	Observer	Instrument
1903	Nov. 1	12 43 p.m.	1	61° 56′·0 S.	61° 53′·3 S.	Beattie	142
		12 43 p.m.	4	61 55 ·2			

Horizontal Intensity. H.

Date		G.M.T.	H (observed)	H	Observer	Instrument
1903	Nov. 1	8 24 a.m. V.				
		9 8 a.m. D.	H_{30} ·17775 H_{40} ·17779	·17801	Beattie	73
		9 58 a.m. V				

260. PORT ST JOHNS. Lat. 31° 37′·8 S.; Long. 29° 33′·0 E. In a hollow about three-quarters of a mile from the Needles Hotel, between the road that runs south west along the coast and the sea.

Declination. D.

Date		G.M.T.	D (observed)	D	Observer	Instrument
1906	Jan. 23	6 37 a.m.	24° 54′·5 W.	25° 18′·2 W.	Brown	31
		6 46 a.m.	24 55 ·3		Morrison	

Dip. θ.

Date		G.M.T.	Needle	θ (observed)	θ	Observer	Instrument
1906	Jan. 23	11 5 a.m.	1	62° 21′·9 S.	62° 1′·5 S.	Morrison	9
		11 6 a.m.	2	62 22 ·2			

Horizontal Intensity. H.

Date		G.M.T.	H (observed)	H	Observer	Instrument
1906	Jan. 23	8 55 a.m. V.				
		9 30 a.m. D.	H_{30} ·17300 H_{40} ·17302	·17547	Morrison	31
		9 57 a.m. V.				

261. Potchefstroom. Lat. 26° 42′·8 S.; Long. 27° 5′·0 E. At point half-way between railway station and gaol.

Declination. D.

Date	G.M.T.	D (observed)	D	Observer	Instrument
1903 June 20	6 32 a.m.	23° 0′·9 W.	23° 0′·4 W.	Beattie Löwinger	73
	6 44 a.m.	23 1 ·2			
	12 50 p.m.	22 59 ·9			
	1 5 p.m.	23 0 ·3			

Dip. θ.

Date	G.M.T.	Needle	θ (observed)	θ	Observer	Instrument
1903 June 20	8 58 a.m.	3	58° 20′·7 S.	58° 21′·7 S.	Beattie	142
	8 58 a.m.	4	58 22 ·8			
	8 58 a.m.	1	58 21 ·4			
	8 54 a.m.	4_9	58 19 ·6			

Horizontal Intensity. H.

Date	G.M.T.	H (observed)	H	Observer	Instrument
1903 June 20	9 53 a.m. V.		·19079	Beattie	73
	10 29 a.m. D.	H_{30} ·19082			
	12 0 noon D.	H_{40} ·19077			
	12 35 p.m. V.				

262. Potfontein. Lat. 30° 12′·2 S.; Long. 24° 7′·0 E.

Declination. D.

Date	G.M.T	D (observed)	D	Observer	Instrument
1903 March 15	5 46 a.m.	25° 44′·0 W.	25° 38 ·5 W.	Beattie	73
	6 0 a.m.	25 43 ·8			
	3 20 p.m.	25 39 ·6			
	3 40 p.m.	25 39 ·5			

Dip θ.

Date	G.M.T.	Needle	θ (observed)	θ	Observer	Instrument
1903 March 15	1 32 p.m.	3	59° 19′·8 S.	59° 21′·4 S.	Beattie	142
	1 35 p.m.	4	59 18 ·4			

Horizontal Intensity. H.

Date	G.M.T.	H (observed)	H	Observer	Instrument
1903 March 15	8 43 a.m. V.	H_{30} ·18454	·18433	Beattie	73
	9 40 a.m. D.	H_{40} ·18461			
	10 16 a.m. V.				

263. PRETORIA. Lat. 25° 45′·3 S.; Long. 28° 12′·0 E. In open space between end of Schoeman Street and the racecourse. About one-third of the distance between end of street and the racecourse.

Declination. D.

Date	G.M.T.	D (observed)	D	Observer	Instrument
1903 June 25	12 39 p.m.	22° 16′·4 W.			
	12 53 p.m.	22 16 ·7	22° 16′·4 W	Beattie	73
	1 4 p.m.	22 16 ·8			

Dip. θ.

Date	G.M.T.	Needle	θ (observed)	θ	Observer	Instrument
1903 June 25	11 0 a.m.	3	58° 3′·4 S.			
	11 0 a.m.	4	58 3 ·0	58° 2′·7 S.	Beattie	142
	10 57 a.m.	1	58 1 ·8			
	10 56 a.m.	4_9	58 2 ·7			

Horizontal Intensity. H.

Date	G.M.T.	H (observed)	H	Observer	Instrument
1903 June 25	8 15 a.m. V.				
	8 47 a.m. D.	H_{30} ·19370 H_{40} ·19384	·19377	Beattie	73
	9 34 a.m. V.				

264. PRINCE ALBERT. Lat. 33° 13′·2 S.; Long. 22° 3′·0 E. In field belonging to Mrs Haak, behind hotel.

Declination. D.

Date	G.M.T.	D (observed)	D	Observer	Instrument
1903 Jan 2	2 26 p.m.	28° 4′·8 W.			
	2 43 p.m.	28 5 ·7	28° 4′·3 W.	Beattie	31
1903 Jan. 3	4 58 a.m.	28 9 ·6			
	5 9 a.m.	28 8 ·6			

Dip. θ.

Date	G.M.T.	Needle	θ (observed)	θ	Observer	Instrument
1903 Jan. 2	9 0 a.m.	1	60° 19′·1 S.			
	9 0 a.m.	2	60 16 ·0	60° 20′·5 S.	Beattie	142
	9 0 a.m.	3	60 14 ·4			

265. PRINCE ALBERT ROAD. Lat. 32° 58′·7 S.; Long. 21° 42′·0 E. In a field adjoining hotel, on the same side of the railway as the hotel on the side of the hotel distant from Cape Town, and about 150 yards from hotel.

Declination. D.

Date		G.M.T.	D (observed)	D	Observer	Instrument
1900	Jan. 5	11 26 a.m.	28° 10′·9 W.			
		11 46 a.m.	28 9·9	27° 48′·4 W.	Beattie Morrison	31
		12 15 p.m	28 10′·2			
1902	Dec. 31	2 30 p.m.	27 55·2			
		2 42 p.m.	27 55·9			
		2 46 p.m.	27 54·9			
		4 20 p.m.	27 56·8	27 53·9	Beattie	31
		4 30 p.m.	27 57·6			
1903	Jan. 1	4 51 a.m.	27 58·2			
		5 4 a.m.	27 58·3			
				27 51·2 (mean adopted)		

Dip. θ.

Date		G.M.T.	Needle	θ (observed)	θ	Observer	Instrument
1899	July 14	10 55 a.m.	1	59° 42′·5 S.	60° 13′·7 S.	Beattie	9
		10 55 a.m.	2	59 40·8			
1900	Jan. 6	7 30 a.m.	1	59 43·9	60 12·0	Beattie	9
		7 30 a.m.	2	59 44·0			
1903	Jan. 1	9 0 a.m.	1	60 10·8			
		9 0 a.m.	2	60 14·2	60 15·4	Beattie	142
		8 59 a.m.	3	60 9·2			
					60 13·7 (mean adopted)		

Horizontal Intensity. H.

Date		G.M.T.	H (observed)	H	Observer	Instrument
1899	July 14	11 44 a.m. D.	H_{30} ·18312			
		12 54 p.m. V.	H_{40} ·18316	·17994		
		1 16 p.m. V.			Morrison	31
1899	July 15	8 0 a.m. V.	H_{30} ·18342	·18015		
		9 11 a.m. D.	H_{40} ·18332			
1900	Jan. 6	9 37 a.m. V.	H_{30} ·18306	·18020	Morrison	31
		10 26 a.m. D.	H_{40} ·18294			
				·18010 (mean adopted)		

266. QUEENSTOWN. Lat. 31° 54′·0 S.; Long. 26° 52′·0 E.

Declination. D.

Date		G.M.T.	D (observed)	D	Observer	Instrument
1902	Jan. 1	5 46 a.m.	26° 32′·7 W.			
		6 56 a.m.	26 32·4	26° 16′·0 W.	Beattie	31
		7 22 a.m.	26 33·1			
		2 2 p.m.	26 30·2			

Dip. θ.

Date		G.M.T.	Needle	θ (observed)	θ	Observer	Instrument
1901	Dec. 31	1 57 p.m.	2	60° 47′·6 S.			
		1 56 p.m.	1	60 51·1	61° 1′·6 S.	Beattie	142
		4 2 p.m.	2	60 47·9			
		4 2 p.m.	1	60 51·9			

Horizontal Intensity. H.

Date		G.M.T.	H (observed)	H	Observer	Instrument
1902	Jan. 1	8 15 a.m. V.	H_{30} ·18180	·18049	Beattie	31
		9 45 a.m. D.	H_{40} ·18188			

267. RANDFONTEIN. Lat. 26° 10′·7 S.; Long. 27° 42′·0 E. Left-hand side of railway, Johannesburg to Randfontein. Place reached by going from Johannesburg end of platform, 255 paces towards Johannesburg along railway, then 266 paces at right angles to railway.

Declination. D.

Date	G.M.T.	D (observed)	D	Observer	Instrument
1903 June 19	6 11 a.m.	22° 46′·3 W.	22° 46′·2 W.	Beattie Löwinger	73
	6 27 a.m.	22 47 ·4			
	6 40 a.m.	22 47 ·3			
	1 47 p.m.	22 45 ·1			
	1 58 p.m.	22 46 ·4			
	2 9 p.m.	22 46 ·3			

Dip. θ.

Date	G M.T.	Needle	θ (observed)	θ	Observer	Instrument
1903 June 19	8 46 a.m.	3	58° 19′·0 S.	58° 19′·2 S.	Beattie	142
	8 47 a.m.	4	58 19 ·6			
	8 50 a.m.	4_9	58 18 ·0			

Horizontal Intensity. H.

Date	G.M.T.	H (observed)	H	Observer	Instrument
1903 June 19	9 40 a.m. V.	H_{30} ·19150	·19151	Beattie	73
	12 11 p.m. D.	H_{40} ·19153			
	2 36 p.m. V.				

268. RATELDRAAI. Lat. 28° 45′·7 S.; Long. 21° 17′·9 E. North of house, east of dam, about 300 yards from the dam.

Declination. D.

Date	G.M.T.	D (observed)	D	Observer	Instrument
1904 Dec. 20	5 25 a.m.	24° 41′·5 W.	24° 53′·5 W.	Beattie Hough	73
	5 38 a.m.	24 41 ·5			
	6 18 a.m.	24 43 ·6			

Dip. θ.

Date	G.M.T.	Needle	θ (observed)	θ	Observer	Instrument
1904 Dec. 20	4 50 a.m.	3_9	57° 58′·4 S.	57° 45′·5 S.	Beattie	142
	4 50 a.m.	4_9	57 56 ·1			

Horizontal Intensity. H.

Date	G.M.T.	H (observed)	H	Observer	Instrument
1904 Dec. 20	6 48 a.m. V.	H_{30} ·19077	·19194	Beattie	73
	8 15 a.m. D.	H_{40} ·19076			

269. RATELDRIFT. Lat. 31° 31′·6 S.; Long. 20° 17′·6 E. Right-hand side of road, Brandwacht to Rateldrift, about one mile from farm on Brandwacht side, and on opposite side of river to farm.

Declination. D.

Date	G.M.T.	D (observed)	D	Observer	Instrument
1905 Jan. 27	6 18 a.m. 6 37 a.m.	27° 24′·1 W. 27 23 ·3	27° 32′·3 W.	Beattie Brown	73

Dip. θ.

Date	G.M.T.	Needle	θ (observed)	θ	Observer	Instrument
1905 Jan. 27	9 34 a.m. 9 34 a.m.	3_9 4_9	59° 1′·5 S. 59 4 ·9	58° 50′·5 S.	Beattie	142

Horizontal Intensity. H.

Date	G.M.T.	H (observed)	H	Observer	Instrument
1905 Jan. 27	7 41 a.m. V. 8 37 a.m. D.	H_{30} ·18312	·18439	Beattie	73

270. RICHMOND (NATAL). Lat. 29° 54′·0 S.; Long. 30° 20′·0 E. Left-hand side of road leaving village towards the north and past the railway station. Two-thirds of distance between road and wood. About 600 paces from the station.

Dip. θ.

Date	G.M.T.	Needle	θ (observed)	θ	Observer	Instrument
1903 Oct. 15	9 7 a.m. 9 8 a.m. 9 7 a.m.	1 4_9 4	61° 7′·3 S. 61 6 ·2 61 8 ·0	61° 5′·4 S.	Beattie	142

Horizontal Intensity. H.

Date	G.M.T.	H (observed)	H	Observer	Instrument
1903 Oct. 15	5 48 a.m. V. 6 30 a.m. D. 7 35 a.m. D. 8 5 a.m. V.	H_{30} ·18070 H_{40} ·18070	·18091	Beattie	73

271. Richmond Road. Lat. 31° 13′·0 S.; Long. 23° 38′·0 E. Left-hand side of railway from Cape Town to De Aar. 300 paces from railway starting from De Aar end of platform.

Declination. D.

Date	G.M.T.	D (observed)	D	Observer	Instrument
1902 July 18	12 46 p.m. 1 3 p.m. 1 13 p.m.	27° 20′·5 W. 27 19 ·9 27 20 ·5	27° 11′·5 W.	Beattie Morrison	31

Dip. θ.

Date	G.M.T.	Needle	θ (observed)	θ	Observer	Instrument
1902 July 18	8 23 a.m. 8 23 a.m. 8 22 a.m. 8 23 a.m.	1 2 3 4	59° 53′·1 S. 59 57 ·9 59 53 ·3 59 53 ·3	59° 54′·4 S.	Beattie	142

Horizontal Intensity. H.

Date	G.M.T.	H (observed)	H	Observer	Instrument
1902 July 18	10 5 a.m. V. 10 42 a.m. D. 11 11 a.m. V.	H_{30} ·18322 H_{40} ·18320	·18231	Morrison	31

272. Rietkuil Farm. (Zwartberg Store.) Lat. 30° 14′·4 S.; Long. 29° 22′·0 E. About 200 yards east of store.

Declination. D.

Date	G.M.T.	D (observed)	D	Observer	Instrument
1903 Nov. 18	8 18 a.m.	24° 10′·0 W.	24° 12′·2 W.	Beattie	73

Dip. θ.

Date	G.M.T.	Needle	θ (observed)	θ	Observer	Instrument
1903 Nov. 18	11 13 a.m. 11 12 a.m.	1 4_9	61° 8′·2 S. 61 7 ·6	61° 5′·3 S.	Beattie	142

Horizontal Intensity. H.

Date	G.M.T.	H (observed)	H	Observer	Instrument
1903 Nov. 18	12 4 p.m. V.	·17951	·17978	Beattie	73

273. Rietpoort. Lat. 31° 4′·4 S.; Long. 20° 55′·1 E. On left-hand side of road, Williston to Rietpoort. On side of farm distant from Williston, between two furrows leading from dam. Opposite pool more distant from house.

Declination. D.

Date	G.M.T.	D (observed)	D	Observer	Instrument
1905 Jan. 22	7 31 a.m. 7 40 a.m.	26° 44′·5 W. 26 45 ·7	26° 53′·5 W.	Beattie Brown	73

Dip. θ.

Date	G.M.T.	Needle	θ (observed)	θ	Observer	Instrument
1905 Jan. 22	9 36 a.m. 9 37 a.m.	3_9 4_9	59° 7′·6 S. 59 6 ·5	58° 54′·5 S.	Beattie	142

Horizontal Intensity. H.

Date	G.M.T.	H (observed)	H	Observer	Instrument
1905 Jan. 22	7 55 a.m. V. 8 43 a.m. D.	H_{30} ·18398 H_{40} ·18397	·18523	Beattie	73

274. Rietvlei. Lat. 24° 35′·0 S.; Long. 30° 40′·0 E. Instead of crossing the drift leading to Oliphant's Berg keep to the right-hand side of the river, continue on this path beyond the farmhouse. Half a mile beyond house at foot of hill where path leads to native kraal.

Declination. D.

Date	G.M.T.	D (observed)	D	Observer	Instrument
1903 Sept. 4	9 9 a.m.	20° 23′·3 W.	20° 24′·2 W.	Beattie	73

Dip. θ.

Date	G.M.T.	Needle	θ (observed)	θ	Observer	Instrument
1903 Sept. 4	11 28 a.m. 11 28 a.m. 11 28 a.m.	1 4 4_9	57° 32′·4 S. 57 29 ·8 57 32 ·2	57° 30′·2 S.	Beattie	142

Horizontal Intensity. H.

Date	G.M.T.	H (observed)	H	Observer	Instrument
1903 Sept. 4	6 46 a.m. V. 8 5 a.m. D. 8 40 a.m. D. 9 29 a.m. V.	H_{30} ·19779 H_{40} ·19783	·19793	Beattie	73

275. Rietvlei, C. C. Lat. 33° 32′·0 S.; Long. 22° 29′·0 E. In camp adjoining the hotel, and 100 yards east of the hotel.

Declination. D.

Date	G.M.T.	D (observed)	D	Observer	Instrument
1903 Jan. 7	4 34 a.m.	28° 3′·3 W.			
	4 46 a.m.	28 4 ·0	27° 58′·8 W.	Beattie	31
	6 54 a.m.	28 4 ·9			

Dip. θ.

Date	G.M.T.	Needle	θ (observed)	θ	Observer	Instrument
1903 Jan. 7	9 24 a.m.	3	60° 25′·4 S.	60° 29′·0 S.	Beattie	142
	9 24 a.m.	4	60 24 ·7			

276. Riversdale. Lat. 34° 5′·0 S.; Long. 21° 16′·0 E. On open grass square on south west of town in front of graveyard.

Declination. D.

Date	G.M.T.	D (observed)	D	Observer	Instrument
1903 Jan. 16	6 34 a.m.	28° 25′·2 W.			
	6 46 a.m.	28 25 ·5			
	6 58 a.m.	28 25 ·3	28° 17′·2 W.	Beattie Morrison	31
	3 36 p.m.	28 18 ·0			
	3 56 p.m.	28 18 ·9			
	4 17 p.m.	28 18 ·5			

Dip. θ.

Date	G.M.T.	Needle	θ (observed)	θ	Observer	Instrument
1903 Jan. 16	9 29 a.m.	1	60° 8′·9 S.			
	9 28 a.m.	3	60 5 ·6	60° 11′·0 S.	Beattie	142
	9 28 a.m.	4	60 7 ·3			

Horizontal Intensity. H.

Date	G.M.T.	H (observed)	H	Observer	Instrument
1903 Jan. 16	2 12 p.m. V.				
	2 51 p.m. D.	H_{30} ·18042 H_{40} ·18042	·17997	Morrison	31
	4 31 p.m. V.				

277. RIVIER PLAATS. Lat. 32° 8′·5 S.; Long. 20° 24′·0 E. Left-hand side of road coming from Middlepost to Sutherland. About 400 paces from house and on Sutherland side of it.

Declination. D.

Date	G.M.T.	D (observed)	D	Observer	Instrument
1905 Jan. 29	8 0 a.m.	27° 9′·0 W.	27° 17′·1 W.	Beattie Brown	73

Dip. θ.

Date	G.M.T.	Needle	θ (observed)	θ	Observer	Instrument
1905 Jan. 29	5 45 a.m.	4_9	58° 59′·3 S.			
	9 18 a.m.	4_9	59 0 ·4			
	9 16 a.m.	3_9	58 59 ·1	58° 49′·1 S.	Beattie	142
	9 14 a.m.	5	59 5 ·5			
	9 14 a.m.	6	59 4 ·4			

Horizontal Intensity. H.

Date	G.M.T.	H (observed)	H	Observer	Instrument
1905 Jan. 29	8 18 a.m. V.	·18329	·18456	Beattie	73

278. ROADSIDE. Lat. 30° 44′·3 S.; Long. 20° 25′·5 E. On road from Brandvlei to Tontelboschkalk, 3½ hours by cart from the latter. At a point where the road is nearest the river and between river and road.

Declination. D.

Date	G.M.T.	D (observed)	D	Observer	Instrument
1905 Jan. 25	7 28 a.m.	26° 43′·1 W.	26° 51′·5 W.	Beattie Brown	73

Dip. θ.

Date	G.M.T.	Needle	θ (observed)	θ	Observer	Instrument
1905 Jan. 25	9 25 a.m.	3_9	58° 30′·0 S.	58° 17′·5 S.	Beattie	142
	9 25 a.m.	4_9	58 30 ·3			

Horizontal Intensity. H.

Date	G.M.T.	H (observed)	H	Observer	Instrument
1905 Jan. 25	6 53 a.m. V.	H_{30} ·18608	·18734	Beattie	73
	8 40 a.m. D.	H_{40} ·18607			

279. ROBERTSON. Lat. 33° 48′·8 S.; Long. 19° 53′·0 E. In municipal field to S.S.W. of dorp. Opposite side of river to village.

Declination. D.

Date	G.M.T.	D (observed)	D	Observer	Instrument
1901 Feb. 7	8 43 a.m.	28° 40′·8 W.			
	8 55 a.m.	28 41 ·0			
	9 5 a.m.	28 40 ·5			
	9 17 a.m.	28 39 ·6	28° 24′·2 W.	Beattie Morrison	31
1901 Feb. 8	2 27 p.m.	28 32 ·8			
	2 35 p.m.	28 32 ·8			
	2 44 p.m.	28 33 ·7			
	4 12 p.m.	28 36 ·1			

Dip. θ.

Date	G.M.T.	Needle	θ (observed)	θ	Observer	Instrument
1901 Feb. 7	10 32 a.m.	1	59° 17′·7 S.	59° 40′·7 S.	Beattie	142
	10 36 a.m.	2	59 20 ·3			

Horizontal Intensity. H.

Date	G.M.T.	H (observed)	H	Observer	Instrument
1901 Feb. 7	9 51 a.m. V.				
	10 29 a.m. D.	H_{30} ·18357 H_{40} ·18354	·18145	Morrison	31
	11 1 a.m. V.				

280. RODEKRANTZ. Lat. 24° 38′·0 S.; Long. 30° 35′·0 E. Left-hand side of road coming from Ohrigstad and going north. About a mile and a half beyond point where road turns off for Kasper's Nek.

Declination. D.

Date	G.M.T.	D (observed)	D	Observer	Instrument
1903 Sept. 3	12 39 p.m.	20° 50′·1 W.	20° 52′·9 W.	Beattie	73

Dip. θ.

Date	G.M.T.	Needle	θ (observed)	θ	Observer	Instrument
1903 Sept. 3	10 7 a.m.	1	57° 45′·8 S.			
	10 7 a.m.	4	57 44 ·0	57° 43′·4 S.	Beattie	142
	10 7 a.m.	4_9	57 44 ·4			

Horizontal Intensity. H.

Date	G.M.T.	H (observed)	H	Observer	Instrument
1903 Sept. 3	11 15 a.m. V.				
	11 45 a.m. D.	H_{30} ·19722 H_{40} ·19722	·19734	Beattie	73
	12 15 p.m. V.				

281. Roodepoort. Lat. 30° 13′·0 S.; Long. 23° 22′·0 E.

Declination. D.

Date	G.M.T.	D (observed)	D	Observer	Instrument
1904 Dec. 26	6 4 a.m. 6 16 a.m.	25° 2′·4 W. 25 2 ·8	25° 14′·1 W.	Beattie Hough	73

Dip. θ.

Date	G.M.T.	Needle	θ (observed)	θ	Observer	Instrument
1904 Dec. 26	5 26 a.m. 5 26 a.m.	3_9 4_9	58° 11′·3 S. 58 9 ·4	57° 58′·3 S.	Beattie	142

Horizontal Intensity. H.

Date	G.M.T.	H (observed)	H	Observer	Instrument
1904 Dec. 26	7 22 a.m. V. 7 34 a.m. D.	H_{30} ·18413 H_{40} ·18417	·18534	Beattie	73

282. Rooidam. Lat. 29° 50′·7 S.; Long. 23° 11′·8 E.

Declination. D.

Date	G.M.T.	D (observed)	D	Observer	Instrument
1904 Dec. 25	5 42 a.m. 5 52 a.m.	24° 47′·3 W. 24 48 ·8	24° 59′·6 W.	Beattie Hough	73

Dip. θ.

Date	G.M.T.	Needle	θ (observed)	θ	Observer	Instrument
1904 Dec. 25	5 13 a.m. 5 13 a.m.	3_9 4_9	58° 52′·4 S. 58 52 ·6	58° 40′·5 S.	Beattie	142

Horizontal Intensity. H.

Date	G.M.T.	H (observed)	H	Observer	Instrument
1904 Dec. 25	6 54 a.m. V. 7 30 a.m. D. 8 0 a.m. V.	H_{30} ·18541 H_{40} ·18536	·18657	Beattie	73

283. Revué. Lat. 18° 59′·0 S.; Long. 33° 3′·0 E. Left-hand side of railway, Umtali to Beira. 250 paces at right angles to railway, starting from mid-point of the loop.

Declination. D.

Date	G.M.T.	D (observed)	D	Observer	Instrument
1903 April 19	8 25 a.m. 8 42 a.m.	16° 7′·9 W. 16 7·9	16° 5′·8 W.	Beattie	73

Horizontal Intensity. H.

Date	G.M.T.	H (observed)	H	Observer	Instrument
1903 April 19	9 12 a.m. V. 10 23 a.m. D. 11 15 a.m. V.	H_{30} ·22009 H_{40} ·22005 H_{25} ·22011 H_{35} ·22001	·21995	Beattie	73

284. Rooipüts. Lat. 29° 17′·4 S.; Long. 21° 38′·6 E. North north east of well, about 200 paces from it. The well is that one to left of main road, Kenhardt to Prieska, with the better water.

Declination. D.

Date	G.M.T.	D (observed)	D	Observer	Instrument
1904 Dec. 22	5 41 a.m. 5 56 a.m.	26° 48′·9 W. 26 48·4	27° 0′·2 W.	Beattie Hough	73

Dip. θ.

Date	G.M.T.	Needle	θ (observed)	θ	Observer	Instrument
1904 Dec. 22	4 58 a.m. 4 58 a.m.	3_9 4_9	58° 29′·8 S. 58 26·8	58° 16′·4 S.	Beattie	142

Horizontal Intensity. H.

Date	G.M.T.	H (observed)	H	Observer	Instrument
1904 Dec. 22	7 12 a.m. V. 7 32 a.m. D. 8 49 a.m. V.	H_{30} ·18428 H_{40} ·18425	·18545	Beattie	73

285. ROOIVAL. Lat. 32° 12′·0 S.; Long. 21° 58′·3 E. At a place 7½ hours by cart from Beaufort West. Right-hand side of road opposite sheep kraal. Farm-house 300 paces nearer Fraserburg.

Declination. D.

Date	G.M.T.	D (observed)	D	Observer	Instrument
1905 Jan. 16	9 17 a.m. 9 23 a.m.	27° 10′·6 W. 27 10 ·0	27° 19′·0 W.	Beattie Brown	73

Dip. θ.

Date	G.M.T.	Needle	θ (observed)	θ	Observer	Instrument
1905 Jan. 16	5 5 a.m. 5 5 a.m.	3_9 4_9	59° 38′·2 S. 59 40 ·3	59° 27′·0 S.	Beattie	142

Horizontal Intensity. H.

Date	G.M.T.	H (observed)	H	Observer	Instrument
1905 Jan. 16	8 43 a.m. V. 9 20 a.m. D.	H_{40} ·18151	·18274	Beattie	73

286. ROSMEAD JUNCTION. Lat. 31° 39′·6 S.; Long. 25° 5′·0 E. On football field.

Declination. D.

Date	G.M.T.	D (observed)	D	Observer	Instrument
1900 July 2	11 9 a.m. 11 22 a.m. 1 28 p.m. 1 41 p.m.	26° 45′·0 W. 26 45 ·4 26 43 ·9 26 43 ·9	26° 14′·3 W.	Beattie Morrison	31
1902 July 17	1 54 p.m. 2 4 p.m. 2 7 p.m. 2 18 p.m.	26 22 ·0 26 22 ·5 26 22 ·1 26 22 ·8	26 13 ·1	Beattie Morrison	31
			26 13 ·7 (mean adopted)		

Dip. θ.

Date	G.M.T.	Needle	θ (observed)	θ	Observer	Instrument
1900 July 2	9 0 a.m. 9 0 a.m.	1 2	59° 57′·5 S. 59 56 ·9	60° 21′·2 S.	Beattie	9
1904 July 2	12 48 p.m. 12 51 p.m.	1 2	60 30 ·6 60 27 ·1	60 20 ·8	Morrison	9
				60 21 ·0 (mean adopted)		

Horizontal Intensity. H.

Date	G.M.T.	H (observed)	H	Observer	Instrument
1900 July 3	9 6 a.m. V. 9 54 a.m. D. 10 42 a.m. V.	H_{30} ·18428 H_{40} ·18425	·18175	Morrison	31
1902 July 17	3 46 p.m. V.	·18254	·18164	Morrison	31
1904 July 2	8 58 a.m. V. 9 55 a.m. D. 10 22 a.m. V.	H_{30} ·18084 H_{40} ·18084	·18172	Morrison	31
			·18170 (mean adopted)		

287. ROUXVILLE. Lat. 30° 31′·6 S.; Long. 26° 47′·3 E. Place of observation between Aliwal North and Rouxville. On farm Vollbank.

Declination. D.

Date	G.M.T.	D (observed)	D	Observer	Instrument
1904 Jan. 7	2 46 p.m. 2 57 p.m.	25° 0′·3 W. 25 0 ·6	25° 6′·8 W.	Beattie	73

Dip. θ.

Date	G.M.T.	Needle	θ (observed)	θ	Observer	Instrument
1904 Jan. 7	10 57 a.m. 10 58 a.m.	1 4_9	60° 15′·0 S. 60 17 ·0	60° 12′·5 S.	Beattie	142

Horizontal Intensity. H.

Date	G.M.T.	H (observed)	H	Observer	Instrument
1904 Jan. 7	3 35 p.m. V.	·18243	·18286	Beattie	73

288. RUSAPI. Lat. 18° 32′·0 S.; Long. 32° 8′·0 E. Right-hand side of railway, Salisbury to Umtali. 165 paces at right angles to line, starting at the Salisbury end of the siding.

Declination. D.

Date	G.M.T.	D (observed)	D	Observer	Instrument
1903 April 25	2 55 p.m. 3 5 p.m.	16° 4′·5 W. 16 5 ·9	16° 3′·4 W.	Beattie	73

Dip. θ.

Date	G.M.T.	Needle	θ (observed)	θ	Observer	Instrument
1903 April 25	10 30 a.m. 10 31 a.m. 10 31 a.m.	3 4 1	52° 31′·5 S. 52 35 ·6 52 34 ·2	52° 35′·2 S.	Beattie	142

Horizontal Intensity. H.

Date	G.M.T.	H (observed)	H	Observer	Instrument
1903 April 25	12 35 p.m. V. 1 5 p.m. D. 1 47 p.m. V.	H_{30} ·22192 H_{40} ·22192 H_{25} ·22192 H_{35} ·22187	·22179	Beattie	73

289. RUSTPLAATS. Lat. 24° 50′·6 S.; Long. 30° 38′·0 E. Left-hand side of road coming from Krugerspost. About half a mile on Ohrigstad side of Rustplaats post office.

Declination. *D.*

Date	G.M.T.	D (observed)	D	Observer	Instrument
1903 Sept. 2	12 37 p.m. 12 55 p.m.	20° 49′·6 W. 20 49 ·9	20° 52′·7 W.	Beattie	73

Dip. *θ.*

Date	G.M.T.	Needle	θ (observed)	θ	Observer	Instrument
1903 Sept. 2	10 5 a.m. 10 6 a.m.	1 4_9	57° 51′·6 S. 57 51 ·0	57° 50′·0 S.	Beattie	142

Horizontal Intensity. *H.*

Date	G.M.T.	H (observed)	H	Observer	Instrument
1903 Sept. 2	11 12 a.m. V. 11 43 a.m. D. 12 20 p.m. V.	H_{30} ·19628 H_{40} ·19626	·19639	Beattie	73

290. RUYTERBOSCH. Lat. 33° 55′·7 S.; Long. 22° 2′·0 E. On public outspan about three and a half hours by cart from Mossel Bay. Half-way between hotel and boarding house, right-hand side of road coming from Mossel Bay.

Declination. *D.*

Date	G.M.T.	D (observed)	D	Observer	Instrument
1903 Jan. 22	8 22 a.m. 8 36 a.m.	28° 7′·0 W. 28 6 ·9	28° 0′·3 W.	Beattie	31

Dip. *θ.*

Date	G.M.T.	Needle	θ (observed)	θ	Observer	Instrument
1903 Jan. 22	11 55 a.m. 11 56 a.m.	3 4	60° 14′·6 S. 60 16 ·9	60° 19′·3 S.	Beattie	142

Horizontal Intensity. *H.*

Date	G.M.T.	H (observed)	H	Observer	Instrument
1903 Jan. 22	9 6 a.m. V. 9 57 a.m. D.	H_{30} ·18049 H_{40} ·18048	·18004	Beattie	31

291. SABIE RIVER. Lat. 25° 6′·1 S.; Long. 30° 45′·0 E. On Nelspruit side of spruit near police camp. Left-hand side of road and half-way up the hill.

Declination. D.

Date	G.M.T.	D (observed)	D	Observer	Instrument
1903 Sept. 7	1 39 p.m.	21° 0′·9 W.			
	1 54 p.m.	21 2·5	21° 4′·0 W.	Beattie	73
	2 14 p.m.	21 0·8			

Dip. θ.

Date	G.M.T.	Needle	θ (observed)	θ	Observer	Instrument
1903 Sept. 7	10 58 a.m.	1	58° 17′·8 S.	58° 16′·6 S.	Beattie	142
	10 59 a.m.	4_9	58 18·0			

Horizontal Intensity. H.

Date	G.M.T.	H (observed)	H	Observer	Instrument
1903 Sept. 7	11 48 a.m. V.				
	1 15 p.m. D.	H_{30} ·19681	·19694	Beattie	73
	2 31 p.m. V.	H_{40} ·19683			

292. SALISBURY. Lat. 17° 50′·3 S.; Long. 31° 3′·0 E. Right-hand side of railway, Salisbury to Umtali. 330 yards from railway, starting from a point on the railway opposite to the cold storage building.

Declination. D.

Date	G.M.T.	D (observed)	D	Observer	Instrument
1903 April 12	2 38 p.m.	16° 9′·4 W.			
	2 50 p.m.	16 8·3			
1903 April 13	5 27 a.m.	16 10·5	16° 7′·5 W.	Beattie	73
	5 44 a.m.	16 10·6			

Dip. θ.

Date	G.M.T.	Needle	θ (observed)	θ	Observer	Instrument
1903 April 13	8 26 a.m.	3	51° 41′·3 S.			
	8 27 a.m.	4	51 40·4	51° 42′·7 S.	Beattie	142
	8 27 a.m.	1	51 41·5			

Horizontal Intensity. H.

Date	G.M.T.	H (observed)	H	Observer	Instrument
1903 April 12	11 2 a.m. V.	H_{30} ·22152			
	12 2 p.m. D.	H_{40} ·22149	·22132	Beattie	73
	1 56 p.m. V.	H_{25} ·22149			
		H_{35} ·22152			

293. SAXONY. Lat. 28° 44′·1 S.; Long. 27° 44′·4 E. On road from Winburg to Senekal. In hollow on right-hand side of road, and about 100 yards from it just before coming to spruit.

Declination. D.

Date	G.M.T.	D (observed)	D	Observer	Instrument
1904 Jan. 23	8 13 a.m.	24° 10′·8 W.	24° 11′·9 W.	Beattie	73
	8 23 a.m.	24 11 ·3			
	1 24 p.m.	24 1 ·3			
	1 31 p.m.	24 2 ·2			

Dip. θ.

Date	G.M.T.	Needle	θ (observed)	θ	Observer	Instrument
1904 Jan. 23	11 50 a.m.	1	59° 49′·8 S.	59° 47′·1 S.	Beattie	142
	11 50 a.m.	4_9	59 52 ·4			

Horizontal Intensity. H.

Date	G.M.T.	H (observed)	H	Observer	Instrument
1904 Jan. 23	8 52 a.m. V.		·18562	Beattie	73
	9 30 a.m. D.	H_{30} ·18519			
	9 56 a.m. V.	H_{40} ·18519			
	12 38 p.m. D.				

294. SCHIETFONTEIN. Lat. 32° 41′·7 S.; Long. 20° 46′·6 E. Left-hand side of road, Sutherland to Matjesfontein, about 400 paces from house, and on Matjesfontein side of it.

Declination. D.

Date	G.M.T.	D (observed)	D	Observer	Instrument
1905 Feb. 1	6 9 a.m.	27° 44′·3 W.	27° 52′·5 W.	Beattie	73
	6 18 a.m.	27 43 ·7		Brown	

Dip. θ.

Date	G.M.T.	Needle	θ (observed)	θ	Observer	Instrument
1905 Feb. 1	9 36 a.m.	4_9	59° 43′·5 S.	59° 32′·0 S.	Beattie	142
	9 38 a.m.	4	59 45 ·3			
	9 35 a.m.	5	59 45 ·2			

Horizontal Intensity. H.

Date	G.M.T.	H (observed)	H	Observer	Instrument
1905 Feb. 1	7 28 a.m. V.	H_{30} ·18188	·18314	Beattie	73
	8 15 a.m. D.	H_{40} ·18187			

295. Schikhoek. Lat. 27° 24′·6 S.; Long. 30° 34′·0 E. Right-hand side of road, Utrecht to Piet Retief, on Utrecht side of river.

Declination. D.

Date	G.M.T.	D (observed)	D	Observer	Instrument
1903 Aug. 14	7 53 a.m. 8 8 a.m.	23° 15′·2 W. 23 15 ·3	23° 16′·0 W.	Beattie	73

Dip. θ.

Date	G.M.T.	Needle	θ (observed)	θ	Observer	Instrument
1903 Aug. 14	11 21 a.m. 11 26 a.m. 11 26 a.m.	1 4 4_9	59° 51′·2 S. 59 48 ·1 59 48 ·8	59° 48′·5 S.	Beattie	142

Horizontal Intensity. H.

Date	G.M.T.	H (observed)	H	Observer	Instrument
1903 Aug. 14	8 29 a.m. V. 9 18 a.m. D. 9 51 a.m. V.	H_{30} ·18651 H_{40} ·18648	·18656	Beattie	73

296. Schoemanshoek. Lat. 25° 27′·9 S.; Long. 30° 21′·0 E. On road from Machadodorp to Lydenburg beyond the undulating part at the foot of the last hill before the level stretch. Tent on right-hand side of road going to Lydenburg just in front of outspan.

Declination. D.

Date	G.M.T.	D (observed)	D	Observer	Instrument
1903 Aug. 30	12 54 p.m. 1 4 p.m.	21° 6′·3 W. 21 6 ·0	21° 8′·9 W.	Beattie	73

Dip. θ.

Date	G.M.T.	Needle	θ (observed)	θ	Observer	Instrument
1903 Aug. 30	11 26 a.m. 11 27 a.m.	1 4_9	58° 30′·6 S. 58 30 ·9	58° 29′·6 S.	Beattie	142

Horizontal Intensity. H.

Date	G.M.T.	H (observed)	H	Observer	Instrument
1903 Aug. 30	12 31 p.m. V.	·19281	·19293	Beattie	73

297. SCHUILPLAATS. Lat. 26° 54′·2 S.; Long. 29° 47′·0 E. 9 miles from Amersfoort, right-hand side of road coming from Vaal River Drift.

Declination. D.

Date	G.M.T.	D (observed)	D	Observer	Instrument
1903 Aug. 8	2 17 p.m. 2 28 p.m.	22° 14′·7 W. 22 15 ·3	22° 16′·1 W.	Beattie	73

Dip. θ.

Date	G.M.T.	Needle	θ (observed)	θ	Observer	Instrument
1903 Aug. 8	10 57 a.m. 10 57 a.m.	1 4	59° 9′·2 S. 59 7 ·1	59° 7′·5 S.	Beattie	142

Horizontal Intensity. H.

Date	G.M.T.	H (observed)	H	Observer	Instrument
1903 Aug. 8	11 55 a.m. V. 1 4 p.m. D.	H_{30} ·18918 H_{40} ·18913	·18922	Beattie	73

298. SECOCOENI'S STAD. Lat. 24° 28′·3 S.; Long. 29° 52′·0 W. Two hours by cart beyond Oliphant's River just in front of drift, and on left-hand side of road.

Declination. D.

Date	G.M.T.	D (observed)	D	Observer	Instrument
1903 July 26	7 30 a.m. 7 43 a.m. 7 56 a.m.	19° 16′·4 W. 19 17 ·4 19 18 ·3	19° 17′·9 W.	Beattie Löwinger	73

Dip. θ.

Date	G.M.T.	Needle	θ (observed)	θ	Observer	Instrument
1903 July 26	10 0 a.m. 9 59 a.m.	1 4	56° 27′·8 S. 56 27 ·7	56° 27′·3 S.	Beattie	142

Horizontal Intensity. H.

Date	G.M.T.	H (observed)	H	Observer	Instrument
1903 July 26	8 10 a.m. V. 9 2 a.m. D.	H_{30} ·20071 H_{40} ·20068	·20070	Beattie	73

299. SERULI. Lat. 21° 55′·7 S.; Long. 27° 19′·0 E. Left-hand side of railway, Mafeking to Bulawayo, on Bulawayo side of tank. 200 paces at right angles to railway, starting from a point 80 paces from tank.

Declination. D.

Date	G.M.T.	D (observed)	D	Observer	Instrument
1903 March 29	8 20 a.m. 9 10 a.m.	20° 30′·7 W. 20 28 ·7	20° 26′·2 W.	Beattie	73

Dip. θ.

Date	G.M.T.	Needle	θ (observed)	θ	Observer	Instrument
1903 March 29	2 29 p.m. 2 28 p.m.	4 1	54° 54′·9 S. 54 53 ·0	54° 55′·7 S.	Beattie	142

Horizontal Intensity. H.

Date	G.M.T.	H (observed)	H	Observer	Instrument
1903 March 29	9 43 a.m. V. 10 52 a.m. D. 11 38 a.m. V.	H_{30} ·20676 H_{40} ·20676 H_{35} ·20675	·20658	Beattie	73

300. SHANGANI. Lat. 19° 45′·8 S.; Long. 29° 24′·0 E. On right-hand side of railway, Bulawayo to Gwelo. 180 paces at right angles to railway, starting from Gwelo end of siding.

Declination. D.

Date	G.M.T.	D (observed)	D	Observer	Instrument
1903 April 5	7 12 a.m. 7 24 a.m. 2 5 p.m. 3 26 p.m. 3 37 p.m.	18° 41′·3 W. 18 41 ·5 18 39 ·9 18 39 ·5 18 39 ·0	18° 37′·9 W.	Beattie	73

Dip. θ.

Date	G.M.T.	Needle	θ (observed)	θ	Observer	Instrument
1903 April 5	12 11 p.m. 12 12 p.m. 12 12 p.m.	3 4 1	54° 38′·0 S. 54 40 ·5 54 39 ·8	54° 41′·1 S.	Beattie	142

Horizontal Intensity. H.

Date	G.M.T.	H (observed)	H	Observer	Instrument
1903 April 5	6 18 a.m. V. 9 13 a.m. D. 10 25 a.m. V.	H_{30} ·22447 H_{40} ·22449 H_{25} ·22451 H_{35} ·22449	·22431	Beattie	73

301. SHASHI. Lat. 21° 23′·2 S.; Long. 27° 27′·0 E. Left-hand side of railway, Mafeking to Bulawayo, on side of tank nearer Bulawayo. 320 paces at right angles to railway from a point 330 paces from tank.

Declination. D.

Date	G.M.T.	D (observed)	D	Observer	Instrument
1903 March 30	2 50 p.m.	19° 31′·6 W.			
	3 17 p.m.	19 30 ·7	19° 28′·6 W.	Beattie	73
	3 33 p.m.	19 31 ·5			

Dip. θ.

Date	G.M.T.	Needle	θ (observed)	θ	Observer	Instrument
1903 March 30	8 25 a.m.	3	56° 57′·1 S.			
	8 25 a.m.	4	56 59 ·9	57° 0′·4 S.	Beattie	142
	8 24 a.m.	1	56 59 ·2			

Horizontal Intensity. H.

Date	G.M.T.	H (observed)	H	Observer	Instrument
1903 March 30	9 27 a.m. V.	H_{30} ·21212			
	10 53 a.m. D.	H_{40} ·21213	·21195	Beattie	31
	11 36 a.m. V.	H_{25} ·21215			

302. SHELA RIVER. Lat. 26° 51′·0 S.; Long. 30° 43′·0 E. On road from Piet Retief to Amsterdam passing through Krom River farm, just over first spruit after passing Shela River. Left-hand side of road.

Declination. D.

Date	G.M.T.	D (observed)	D	Observer	Instrument
1903 Aug. 23	2 25 p.m.	21° 42′·5 W.	21° 43′·8 W.	Beattie	73
	2 38 p.m.	21 42 ·4			

Dip. θ.

Date	G.M.T.	Needle	θ (observed)	θ	Observer	Instrument
1903 Aug. 23	11 35 a.m.	1	59° 12′·2 S.			
	11 35 a.m.	4	59 9 ·0	59° 10′·1 S.	Beattie	142
	11 35 a.m.	4_9	59 12 ·4			

Horizontal Intensity. H.

Date	G.M.T.	H (observed)	H	Observer	Instrument
1903 Aug. 23	12 28 p.m. V.	H_{30} ·18888			
	1 1 p.m. D.	H_{40} ·18892	·18896	Beattie	73
	1 31 p.m. V.				

303. SHOSHONG ROAD. Lat. 23° 34′·8 S.; Long. 26° 34′·0 E. Left-hand side of railway, Mafeking to Bulawayo. 10 paces from blind end of siding and towards Bulawayo. 200 paces perpendicular to railway from that point.

Declination. D.

Date	G.M.T.	D (observed)	D	Observer	Instrument
1903 March 25	8 31 a.m.	22° 0′·2 W.			
	8 47 a.m.	21 59 ·4	21° 54′·3 W.	Beattie	73
	1 46 p.m.	21 53 ·8			
	1 58 p.m.	21 54 ·3			

Dip. θ.

Date	G.M.T.	Needle	θ (observed)	θ	Observer	Instrument
1903 March 26	6 19 p.m.	3	56° 36′·0 S.			
	6 21 p.m.	4	56 39 ·6	56° 40′·0 S.	Beattie	142
	6 22 p.m.	1	56 39 ·2			

Horizontal Intensity. H.

Date	G.M.T.	H (observed)	H	Observer	Instrument
1903 March 25	9 12 a.m. V.	H_{30} ·19764			
	10 25 a.m. D.	H_{35} ·19766			
	11 32 a.m. V.	H_{25} ·19765	·19750	Beattie	73
1903 March 26	10 15 a.m. V.	H_{30} ·19774			
	11 4 a.m. D.	H_{35} ·19780			
	11 55 a.m. V.	H_{25} ·19778			

304. SIGNAL HILL. Lat. 33° 55′·0 S.; Long. 18° 24′·3 E. Near signal station and to the west of it.

Declination. D.

Date	G.M.T.	D (observed)	D	Observer	Instrument
1901 Nov. 28	4 0 p.m.	28° 55′·2 W.			
	4 9 p.m.	28 56 ·1			
	4 19 p.m.	28 55 ·5	28° 50′·3 W.	Beattie Morrison	31
1901 Nov. 29	6 47 a.m.	28 59 ·8			
	6 53 a.m.	28 59 ·8			

Dip. θ.

Date	G.M.T.	Needle	θ (observed)	θ	Observer	Instrument
1901 Nov. 28	6 39 a.m.	2	58° 53′·0 S.			
	6 37 a.m.	1	58 50 ·3			
1901 Nov. 29	10 9 a.m.	2	58 56 ·4	59° 7′·4 S.	Beattie	142
	10 10 a.m.	1	58 54 ·6			
1901 Nov. 30	5 38 a.m.	2	58 53 ·4			
	5 33 a.m.	1	58 51 ·6			

Horizontal Intensity. H.

Date	G.M.T.	H (observed)	H	Observer	Instrument
1901 Nov. 28	1 29 p.m. V.	H_{30} ·18417			
	2 20 p.m. D.	H_{40} ·18418			
	3 3 p.m. V.				
1901 Nov. 30	8 46 a.m. V.	H_{30} ·18440			
	9 30 a.m. D.	H_{40} ·18436	·18271	Morrison	31
	10 20 a.m. V.				
	1 9 p.m. V.				
	1 57 p.m. D.	H_{30} ·18414			
	2 26 p.m. V.	H_{40} ·18418			

305. Rifle Range, Simonstown. Lat. 34° 12′ S.; Long. 18° 26′ E.

Declination. D.

Date		G.M.T.	D (observed)	D	Observer	Instrument
1901	Oct. 11	6 32 a.m.	28° 48′·5 W.			
		7 47 a.m.	28 52 ·2			
		10 10 a.m.	28 47 ·9	28° 42′·1 W.	Beattie Löwinger	31
		3 40 p.m.	28 48 ·4			
1901	Oct. 12	8 56 a.m.	28 51 ·6			
		3 10 p.m.	28 51 ·1			
1901	Oct. 11	10 57 a.m.	28 46 ·4			
		12 0 noon	28 47 ·0	28 41 ·3	Beattie Löwinger	25 (of 'Discovery' Expedition)
1901	Oct. 12	8 12 a.m.	28 53 ·5			
		3 58 p.m.	28 48 ·1			
1901	Oct. 11	7 21 a.m.	28 51 ·2			
		9 27 a.m.	28 51 ·6	28° 42′·4 W.	Beattie Löwinger	36 (of 'Discovery' Expedition)
		1 30 p.m.	28 48 ·7			
		3 12 p.m.	28 48 ·4			
				28 41 ·9 (mean adopted)		

Dip. θ.

Date		G.M.T.	Needle	θ (observed)	θ	Observer	Instrument
1901	Oct. 6	12 48 p.m.	1	59° 1′·2 S.		Beattie	
		12 46 p.m.	2	58 59 ·1			
		1 11 p.m.	1	58 59 ·4		Morrison	27 (of 'Discovery' Expedition)
		1 11 p.m.	2	58 55 ·9			
1901	Oct. 7	12 39 p.m.	1	58 58 ·9		Barne	
		12 39 p.m.	2	58 59 ·8			
1901	Oct. 10	1 34 p.m.	1	59 3 ·3		Beattie	
		1 31 p.m.	2	58 58 ·5			
1901	Oct. 6	10 0 a.m.	2	58 57 ·1		Armitage	
1901	Oct. 7	12 4 p.m.	2	58 55 ·2		Armitage	26 (of 'Discovery' Expedition)
		12 27 p.m.	1_{27}	58 54 ·0	59° 15′·8 S.		
1901	Oct. 9	11 30 a.m.	2	59 2 ·0		Morrison	
		11 44 a.m.	1_{27}	59 2 ·2			
1901	Oct. 10	1 9 p.m.	2	58 57 ·3		Armitage	
1901	Oct. 6	1 53 p.m.	2	59 2 ·4		Beattie	
		1 54 p.m.	1	58 58 ·7			
		1 22 p.m.	2	59 2 ·6		Morrison	
		1 15 p.m.	1	59 0 ·8			
1901	Oct. 7	12 5 p.m.	2	59 1 ·4		Beattie	142
		12 5 p.m.	1	58 57 ·5			
1901	Oct. 9	11 4 a.m.	2	59 4 ·1		Morrison	
		11 4 a.m.	1	59 1 ·0			
1901	Oct. 10	1 28 p.m.	2	59 3 ·9		Beattie	
		1 26 p.m.	1	58 59 ·6			

Horizontal Intensity. H.

Date	G.M.T.	H (observed)	H	Observer	Instrument
1901 Oct. 7	10 6 a.m. V. 11 15 a.m. D. 2 57 p.m. V.	H_{30} ·18379 H_{40} ·18377	·18210	Morrison	25 (of 'Discovery' Expedition)
1901 Oct. 8	7 57 a.m. V. 8 40 a.m. D. 4 18 p.m. V.	H_{30} ·18392 H_{40} ·18394	·18225	Morrison	25 (of 'Discovery' Expedition)
	12 36 p.m. V. 1 31 p.m. D. 2 7 p.m. V.	H_{30} ·18359 H_{40} ·18364	·18194	Morrison	31
1901 Oct. 12	9 46 a.m. V. 12 18 p.m. D. 2 46 p.m. V.	H_{30} ·18389 H_{40} ·18387	·18220	Beattie	31
1901 Oct. 9	2 15 p.m. V. 3 2 p.m. D.	H_{30} ·18391 H_{40} ·18394	·18225	Beattie	
1901 Oct. 10	1 15 p.m. V. 2 0 p.m. D.	H_{30} ·18355 H_{40} ·18352	·18186	Beattie	
1901 Oct. 12	11 24 a.m. V. 12 44 p.m. D. 1 7 p.m. V.	H_{30} ·18379 H_{40} ·18381	·18212	Beattie	25 (of 'Discovery' Expedition)
	11 6 a.m. V. 12 45 p.m. D. 1 51 p.m. V.	H_{30} ·18371 H_{40} ·18367	·18201	Beattie	
			·18209	(mean adopted)	

305 A. GLENCAIRN, SIMONSTOWN. Lat. 34° 10′·8 S.; Long. 18° 26′·0 E. On raised patch in middle of river bed.

Declination. D.

Date	G.M.T.	D (observed)	D	Observer	Instrument
1901 Jan. 5	3 33 p.m.	28° 44′·6 W.			
	3 45 p.m.	28 44 ·0			
	3 58 p.m.	28 42 ·4	28° 37′·2 W.	Beattie Morrison	31
1901 Jan. 7	7 3 a.m.	28 53 ·7			
	3 47 p.m.	28 45 ·2			
	4 1 p.m.	28 45 ·1			

Dip. θ.

Date	G.M.T.	Needle	θ (observed)	θ	Observer	Instrument
1901 Jan. 5	2 4 p.m. 2 4 p.m.	1 2	58° 48′·0 S. 58 50 ·3	59° 11′·7 S.	Beattie	142

Horizontal Intensity. H.

Date	G.M.T.	H (observed)	H	Observer	Instrument
1901 Jan. 5	9 51 a.m. V. 10 40 a.m. D. 11 20 a.m. V.	H_{30} ·18493 H_{40} ·18492	·18252	Morrison	31

306. Sir Lowry's Pass. Lat. 34° 7′·3 S.; Long. 18° 55′·0 E. Field south of railway station, and about 400 yards distant.

Declination. D.

Date		G.M.T.	D (observed)	D	Observer	Instrument
1901	Jan. 10	6 9 a.m.	29° 10′·5 W.	28° 58′·0 W.	Beattie Morrison	31
		6 24 a.m.	29 10·1			
		6 38 a.m.	29 9·5			
		3 58 p.m.	29 7·5			
		4 12 p.m.	29 7·8			
		4 22 p.m.	29 7·8			

Dip. θ.

Date		G.M.T.	Needle	θ (observed)	θ	Observer	Instrument
1898	May 14	11 30 a.m.	1	58° 40′·0 S.	59° 24′·5 S.	Beattie	9
		11 30 a.m.	2	58 37·4			
1898	May 15	9 30 a.m.	1	58 40·2			
		10 30 a.m.	2	58 36·4			
1901	Jan. 11	10 3 a.m.	1	59 1·5	59 25·4	Beattie	142
		10 6 a.m.	2	59 4·7			
					59 25·0 (mean adopted)		

Horizontal Intensity. H.

Date		G.M.T.	H (observed)	H	Observer	Instrument
1898	May 15	3 12 p.m. V.	H_{30} ·18572	·18086	Morrison	31
		3 0 p.m. D.	H_{40} ·18584			
1901	Jan. 10	1 36 p.m. V.				
		2 15 p.m. D.	H_{30} ·18370	·18130	Morrison	31
		3 16 p.m. V.	H_{40} ·18369			
				·18108 (mean adopted)		

307. Smaldeel. Lat. 28° 24′·3 S.; Long. 26° 44′·0 E. Right-hand side of railway going to Kroonstad. 374 paces at right angles to railway, starting from Bloemfontein end of platform.

Declination. D.

Date		G.M.T.	D (observed)	D	Observer	Instrument
1903	June 7	7 18 a.m.	24° 23′·2 W.	24° 21′·4 W.	Beattie	73
		7 32 a.m.	24 24·5			
		12 35 p.m.	24 19·7			
		12 48 p.m.	24 20·6			

Dip. θ.

Date		G.M.T.	Needle	θ (observed)	θ	Observer	Instrument
1903	June 7	8 54 a.m.	3	59° 9′·0 S.	59° 7′·0 S.	Beattie	142
		8 56 a.m.	4	59 4·9			
		8 54 a.m.	1	59 5·6			

Horizontal Intensity. H.

Date		G.M.T.	H (observed)	H	Observer	Instrument
1903	June 7	9 44 a.m. V.				
		10 35 a.m. D.	H_{30} ·18680	·18680	Beattie	73
		11 14 a.m. D.	H_{40} ·18680			
		12 16 p.m. V.				

308. SPITZKOPJE. Lat. 25° 18′·2 S.; Long. 30° 49′·0 E. Left-hand side of road coming from Pilgrim's Rest. Half a mile from store. On Nelspruit side of store and of spruit.

Declination. D.

Date	G.M.T.	D (observed)	D	Observer	Instrument
1903 Sept. 8	11 56 a.m.	20° 33′·1 W.			
	12 7 p.m.	20 33 ·1	20° 36′·4 W.	Beattie	73
	1 3 p.m.	20 33 ·2			

Dip. θ.

Date	G.M.T.	Needle	θ (observed)	θ	Observer	Instrument
1903 Sept. 8	10 38 a.m.	1	58° 49′·3 S.	58° 48′·0 S.	Beattie	142

Horizontal Intensity. H.

Date	G.M.T.	H (observed)	H	Observer	Instrument
1903 Sept. 8	11 31 a.m. V.	·19212	·19224	Beattie	73

309. SPRINGFONTEIN. Lat. 30° 16′·7 S.; Long. 25° 44′·0 E. Right-hand side of railway, Springfontein to Bloemfontein. 365 paces from Bloemfontein end of platform towards Bloemfontein, then 397 paces perpendicular to railway.

Declination. D.

Date	G.M.T.	D (observed)	D	Observer	Instrument
1903 May 27	2 9 p.m.	25° 55′·8 W.			
	2 25 p.m.	25 56 ·2	25° 57′·3 W.	Beattie	73
1903 May 28	7 19 a.m.	26 0 ·4			
	7 53 a.m.	25 59 ·8			

Dip. θ.

Date	G.M.T.	Needle	θ (observed)	θ	Observer	Instrument
1903 May 28	9 13 a.m.	3	60° 14′·5 S.			
	9 14 a.m.	4	60 12 ·9	60° 15′·0 S.	Beattie	142
	9 14 a.m.	1	60 15 ·9			

Horizontal Intensity. H.

Date	G.M.T.	H (observed)	H	Observer	Instrument
1903 May 26	12 55 p.m. V.				
	1 53 p.m. D.	H_{30} ·18104 H_{40} ·18101			
	2 39 p.m. V.		·18102	Beattie	73
1903 May 27	10 53 a.m. V.				
	11 35 a.m. D.	H_{30} ·18104 H_{40} ·18101			
	12 46 p.m. V.				

310. SPRINGS. Lat. 26° 13′·0 S.; Long. 28° 27′·0 E. Left-hand side of railway, Germiston to Springs. Place reached by going 566 paces at right angles to railway from end of island platform away from Germiston, then going 300 paces parallel to railway towards Germiston.

Declination. D.

Date	G.M.T.	D (observed)	D	Observer	Instrument
1903 June 18	2 21 p.m.	22° 20′·0 W.	22° 19′·7 W.	Beattie	73
	2 42 p.m.	22 20 ·2			

Dip. θ.

Date	G.M.T.	Needle	θ (observed)	θ	Observer	Instrument
1903 June 18	11 1 a.m.	3	58° 21′·1 S.	58° 21′·7 S.	Beattie	142
	11 7 a.m.	1	58 21 ·6			

Horizontal Intensity. H.

Date	G.M.T.	H (observed)	H	Observer	Instrument
1903 June 18	12 17 p.m. V.	H_{30} ·19119	·19116	Beattie	73
	12 50 p.m. D.	H_{40} ·19113			

311. STANFORD. Lat. 34° 26′·7 S.; Long. 19° 28′·0 E. In a field on opposite side of river to village.

Declination. D.

Date	G.M.T.	D (observed)	D	Observer	Instrument
1901 Jan. 21	5 42 a.m.	28° 54′·6 W.	28° 43′·0 W.	Beattie Morrison	31
	5 57 a.m.	28 56 ·1			
	6 7 a.m.	28 57 ·0			
	6 16 a.m.	28 57 ·3			
	4 22 p.m.	28 49 ·9			
	4 32 p.m.	28 50 ·2			

Dip. θ.

Date	G.M.T.	Needle	θ (observed)	θ	Observer	Instrument
1901 Jan. 22	8 21 a.m.	1	59° 18′·5 S.	59° 41′·2 S.	Beattie	142
	8 21 a.m.	2	59 19 ·9			

Horizontal Intensity. H.

Date	G.M.T.	H (observed)	H	Observer	Instrument
1901 Jan. 21	9 34 a.m. V.	H_{30} ·18365	·18127	Morrison	31
	10 14 a.m. D.	H_{40} ·18359			
	10 52 a.m. V.				

312. Stanger. Lat. 29° 21′·1 S.; Long. 31° 15′·0 E. Right-hand side of railway, Durban to Hlabisa. 180 paces at right angles to railway, starting from middle of goods shed.

Declination. D.

Date	G.M.T.	D (observed)	D	Observer	Instrument
1903 Oct. 20	6 52 a.m.	23° 0′·5 W.			
	7 2 a.m.	23 1 ·1			
	11 29 a.m.	22 52 ·7	22° 59′·3 W.	Beattie	73
	11 38 a.m.	22 51 ·4			
	11 51 a.m.	22 51 ·4			

Dip. θ.

Date	G.M.T.	Needle	θ (observed)	θ	Observer	Instrument
1903 Oct. 20	9 54 a.m.	1	61° 18′·8 S.			
	9 55 a.m.	4_9	61 17 ·5	61° 16′·3 S.	Beattie	142
	9 59 a.m.	4	61 18 ·2			

Horizontal Intensity. H.

Date	G.M.T.	H (observed)	H	Observer	Instrument
1903 Oct. 20	7 32 a.m. V.				
	8 21 a.m. D.	H_{30} ·17991 H_{40} ·17988	·18011	Beattie	73
	8 53 a.m. V.				

313. Steenkampspoort. Lat. 32° 6′·3 S.; Long. 21° 44′·1 E. On road from Fraserburg Road to Fraserburg, 3 hours from latter. Alongside ditch on left-hand side of road half-way between dam and farmhouse.

Declination. D.

Date	G.M.T.	D (observed)	D	Observer	Instrument
1905 Jan. 17	8 4 a.m.	27° 15′·6 W.	27° 23′·4 W.	Beattie Brown	73

Dip. θ.

Date	G.M.T.	Needle	θ (observed)	θ	Observer	Instrument
1905 Jan. 17	5 24 a.m.	3_9	59° 35′·8 S.	59° 23′·6 S.	Beattie	142
	5 24 a.m.	4_9	59 36 ·1			

Horizontal Intensity. H.

Date	G.M.T.	H (observed)	H	Observer	Instrument
1905 Jan. 17	8 22 a.m. V.	·18211	·18335	Beattie	73

314. STELLENBOSCH. Lat. 33° 56′·0 S.; Long. 18° 50′·0 E. In centre of college field.

Declination. D.

Date	G.M.T.	D (observed)	D	Observer	Instrument
1900 June 4	9 48 a.m.	28° 59′·5 W.			
	10 2 a.m.	28 58 ·5			
	12 59 p.m.	28 59 ·3	28° 44′·9 W.	Beattie Morrison	31
	1 14 p.m.	28 57 ·4			
	1 25 p.m.	28 57 ·1			

Dip. θ.

Date	G.M.T.	Needle	θ (observed)	θ	Observer	Instrument
1899 Sept. 16	Forenoon	1	58° 47′·7 S.		Beattie	
	Forenoon	2	58 46 ·7	59° 18′·6 S.		9
	3 5 p.m.	1	58 49 ·0		Morrison	
	3 5 p.m.	2	58 46 ·8			
1901 Aug. 7	3 30 p.m.	1	59 4 ·8			
	3 40 p.m.	2	59 8 ·0			
1901 Aug. 14	3 24 p.m.	1	59 8 ·5	59 25 ·4	Morrison	142
	3 25 p.m.	2	59 12 ·1			
1901 Aug. 27	3 43 p.m.	1	59 7 ·3			
	3 44 p.m.	2	59 10 ·2			
1906 Feb. 22	4 48 p.m.	1	59 40 ·8	59 17 ·2	Morrison	9
	4 40 p.m.	2	59 41 ·2			
				59 20 ·4	(mean adopted)	

Horizontal Intensity. H.

Date	G.M.T.	H (observed)	H	Observer	Instrument
1899. Sept. 2	9 31 a.m. V.	H_{30} ·18553	·18183	Morrison	31
	10 32 a.m. D.	H_{40} ·18552			
1900 June 16	10 12 a.m. V.				
	11 2 a.m. D.	H_{30} ·18486	·18194	Morrison	31
	11 40 a.m. V.	H_{40} ·18485			
1906 March 1	3 41 p.m. V.				
	4 25 p.m. D.	H_{30} ·17987	·18199	Morrison	31
	4 56 p.m. V.	H_{40} ·17982			
1901 Aug. 8	1 34 p.m. V.				
	2 13 p.m. D.	H_{30} ·18355		Morrison	31
	2 49 p.m. V.	H_{40} ·18355			
1901 Aug. 15	1 23 p.m. V.				
	2 6 p.m. D.	H_{30} ·18343	·18194	Morrison	31
	2 42 p.m. V.	H_{40} ·18347			
1901 Aug. 29	2 57 p.m. V.				
	3 31 p.m. D.	H_{30} ·18361		Morrison	31
	4 4 p.m. V.	H_{40} ·18362			
			·18192 (mean adopted)		

315. Sterkstroom. Lat. 31° 34′·5 S.; Long. 26° 33′·0 E. In front of railway station, and about 220 yards from it.

Declination. D.

Date	G.M.T.	D (observed)	D	Observer	Instrument
1901 Dec. 30	5 2 a.m.	26° 14′·0 W.	25° 58′·2 W.	Beattie	31
	6 21 a.m.	26 16 ·6			

Dip. θ.

Date	G.M.T.	Needle	θ (observed)	θ	Observer	Instrument
1901 Dec. 30	8 41 a.m.	2	60° 37′·6 S.	60° 49′·3 S.	Beattie	142
	8 43 a.m.	1	60 37 ·0			

Horizontal Intensity. H.

Date	G.M.T.	H (observed)	H	Observer	Instrument
1901 Dec. 30	6 41 a.m. V.	·18230	·18095	Beattie	31

316. Steynsburg. Lat. 31° 18′·5 S.; Long. 25° 48′·0 E. On open ground above market-place.

Declination. D.

Date	G.M.T.	D (observed)	D	Observer	Instrument
1901 Dec. 23	5 4 a.m.	26° 23′·4 W.	26° 6′·1 W.	Beattie Morrison	31
	5 12 a.m.	26 22 ·9			
	5 20 a.m.	26 23 ·4			

Dip. θ.

Date	G.M.T.	Needle	θ (observed)	θ	Observer	Instrument
1901 Dec. 23	6 24 a.m.	2	60° 3′·8 S.	60° 15′·1 S.	Beattie	142
	6 26 a.m.	1	60 2 ·3			

Horizontal Intensity. H.

Date	G.M.T.	H (observed)	H	Observer	Instrument
1901 Dec. 23	8 11 a.m. V.	H_{30} ·18394	·18258	Morrison	31
	8 50 a.m. D.	H_{40} ·18391			
	9 22 a.m. V.				

317. Still Bay. Lat. 34° 22′·0 S.; Long. 21° 25′·0 E. In field adjoining Samuel's winkel.

Declination. D.

Date	G.M.T.	D (observed)	D	Observer	Instrument
1903 Jan. 17	8 15 a.m.	28° 28′·7 W.	28° 21′·7 W.	Beattie Morrison	31
	8 27 a.m.	28 29·6			
	8 36 a.m.	28 29·6			
	2 2 p.m.	28 23·1			
	2 10 p.m.	28 22·3			

Dip. θ.

Date	G.M.T.	Needle	θ (observed)	θ	Observer	Instrument
1903 Jan. 17	1 16 p.m.	3	60° 15′·0 S.	60° 18′·3 S.	Beattie	142
	1 16 p.m.	4	60 14·4			

Horizontal Intensity. H.

Date	G.M.T.	H (observed)	H	Observer	Instrument
1903 Jan. 17	9 0 a.m. V.	H_{30} ·18010	·17967	Morrison	31
	9 39 a.m. D.	H_{40} ·18003			
	10 0 a.m. V.				

318. Stormberg Junction. Lat. 31° 17′·5 S.; Long. 26° 16′·0 E.

Declination. D.

Date	G.M.T.	D (observed)	D	Observer	Instrument
1901 Dec. 24	6 32 a.m.	26° 16′·9 W.	25° 58′·5 W.	Beattie Morrison	31
	6 42 a.m.	26 17·5			
	6 52 a.m.	26 18·3			
	8 9 a.m.	26 15·0			
	8 18 a.m.	26 14·7			
	8 26 a.m.	26 14·4			

Horizontal Intensity. H.

Date	G.M.T.	H (observed)	H	Observer	Instrument
1901 Dec. 24	8 46 a.m. V.	H_{30} ·18272	·18138	Morrison	31
	9 24 a.m. D.	H_{40} ·18273			

319. Storms River. Lat. 33° 58′·0 S.; Long. 23° 49′·5 E. In field behind the boarding house.

Declination. D.

Date	G.M.T.	D (observed)	D	Observer	Instrument
1903 Feb. 2	1 40 p.m.	27° 43′·7 W.	27° 41′·5 W.	Beattie	31

Dip. θ.

Date	G.M.T.	Needle	θ (observed)	θ	Observer	Instrument
1903 Feb. 2	12 58 p.m.	3	60° 56′·4 S.	61° 0′·8 S.	Beattie	142
	12 59 p.m.	4	60 58·6			

Horizontal Intensity. H.

Date	G.M.T.	H (observed)	H	Observer	Instrument
1903 Feb. 2	8 34 a.m. V.		·17790	Beattie	31
	9 27 a.m. D.	H_{30} ·17832			
	10 57 a.m. V.	H_{40} ·17823			

320. Strandfontein. Lat. 34° 5′·3 S.; Long. 18° 34′·0 E. On side of forester's house distant from Muizenberg, and about 50 yards from the house.

Declination. D.

Date	G.M.T.	D (observed)	D	Observer	Instrument
1901 Feb. 22	8 49 a.m.	29° 4′·9 W.	28° 52′·5 W.	Beattie	31
	9 6 a.m.	29 5·8			
	9 20 a.m.	29 5·3			
	1 37 p.m.	29 0·5			
	1 48 p.m.	28 59·7			
	1 59 p.m.	28 59·2			

Dip. θ.

Date	G.M.T.	Needle	θ (observed)	θ	Observer	Instrument
1901 Feb. 21	9 42 a.m.	2	58° 57′·4 S.	59° 14′·2 S.	Beattie	142
	9 46 a.m.	1	58 54·5			
	3 22 p.m.	2	58 57·3			
	3 25 p.m.	1	58 53·6			
1901 Feb. 22	11 42 a.m.	2	58 57·3			
	11 42 a.m.	1	58 53·7			
1901 Feb. 23	11 13 a.m.	2	58 54·8			
	11 13 a.m.	1	58 51·1			
1901 Feb. 21	11 42 a.m.	2	58 50·3		Beattie	9
	11 42 a.m.	1	58 52·9			
	1 46 p.m.	2	58 49·9			
	1 45 p.m.	1	58 54·9			
1901 Feb. 22	9 15 a.m.	2	58 50·2			
	9 3 a.m.	1	58 51·8			
1901 Feb. 23	12 42 p.m.	2	58 48·5			
	12 43 p.m.	1	58 52·6			

321. SUTHERLAND. Lat. 32° 25′·0 S.; Long. 20° 39′·3 E. Left-hand side of road, Middlepost to Sutherland. About 100 paces from cottage used as a hospital, and between it and village, on Middlepost side of village.

Declination. D.

Date	G.M.T.	D (observed)	D	Observer	Instrument
1905 Jan. 30	5 5 a.m.	27° 21′·5 W.	27° 31′·1 W.	Beattie Brown	73

Dip. θ.

Date	G.M.T.	Needle	θ (observed)	θ	Observer	Instrument
1905 Jan. 30	4 18 a.m.	4_9	59° 28′·5 S.	59° 15′·8 S.	Beattie	142

Horizontal Intensity. H.

Date	G.M.T.	H (observed)	H	Observer	Instrument
1905 Jan. 30	4 48 a.m. V.	·18386	·18495	Beattie	73

322. SWELLENDAM. Lat. 34° 2′·0 S.; Long. 20° 27′·0 E. In a field to the west of the central hotel, about 100 yards distant from it, and about 60 yards from the main street of the village.

Declination. D.

Date	G.M.T.	D (observed)	D	Observer	Instrument
1901 Feb. 2	6 32 a.m. 6 42 a.m. 6 51 a.m.	28° 43′·0 W. 28 43 ·0 28 41 ·5			
1901 Feb. 3	2 54 p.m. 3 1 p.m. 3 8 p.m. 3 17 p.m.	28 35 ·5 28 35 ·5 28 36 ·0 28 35 ·9	28° 24′·7 W.	Beattie Morrison	31

Dip. θ.

Date	G.M.T.	Needle	θ (observed)	θ	Observer	Instrument
1901 Feb. 3	9 15 a.m. 9 15 a.m.	1 2	59° 30′·3 S. 59 32 ·3	59° 53′·0 S.	Beattie	142

Horizontal Intensity. H.

Date	G.M.T.	H (observed)	H	Observer	Instrument
1901 Feb. 4	8 32 a.m. V. 8 56 a.m. D. 9 40 a.m. V.	H_{30} ·18318 H_{40} ·18316	·18107	Morrison	31

323. Taungs. Lat. 27° 34′·8 S.; Long. 24° 45′·0 E. Left-hand side of railway, Kimberley to Mafeking. 200 paces perpendicular to railway, from a point 100 paces from Mafeking end of platform and going towards Mafeking.

Declination. D.

Date	G.M.T.	D (observed)	D	Observer	Instrument
1903 March 18	7 43 a.m.	24° 7′·5 W.	24° 5′·3 W.	Beattie	73
	7 53 a.m.	24 7 ·2			

Dip. θ.

Date	G.M.T.	Needle	θ (observed)	θ	Observer	Instrument
1903 March 18	12 48 p.m.	3	58° 19′·3 S.	58° 21′·8 S.	Beattie	142
	12 48 p.m.	4	58 19 ·9			
	12 48 p.m.	1	58 19 ·4			

Horizontal Intensity. H.

Date	G.M.T.	H (observed)	H	Observer	Instrument
1903 March 18	6 24 a.m. V.	H_{30} ·18909	·18887	Beattie	73
	7 27 a.m. D.	H_{40} ·18906			
	10 8 a.m. V.				

324. Thaba 'Nchu. Lat. 29° 10′·7 S.; Long. 26° 49′·0 E. On right-hand side of railway coming from Bloemfontein. 284 paces at right angles to railway, starting from end of loop nearer Bloemfontein.

Declination. D.

Date	G.M.T.	D (observed)	D	Observer	Instrument
1903 June 2	2 33 p.m.	24° 25′·2 W.	24° 27′·3 W.	Beattie	73
	2 54 p.m.	24 24 ·2			
1903 June 3	8 0 a.m.	24 32 ·7			
	8 19 a.m.	24 31 ·8			

Dip. θ.

Date	G.M.T.	Needle	θ (observed)	θ	Observer	Instrument
1903 June 3	9 9 a.m.	3	59° 36′·1 S.	59° 36′·6 S.	Beattie	142
	9 10 a.m.	4	59 35 ·6			
	9 12 a.m.	1	59 36 ·4			

Horizontal Intensity. H.

Date	G.M.T.	H (observed)	H	Observer	Instrument
1903 June 2	12 0 p.m. V.		·18548	Beattie	73
	12 40 p.m. D.	H_{30} ·18529			
	1 25 p.m. V.				
1903 June 3	7 55 a.m. D.	·18566			

325. TINFONTEIN. Lat. 30° 24′·0 S.; Long. 26° 54′·8 E. On outspan near Tinfontein Nek, on Zastron side of the nek.

Declination. D.

Date	G.M.T.	D (observed)	D	Observer	Instrument
1904 Jan 8.	2 58 p.m. 3 7 p.m.	24° 38′·7 W. 24 38 ·5	24° 45′·0 W.	Beattie	73

Dip. θ.

Date	G.M.T.	Needle	θ (observed)	θ	Observer	Instrument
1904 Jan. 8	12 45 p.m. 12 45 p.m.	1 4_9	60° 35′·7 S. 60 37 ·4	60° 33′·0 S.	Beattie	142

Horizontal Intensity. H.

Date	G.M.T.	H (observed)	H	Observer	Instrument
1904 Jan. 9	8 4 a.m. V.	·18039	·18077	Beattie	73

326. TOISE RIVER. Lat. 32° 27′·3 S.; Long. 27° 28′·7 E. About 300 yards from railway station on right of road leading from railway station to wool washery, in a field belonging to the hotel keeper.

Declination. D.

Date	G.M.T.	D (observed)	D	Observer	Instrument
1906 Jan. 10	6 23 a.m. 6 32 a.m. 6 39 a.m.	26° 5′·0 W. 26 5 ·8 26 5 ·6	26° 28′·2 W.	Brown Morrison	31

Dip. θ.

Date	G.M.T.	Needle	θ (observed)	θ	Observer	Instrument
1906 Jan. 10	10 52 a.m. 10 51 a.m.	1 2	62° 7′·0 S. 62 6 ·9	61° 46′·7 S.	Morrison	9

Horizontal Intensity. H.

Date	G.M.T.	H (observed)	H	Observer	Instrument
1906 Jan. 10	8 26 a.m. V. 9 1 a.m. D. 9 30 a.m. V.	H_{30} ·17319 H_{40} ·17314	·17559	Morrison	31

327. TOUWS RIVER. Lat. 33° 21′·0 S.; Long. 20° 3′·0 E. On opposite bank of river to railway station.

Declination. D.

Date	G.M.T.	D (observed)	D	Observer	Instrument
1900 June 3	9 23 a.m. 9 31 a.m. 9 49 a.m.	29° 12′·9 W. 29 12 ·7 29 12 ·9	28° 49′·6 W.	Beattie Morrison	31

Dip. θ.

Date	G.M.T.	Needle	θ (observed)	θ	Observer	Instrument
1899 Oct. 8	3 40 p.m. 3 40 p.m.	1 2	59° 2′·9 S. 58 59 ·9	59° 31′·4 S.	Beattie	9
1900 June 3	2 0 p.m. 2 0 p.m.	1 2	59 6 ·2 59 8 ·5	59 32 ·4	Beattie	9
				59 31 ·9 (mean adopted)		

Horizontal Intensity. H.

Date	G.M.T.	H (observed)	H	Observer	Instrument
1899 Oct. 6	11 37 a.m. V. 12 41 p.m. D.	H_{30} ·18467 H_{40} ·18462	·18149	Morrison	31
1900 June 3	10 22 a.m. V. 11 17 a.m. D. 11 57 a.m. V.	H_{30} ·18412 H_{40} ·18410	·18153	Morrison	31
			·18151 (mean adopted)		

328. TSOLO. Lat. 31° 18′·2 S.; Long. 28° 45′·6 E. Three hours by ox wagon from Mount Frere, on road to Qumbu.

Declination. D.

Date	G.M.T.	D (observed)	D	Observer	Instrument
1906 Feb. 3	6 58 a.m. 7 7 a.m. 7 16 a.m.	24° 26′·0 W. 24 26 ·3 24 26 ·0	24° 49′·6 W.	Brown Morrison	31

Dip. θ.

Date	G.M.T.	Needle	θ (observed)	θ	Observer	Instrument
1906 Feb. 3	11 19 a.m. 11 19 a.m.	1 2	61° 17′·0 S. 61 18 ·9	60° 57′·3 S.	Morrison	9

Horizontal Intensity. H.

Date	G.M.T.	H (observed)	H	Observer	Instrument
1906 Feb. 3	9 7 a.m. V. 9 38 a.m. D. 10 3 a.m. V.	H_{30} ·17751 H_{40} ·17755	·18002	Morrison	31

329. TUGELA. Lat. 29° 12′·0 S.; Long. 31° 25′·0 E. Left-hand side of railway, Durban to Hlabisa. 130 paces at right angles to railway, starting from a point 200 paces along railway from Tugela end of platform.

Dip. θ.

Date	G.M.T.	Needle	θ (observed)	θ	Observer	Instrument
1903 Oct. 21	8 43 a.m.	1	61° 11′·0 S.	61° 9′·0 S.	Beattie	142

Horizontal Intensity. H.

Date	G.M.T.	H (observed)	H	Observer	Instrument
1903 Oct 21	7 56 a.m. V.	·18118	·18139	Beattie	73

330. TULBAGH ROAD. Lat. 33° 19′·3 S.; Long. 19° 10′·0 E.

Declination. D.

Date	G.M.T.	D (observed)	D	Observer	Instrument
1902 Jan. 13	8 0 a.m.	28° 36′·1 W.	28° 29′·1 W.	Morrison	31
	8 12 a.m.	28 36 ·3			
	8 21 a.m.	28 35 ·4			
	4 22 p.m.	28 36 ·2			
	4 31 p.m.	28 37 ·1			
	4 37 p.m.	28 36 ·6			

Dip. θ.

Date	Needle	θ (observed)	θ	Observer	Instrument
1902 Jan. 13	2	58° 56′·0 S.	59° 2′·4 S.	Morrison	142
	1	58 49 ·5			

Horizontal Intensity. H.

Date	G.M.T.	H (observed)	H	Observer	Instrument
1902 Jan. 13	9 28 a.m. V.	H_{30} ·18445	·18313	Morrison	31
	10 11 a.m. D.	H_{40} ·18442			
	11 23 a.m. V.				

331. Twelfelhoek. Lat. 27° 27′·4 S.; Long. 29° 20′·4 E. On hillside above Commando spruit; on Vrede side of spruit.

Declination. *D.*

Date	G.M.T.	D (observed)	D	Observer	Instrument
1904 Feb. 4	1 1 p.m. 1 10 p.m.	21° 53′·4 W. 21 53 ·6	22° 1′·0 W.	Beattie	73

Dip. *θ.*

Date	G.M.T.	Needle	θ (observed)	θ	Observer	Instrument
1904 Feb. 4	11 48 a.m. 11 46 a.m.	1 4_9	59° 32′·2 S. 59 33 ·3	59° 28′·6 S.	Beattie	142

Horizontal Intensity. *H.*

Date	G.M.T.	H (observed)	H	Observer	Instrument
1904 Feb. 4	1 48 p.m. V.	·18750	·18796	Beattie	73

332. Tweepoort. Lat. 26° 36′·7 S.; Long. 30° 43′·0 E. On road Amsterdam to Bremersdorp between third and fourth drifts, about two miles from Amsterdam. Right-hand side of road going from Amsterdam and just at fourth drift.

Declination. *D.*

Date	G.M.T.	D (observed)	D	Observer	Instrument
1903 Aug. 24	11 58 a.m. 12 11 p.m.	22° 1′·9 W. 22 1 ·4	22° 3′·4 W.	Beattie	73

Dip. *θ.*

Date	G.M.T.	Needle	θ (observed)	θ	Observer	Instrument
1903 Aug. 24	10 45 a.m. 10 47 a.m. 10 45 a.m.	1 4 4_9	58° 54′·0 S. 58 50 ·2 58 51 ·9	58° 50′·9 S.	Beattie	142

Horizontal Intensity. *H.*

Date	G.M.T.	H (observed)	H	Observer	Instrument
1903 Aug. 24	11 42 a.m. V.	·19254	19260	Beattie	73

333. TWEE RIVIEREN. Lat. 33° 50′·3 S.; Long. 23° 56′·5 E. In field in front of Schreiber's house, at corner of garden wall nearer house.

Declination. D.

Date	G.M.T.	D (observed)	D	Observer	Instrument
1903 Feb. 11	2 8 p.m.	27° 43′·0 W.	27° 40′·4 W.	Beattie	31
	3 4 p.m.	27 42 ·5			

Dip. θ.

Date	G.M.T.	Needle	θ (observed)	θ	Observer	Instrument
1903 Feb. 11	12 59 p.m.	3	60° 56′·4 S.	60° 59′·3 S.	Beattie	142
	12 59 p.m.	4	60 56 ·1			

Horizontal Intensity. H.

Date	G.M.T.	H (observed)	H	Observer	Instrument
1903 Feb. 11	3 34 p.m. V.	·17821	·17783	Beattie	31

334. TYGERFONTEIN. Lat. 34° 10′·0 S.; Long. 21° 35′·5 E. On opposite side of road to store alongside clump of trees.

Declination. D.

Date	G.M.T.	D (observed)	D	Observer	Instrument
1903 Jan. 18	2 33 p.m.	28° 14′·1 W.			
	2 47 p.m.	28 14 ·9			
	3 0 p.m.	28 14 ·3	28° 13′·0 W.	Beattie	31
1903 Jan. 19	8 5 a.m.	28 20 ·8			
	8 15 a.m.	28 20 ·8			

Dip. θ.

Date	G.M.T.	Needle	θ (observed)	θ	Observer	Instrument
1903 Jan. 19	5 20 a.m.	3	60° 15′·5 S.	60° 18′·4 S.	Beattie	142
	5 20 a.m.	4	60 14 ·1			

Horizontal Intensity. H.

Date	G.M.T.	H (observed)	H	Observer	Instrument
1903 Jan. 19	8 43 a.m. V.	·18012	·17967	Beattie	31

335. TYGERKLOOF DRIFT. Lat. 28° 10′·8 S.; Long. 28° 35′·2 E. On Bethlehem Harrismith road. On Harrismith side of drift, and about half a mile from drift.

Declination. D.

Date	G.M.T.	D (observed)	D	Observer	Instrument
1904 Jan. 30	12 56 p.m. 1 6 p.m.	22° 49′·4 W. 22 49·2	22° 56′·8 W.	Beattie	73

Dip. θ.

Date	G.M.T.	Needle	θ (observed)	θ	Observer	Instrument
1904 Jan. 30	11 14 a.m. 11 14 a.m.	1 4_9	59° 24′·3 S. 59 27·2	59° 21′·5 S.	Beattie	142

Horizontal Intensity. H.

Date	G.M.T.	H (observed)	H	Observer	Instrument
1904 Jan. 30	12 41 p.m. V.	·18793	·18836	Beattie	73

336. THIRTYFIRST. Lat. 25° 40′·6 S.; Long. 29° 37′·6 E. About 12 miles from Middelburg. Left-hand side of road, Roos Senekal to Middelburg. Farmhouse on right-hand side of road, about half a mile off.

Declination. D.

Date	G.M.T.	D (observed)	D	Observer	Instrument
1903 July 31	1 24 p.m. 1 36 p.m. 1 50 p.m.	22° 1′·1 W. 21 59·6 21 59·7	22° 0′·9 W.	Beattie Löwinger	73

Dip. θ.

Date	G.M.T.	Needle	θ (observed)	θ	Observer	Instrument
1903 July 31	10 50 a.m. 10 51 a.m.	1 4	58° 27′·0 S. 58 24·2	58° 25′·0 S.	Beattie	142

Horizontal Intensity. H.

Date	G.M.T.	H (observed)	H	Observer	Instrument
1903 July 31	11 42 a.m. V. 12 13 p.m. D. 12 43 p.m. D. 1 13 p.m. V.	H_{30} ·19192 H_{40} ·19190	·19197	Beattie	73

337. Uitenhage. Lat. 33° 47′·0 S.; Long. 25° 24′·0 E. On Malay football field.

Declination. D.

Date	G.M.T.	D (observed)	D	Observer	Instrument
1900 July 25	8 38 a.m.	27° 53′·2 W.	27° 22′·6 W.	Beattie	31
	8 48 a.m.	27 53·4		Morrison	

Dip. θ.

Date	G.M.T.	Needle	θ (observed)	θ	Observer	Instrument
1900 July 24	9 30 a.m.	1	61° 0′·0 S.			
	9 30 a.m.	2	61 1·1			
1900 July 26	11 40 a.m.	1	61 2·0	61° 24′·5 S.	Beattie	9
	11 40 a.m.	2	61 1·0			
	1 5 p.m.	1	61 1·0			
	1 5 p.m.	2	61 1·7			
1904 July 17	7 28 a.m.	1	61 31·8	61 25·2	Morrison	9
	7 26 a.m.	2	61 34·5			
				61 24·6 (mean adopted)		

Horizontal Intensity. H.

Date	G.M.T.	H (observed)	H	Observer	Instrument
1900 July 24	11 59 a.m. V.				
	1 46 p.m. D.	H_{30} ·17938 H_{40} ·17941	·17695	Morrison	31
	2 25 p.m. V.				
1904 July 17	12 54 p.m. V.				
	1 58 p.m. D.	H_{30} ·17560 H_{40} ·17561	·17651	Morrison	31
	2 18 p.m. V.				
			·17673 (mean adopted)		

338. Uitkyk. Lat. 25° 49′·5 S.; Long. 29° 25′·0 E. Right-hand side of railway, Middelburg to Pretoria. 219 paces perpendicular to railway, starting from a point 36 paces along railway from Pretoria end of platform and towards Pretoria.

Declination. D.

Date	G.M.T.	D (observed)	D	Observer	Instrument
1903 Sept. 20	4 55 a.m.	21° 19′·2 W.			
	5 10 a.m.	21 20·0			
	5 23 a.m.	21 19·3	21° 24′·0 W.	Beattie	73
	9 28 a.m.	21 25·5			
	9 38 a.m.	21 24·6			

Dip. θ.

Date	G.M.T.	Needle	θ (observed)	θ	Observer	Instrument
1903 Sept. 20	7 57 a.m.	1	58° 12′·6 S.	58° 10′·0 S.	Beattie	142
	7 57 a.m.	4_9	58 10·7			

Horizontal Intensity. H.

Date	G.M.T.	H (observed)	H	Observer	Instrument
1903 Sept. 20	7 7 a.m. V.				
	9 4 a.m. D.	H_{30} ·19440 H_{40} ·19448	·19456	Beattie	73
	10 15 a.m. V.				

339. UITSPAN FARM. Lat. 31° 41′·2 S.; Long. 21° 27′·2 E. Right-hand side of road, Fraserburg to Williston. Opposite Williston end of dam on farm.

Declination. D.

Date	G.M.T.	D (observed)	D	Observer	Instrument
1905 Jan. 19	8 30 a.m.	27° 30′·5 W.	27° 38′·5 W.	Beattie Brown	73

Dip. θ.

Date	G.M.T.	Needle	θ (observed)	θ	Observer	Instrument
1905 Jan. 19	10 24 a.m.	3_9	59° 47′·2 S.	59° 34′·3 S.	Beattie	142
	10 24 a.m.	4_9	59 46 ·2			

Horizontal Intensity. H.

Date	G.M.T.	H (observed)	H	Observer	Instrument
1905 Jan. 19	8 48 a.m. V.	H_{30} ·18035	·18157	Beattie	73
	9 28 a.m. D.	H_{40} ·18030			

340. UMHLATUZI. Lat. 28° 51′·7 S.; Long. 31° 54′·0 E. Right-hand side of railway, Durban to Hlabisa. 130 paces at right angles to railway from Tugela end of platform.

Declination. D.

Date	G.M.T.	D (observed)	D	Observer	Instrument
1903 Oct. 26	2 9 p.m.	22° 24′·5 W.	22° 28′·9 W.	Beattie	73
	2 27 p.m.	22 23 ·6			
	2 47 p.m.	22 24 ·5			
1903 Oct. 27	5 0 a.m.	22 27 ·0			
	5 9 a.m.	22 26 ·4			

Dip. θ.

Date	G.M.T.	Needle	θ (observed)	θ	Observer	Instrument
1903 Oct. 26	8 26 a.m.	1	60° 39′·4 S.	60° 38′·4 S.	Beattie	142
	8 26 a.m.	4_9	60 39 ·9			
	8 27 a.m.	4	60 42 ·6			

Horizontal Intensity. H.

Date	G.M.T.	H (observed)	H	Observer	Instrument
1903 Oct. 26	9 10 a.m. V.		·18602	Beattie	73
	11 40 a.m. D.	H_{30} ·18585			
	12 22 p.m. D.	H_{40} ·18577			
	3 11 p.m. V.				

341. UMHLENGANA PASS. Lat. 31° 36′·0 S.; Long. 29° 19′·6 E. About five miles south east of Umhlengana Pass.

Declination. D.

Date	G.M.T.	D (observed)	D	Observer	Instrument
1906 Jan. 22	6 8 a.m.	24° 46′·4 W.	25° 10′·5 W.	Brown Morrison	31
	6 16 a.m.	24 47 ·8			
	6 26 a.m.	24 47 ·4			

Dip. θ.

Date	G.M.T.	Needle	θ (observed)	θ	Observer	Instrument
1906 Jan. 22	10 38 a.m.	1	62° 9′·8 S.	61° 50′·2 S.	Morrison	9
	10 40 a.m.	2	62 11 ·6			

Horizontal Intensity. H.

Date	G.M.T.	H (observed)	H	Observer	Instrument
1906 Jan. 22	8 4 a.m. V.	H_{30} ·17542	·17787	Morrison	31
	8 51 a.m. D.				
	9 27 a.m. V.				

342. UMTALI. Lat. 18° 59′·2 S.; Long. 32° 39′·0 E.

Declination. D.

Date	G.M.T.	D (observed)	D	Observer	Instrument
1903 April 16	2 33 p.m.	15° 52′·2 W.	15° 50′·1 W.	Beattie	73
	2 57 p.m.	15 52 ·7			
1903 April 18	8 34 a.m.	15 51 ·7			
	8 48 a.m.	15 52 ·0			

Dip. θ.

Date	G.M.T.	Needle	θ (observed)	θ	Observer	Instrument
1903 April 18	10 0 a.m.	3	52° 57′·9 S.	53° 0′·0 S.	Beattie	142
	10 1 a.m.	4	52 59 ·0			
	10 1 a.m.	1	52 58 ·5			

Horizontal Intensity. H.

Date	G.M.T.	H (observed)	H	Observer	Instrument
1903 April 16	11 18 a.m. V.	H_{30} ·22017	·22005	Beattie	73
	12 10 p.m. D.	H_{40} ·22016			
	12 58 p.m. V.	H_{25} ·22019			
		H_{35} ·22018			

343. Umtata. Lat. 31° 35′·9 S.; Long. 28° 47′·1 E.

Declination. D.

Date	G.M.T.	D (observed)	D	Observer	Instrument
1906 Jan. 20	6 20 a.m. 6 28 a.m. 6 36 a.m.	24° 50′·1 W. 24 49·4 24 48·1	25° 12′·3 W.	Brown Morrison	31

Dip. θ.

Date	G.M.T.	Needle	θ (observed)	θ	Observer	Instrument
1906 Jan. 19	3 9 p.m. 3 10 p.m.	1 2	61° 54′·5 S. 61 57·0	61° 35′·4 S.	Morrison	9

Horizontal Intensity. H.

Date	G.M.T.	H (observed)	H	Observer	Instrument
1906 Jan. 20	8 40 a.m. V. 9 14 a.m. D. 9 44 a.m. V.	H_{30} ·17631 H_{40} ·17638	·17879	Morrison	31

344. Umtwalumi. Lat. 30° 28′·0 S.; Long. 30° 40′·0 E. Right-hand side of railway coming from Durban. 200 paces at right angles to railway reckoned from a point 300 paces from Shepstone end of platform along railway towards Shepstone.

Dip. θ.

Date	G.M.T.	Needle	θ (observed)	θ	Observer	Instrument
1903 Nov. 2	11 43 a.m. 11 43 a.m.	1 4	61° 46′·8 S. 61 48·5	61° 45′·4 S.	Beattie	142

Horizontal Intensity. H.

Date	G.M.T.	H (observed)	H	Observer	Instrument
1903 Nov. 2	9 15 a.m. V. 10 8 a.m. D. 10 52 a.m. V.	H_{30} ·17710 H_{40} ·17712	·17735	Beattie	73

345. Umzinto. Lat. 30° 19′·4 S.; Long. 30° 39′·0 E. Left-hand side of railway coming from Alexandra Junction. In hollow opposite shed on Junction side of Umzinto.

Declination. D.

Date	G.M.T.	D (observed)	D	Observer	Instrument
1903 Nov. 4	6 32 a.m.	23° 39′·4 W.	23° 40′·1 W.	Beattie	73
	6 43 a.m.	23 38 ·3			

Dip. θ.

Date	G.M.T.	Needle	θ (observed)	θ	Observer	Instrument
1903 Nov. 4	9 7 a.m.	1	61° 25′·4 S.	61° 22′·2 S.	Beattie	142
	9 8 a.m.	4_9	61 23 ·2			
	9 7 a.m.	4	61 25 ·5			

Horizontal Intensity. H.

Date	G.M.T.	H (observed)	H	Observer	Instrument
1903 Nov. 4	7 2 a.m. V.	H_{30} ·18078	·18102	Beattie	73
	7 43 a.m. D.	H_{40} ·18078			
	8 16 a.m. V.				

346. Underberg Hotel. Lat. 29° 47′·9 S.; Long. 29° 30′·5 E. On right-hand side of road coming from Bulwer. Bulwer side of hotel.

Declination. D.

Date	G.M.T.	D (observed)	D	Observer	Instrument
1903 Nov. 13	2 14 p.m.	24° 31′·5 W.	24° 39′·2 W.	Beattie	73
	2 22 p.m.	24 29 ·5			
	2 38 p.m.	24 31 ·4			
1903 Nov. 14	4 46 a.m.	24 43 ·5			
	4 55 a.m.	24 43 ·2			
	6 39 a.m.	24 38 ·7			
	6 50 a.m.	24 38 ·1			
	9 32 a.m.	24 32 ·6			

Dip. θ.

Date	G.M.T.	Needle	θ (observed)	θ	Observer	Instrument
1903 Nov. 13	2 24 p.m.	1	61° 5′·8 S.	61° 1′·8 S.	Beattie	142
	2 27 p.m.	4_9	61 3 ·1			
	2 28 p.m.	4	61 4 ·1			

Horizontal Intensity. H.

Date	G.M.T.	H (observed)	H	Observer	Instrument
1903 Nov. 14	9 39 a.m. V.	H_{30} ·18041	·18068	Beattie	73
	10 18 a.m. D.	H_{40} ·18039			
	10 47 a.m. D.				
	11 12 a.m. V.				

347. Upington. Lat. 28° 27′·7 S.; Long. 21° 14′·9 E. Near Rondhavel on opposite side of Orange River to Upington. In the first small river bed between Kenhardt and the Orange River, about 100 paces to the left of the road.

Declination. D.

Date	G.M.T.	D (observed)	D	Observer	Instrument
1904 Dec. 19	6 54 a.m. 7 7 a.m.	26° 51′·4 W. 26 52 ·9	27° 3′·5 W.	Beattie	73

Dip. θ.

Date	G.M.T.	Needle	θ (observed)	θ	Observer	Instrument
1904 Dec. 19	5 21 a.m. 5 21 a.m.	3_9 4_9	57° 43′·7 S. 57 40 ·7	57° 42′·2 S.	Beattie	142

Horizontal Intensity. H.

Date	G.M.T.	H (observed)	H	Observer	Instrument
1904 Dec. 19	7 24 a.m. V. 8 23 a.m. D. 8 51 a.m. D. 9 17 a.m. V.	H_{30} ·19067 H_{40} ·19064	·19183	Beattie	73

348. Utrecht, West of. Lat. 27° 39′·9 S.; Long. 30° 16′·0 E. About nine miles west of Utrecht on left-hand side of road, Wakkerstroom to Utrecht coming from Wakkerstroom. About 1½ miles on Utrecht side of stables.

Declination. D.

Date	G.M.T.	D (observed)	D	Observer	Instrument
1903 Aug. 11	2 5 p.m. 2 23 p.m.	22° 40′·6 W. 22 39 ·6	22° 41′·4 W.	Beattie	73

Dip. θ.

Date	G.M.T.	Needle	θ (observed)	θ	Observer	Instrument
1903 Aug. 11	11 12 a.m. 11 14 a.m.	1 4	59° 34′·6 S. 59 34 ·8	59° 33′·8 S.	Beattie	142

Horizontal Intensity. H.

Date	G.M.T.	H (observed)	H	Observer	Instrument
1903 Aug. 11	12 12 p.m. V. 12 50 p.m. D.	H_{30} ·18776 H_{40} ·18773	·18780	Beattie	73

349. VAN REENEN. Lat. 28° 22′·2 S.; Long. 29° 24′·5 E. In field opposite hotel, 200 paces from railway on side away from hotel.

Declination. D.

Date	G.M.T.	D (observed)	D	Observer	Instrument
1903 Nov. 27	8 8 a.m. 8 18 a.m. 11 8 a.m.	23° 2′·2 W. 23 1·2 23 1·8 23 1·2	23° 5′·1 W.	Beattie	73

Horizontal Intensity. H.

Date	G.M.T.	H (observed)	H	Observer	Instrument
1903 Nov. 27	10 46 a.m. V. 11 30 a.m. D. 12 14 p.m. D. 12 52 p.m. V.	H_{30} ·18570 H_{40} ·18568	·18596	Beattie	73

350. VAN WYK'S FARM. Lat. 33° 49′·4 S.; Long. 21° 12′·0 E. Half-way between Ladismith and Riversdale. Left-hand side of road Ladismith to Riversdale. About 150 paces from road.

Declination. D.

Date	G.M.T.	D (observed)	D	Observer	Instrument
1903 Jan. 15	6 56 a.m. 7 10 a.m. 7 23 a.m. 2 0 p.m. 2 5 p.m.	28° 16′·0 W. 28 15·9 28 15·8 28 16·6 28 15·8	28° 11′·4 W.	Beattie Morrison	31

Dip. θ.

Date	G.M.T.	Needle	θ (observed)	θ	Observer	Instrument
1903 Jan. 15	12 36 a.m. 12 37 a.m.	3 4	59° 57′·7 S. 60 0·8	60° 3′·0 S.	Beattie	142

Horizontal Intensity. H.

Date	G.M.T.	H (observed)	H	Observer	Instrument
1903 Jan. 15	9 6 a.m. V. 9 49 a.m. D. 10 10 a.m. V.	H_{30} ·18066 H_{40} ·18073	·18025	Morrison	31

351. VAN WYK'S VLEI. Lat. 30° 22′·3 S.; Long. 21° 50′·0 E. Place of observation on Carnarvon side of ditch flowing from dam to irrigated lands; opposite end of village away from Carnarvon.

Declination. D.

Date	G.M.T.	D (observed)	D	Observer	Instrument
1904 Dec. 12	5 53 a.m.	26° 53′·8 W.	27° 4′·8 W.	Beattie	73

Dip. θ.

Date	G.M.T.	Needle	θ (observed)	θ	Observer	Instrument
1904 Dec. 12	5 17 a.m. 5 17 a.m.	3_9 4_9	58° 57′·0 S. 58 59 ·6	58° 46′·6 S.	Beattie	142

Horizontal Intensity. H.

Date	G.M.T.	H (observed)	H	Observer	Instrument
1904 Dec. 12	6 48 a.m. V. 7 17 a.m. D.	H_{30} ·18350 H_{40} ·18341	·18461	Beattie	73

352. VICTORIA FALLS. Lat. 17° 55′·6 S.; Long. 25° 51′·0 E. Bulawayo side of water tank, 83 paces at right angles to the railway line reckoned from a point 79 paces along the railway from the tank.

Declination. D.

Date	G.M.T.	D (observed)	D	Observer	Instrument
1904 July 8	12 40 p.m. 12 46 p.m.	17° 41′·8 W. 17 42 ·8	17° 52′·5 W.	Beattie	73

Dip. θ.

Date	G.M.T.	Needle	θ (observed)	θ	Observer	Instrument
1904 July 8	9 26 a.m. 9 26 a.m.	1 4	51° 32′·7 S. 51 30 ·3	51° 24′·4 S.	Beattie	142

Horizontal Intensity. H.

Date	G.M.T.	H (observed)	H	Observer	Instrument
1904 July 8	10 34 a.m. V. 11 14 a.m. D. 11 34 a.m. D.	H_{30} ·22010 H_{40} ·22011	·22070	Beattie	73

353. VILLIERSDORP. Lat. 33° 59′·5 S.; Long. 19° 19′·0 E. In a field bordering on the main road through the village, to the southward of a small stream almost opposite Hayne's house.

Declination. D.

Date	G.M.T.	D (observed)	D	Observer	Instrument
1901 Jan. 16	6 1 a.m.	28° 51′·6 W.	28° 37′·9	Beattie Morrison	31
	6 16 a.m.	28 52 ·2			
	6 27 a.m.	28 52 ·6			
	3 1 p.m.	28 44 ·5			
	3 11 p.m.	28 44 ·7			
	3 22 p.m.	28 44 ·8			

Dip. θ.

Date	G.M.T.	Needle	θ (observed)	θ	Observer	Instrument
1901 Jan. 16	9 25 a.m.	1	59° 7′·8 S.	59° 30′·8 S.	Beattie	142
	9 35 a.m.	2	59 9 ·4			

Horizontal Intensity. H.

Date	G.M.T.	H (observed)	H	Observer	Instrument
1901 Jan. 16	12 54 p.m. V.	H_{30} ·18385	·18118	Morrison	31
	1 38 p.m. D.	H_{40} ·18359			
	2 17 p.m. V.				

354. VIRGINIA. Lat. 28° 7′·5 S.; Long. 26° 55′·0 E. Right-hand side of railway going to Kroonstad. 238 paces from it, starting from a point 85 paces from dead end towards Kroonstad.

Declination. D.

Date	G.M.T.	D (observed)	D	Observer	Instrument
1903 June 9	7 52 a.m.	24° 6′·4 W.	24° 6′·4 W.	Beattie	73
	8 9 a.m.	24 7 ·4			

Dip. θ.

Date	G.M.T.	Needle	θ (observed)	θ	Observer	Instrument
1903 June 9	9 11 a.m.	3	59° 3′·8 S.	59° 2′·4 S.	Beattie	142
	9 12 a.m.	4	59 1 ·8			
	9 12 a.m.	1	59 0 ·0			

Horizontal Intensity. H.

Date	G.M.T.	H (observed)	H	Observer	Instrument
1903 June 9	10 32 a.m. V.	H_{30} ·18751	·18749	Beattie	73
	11 17 a.m. D.	H_{40} ·18748			
	11 44 a.m. V.				

355. VLAKLAAGTE. Lat. 26° 50′·6 S.; Long. 29° 5′·0 E. Left-hand side of railway, Germiston to Durban. 28 paces along the railway from dead end towards Germiston, then 145 paces at right angles to railway.

Declination. D.

Date	G.M.T.	D (observed)	D	Observer	Instrument
1903 Sept. 28	4 43 a.m.	22° 17′·9 W.	22° 23′·4 W.	Beattie	73
	4 54 a.m.	22 17 ·0			
	6 40 a.m.	22 21 ·6			
	6 50 a.m.	22 21 ·1			
	8 51 a.m.	22 26 ·1			
	9 0 a.m.	22 26 ·4			

Dip. θ.

Date	G.M.T.	Needle	θ (observed)	θ	Observer	Instrument
1903 Sept. 28	9 40 a.m.	1	58° 55′·0 S.	58° 54′·3 S.	Beattie	142
	9 40 a.m.	4_9	58 56 ·9			

Horizontal Intensity. H.

Date	G.M.T.	H (observed)	H	Observer	Instrument
1903 Sept. 28	7 20 a.m. V.	H_{30} ·19043	·19059	Beattie	73
	8 8 a.m. D.	H_{40} ·19045			
	8 37 a.m. V.				

356. VOGELVLEI. Lat. 29° 8′·3 S.; Long. 27° 31′·1 E. 50 paces from road on opposite side of road to farmhouse.

Declination. D.

Date	G.M.T.	D (observed)	D	Observer	Instrument
1904 Jan. 21	12 10 p.m.	23° 57′·3 W.	24° 4′·9 W.	Beattie	73
	12 19 p.m.	23 58 ·0			

Dip. θ.

Date	G.M.T.	Needle	θ (observed)	θ	Observer	Instrument
1904 Jan. 21	8 45 a.m.	1	60° 3′·5 S.	60° 0′·1 S.	Beattie	142
	8 46 a.m.	4_9	60 4 ·8			
	8 47 a.m.	6	60 3 ·3			

Horizontal Intensity. H.

Date	G.M.T.	H (observed)	H	Observer	Instrument
1904 Jan. 21	9 56 a.m. V.	H_{30} ·18377	·18422	Beattie	73
	10 42 a.m. D.	H_{40} ·18382			
	11 8 a.m. D.				

357. VONDELING. Lat. 33° 19′·8 S.; Long. 23° 4′·0 E. 300 yards south of house, 100 yards from graveyard.

Declination. D.

Date	G.M.T.	D (observed)	D	Observer	Instrument
1903 Feb. 21	3 58 p.m.	27° 53′·3 W.			
	4 14 p.m.	27 53 ·2			
	4 30 p.m.	27 52 ·7	27° 46′·3	Beattie	31
1903 Feb. 22	9 10 a.m.	27 46 ·7			
	9 30 a.m.	27 46 ·8			
	9 49 a.m.	27 45 ·7			

Dip. θ.

Date	G.M.T.	Needle	θ (observed)	θ	Observer	Instrument
1903 Feb. 22	5 54 a.m.	3	60° 31′·9 S.	60° 34′·0 S.	Beattie	142
	5 55 a.m.	4	60 30 ·1			

Horizontal Intensity. H.

Date	G.M.T.	H (observed)	H	Observer	Instrument
1903 Feb. 21	12 9 p.m. V.	H_{30} ·17896	·17858	Beattie	31
	2 14 p.m. V.	H_{40} ·17896			

358. VREDEFORT. Lat. 27° 1′·2 S.; Long. 27° 22′·9 E. Right-hand side of road, Vredefort Road to Vredefort, on hillside about half a mile from the dorp, and on railway side of it.

Declination. D.

Date	G.M.T.	D (observed)	D	Observer	Instrument
1904 Feb. 10	7 51 a.m.	22° 30′·1 W.			
	8 0 a.m.	22 29 ·8	22° 35′·6 W.	Beattie	73
	11 59 a.m.	22 29 ·0			
	12 9 p.m.	22 29 ·3			

Dip. θ.

Date	G.M.T.	Needle	θ (observed)	θ	Observer	Instrument
1904 Feb. 10	11 17 a.m.	1	57° 42′·5 S.	57° 37′·4 S.	Beattie	142
	11 26 a.m.	4_9	57° 41 ·4			

Horizontal Intensity. H.

Date	G.M.T.	H (observed)	H	Observer	Instrument
1904 Feb. 10	8 52 a.m. V.				
	9 26 a.m. D.	H_{30} ·19230	·19276	Beattie	73
	10 0 a.m. V.	H_{40} ·19232			

359. Vredefort Road. Lat. 27° 7′·0 S.; Long. 27° 45′·0 E. Right-hand side of road, Heilbron to Vredefort Road. About 600 paces from point where road crosses the railway on the Heilbron side of the railway.

Dip. θ.

Date	G.M.T.	Needle	θ (observed)	θ	Observer	Instrument
1904 Feb. 9	11 24 a.m.	1	59° 6′·5 S.	59° 3′·9 S.	Beattie	142
	11 24 a.m.	4_9	59 9·7			

Horizontal Intensity. H.

Date	G.M.T.	H (observed)	H	Observer	Instrument
1904 Feb. 9	12 38 p.m. V.	·17904	·17948	Beattie	73

360. Vryburg. Lat. 26° 57′·1 S.; Long. 24° 43′·0 E.

Declination. D.

Date	G.M.T.	D (observed)	D	Observer	Instrument
1906 Jan. 29	1 33 p.m.	22° 27′·3 W.	22° 57′·5 W.	Beattie	73
	1 42 p.m.	22 27·1			
	3 8 p.m.	22 30·4			
1906 Jan. 30	7 23 a.m.	22 37·0			
	7 31 a.m.	22 37·4			
	9 44 a.m.	22 30·8			

Dip. θ.

Date	G.M.T.	Needle	θ (observed)	θ	Observer	Instrument
1906 Jan. 30	8 42 a.m.	3_9	58° 21′·6 S.	58° 2′·3 S.	Beattie	142
	8 42 a.m.	5	58 22·4			

Horizontal Intensity. H.

Date	G.M.T.	H (observed)	H	Observer	Instrument
1906 Jan. 29	8 57 a.m. V.	H_{30} ·18992	·19179	Beattie	73
	9 37 a.m. D.	H_{40} ·18993			
	10 19 a.m. V.				
1906 Jan. 30	7 48 a.m. V.	·19019			
	9 30 a.m. V.				

361. WAKKERSTROOM. Lat. 27° 21′·5 S.; Long. 30° 9′·0 E. On south side of town, just over the spruit on the golf course.

Declination. D.

Date	G.M.T.	D (observed)	D	Observer	Instrument
1903 Aug. 10	1 7 p.m. 1 18 p.m.	22° 25′·5 W. 22 25 ·9	22° 27′·0 W.	Beattie	73

Dip. θ.

Date	G.M.T.	Needle	θ (observed)	θ	Observer	Instrument
1903 Aug. 10	11 30 a.m. 11 32 a.m.	1 4	59° 25′·1 S. 59 24 ·3	59° 24′·0 S.	Beattie	142

Horizontal Intensity. H.

Date	G.M.T.	H (observed)	H	Observer	Instrument
1903 Aug. 10	12 30 p.m. V.	·18868	·18874	Beattie	73

362. WANKIE. Lat. 18° 22′·3 S.; Long. 26° 28′·5 E. Right-hand side of railway, Falls to Bulawayo. Right-hand side of road to Falls from Wankie, about 280 yards beyond railway cottages, 50 paces from the road in first hollow with a flat expanse of 300 yards (about).

Declination. D.

Date	D	Observer	Instrument
1904 July 12	16° 6′·1 W.	Beattie	73

Dip. θ.

Date	G.M.T.	Needle	θ (observed)	θ	Observer	Instrument
1904 July 11	4 29 a.m. 4 29 a.m.	1 4	51° 13′·3 S. 51 12 ·1	51° 5′·6 S.	Beattie	142

Horizontal Intensity. H.

Date	G.M.T.	H (observed)	H	Observer	Instrument
1904 July 12	7 39 a.m. V. 8 20 a.m. D. 8 53 a.m. V.	H_{30} ·22133 H_{40} ·22131	·22192	Beattie	73

363. WARMBAD (WATERBERG). Lat. 24° 53′·0 S.; Long. 28° 20′·0 E. Right-hand side of railway, Pretoria to Pietersburg. Along railway 129 paces from Pietersburg end of platform towards Pietersburg, then 203 paces at right angles to the railway.

Declination. D.

Date	G.M.T.	D (observed)	D	Observer	Instrument
1903 Sept. 23	2 30 p.m.	21° 17′·8 W.	21° 21′·1 W.	Beattie	73

Dip. θ.

Date	G.M.T.	Needle	θ (observed)	θ	Observer	Instrument
1903 Sept. 23	3 24 p.m.	1	57° 12′·6 S.	57° 11′·0 S.	Beattie	142

Horizontal Intensity. H.

Date	G.M.T.	H (observed)	H	Observer	Instrument
1903 Sept. 23	2 40 p.m. V.	·19676	·19691	Beattie	73

364. WARMBAD (ZOUTPANSBERG). Lat. 22° 24′·9 S.; Long. 29° 12′·0 E.

Declination. D.

Date	G.M.T.	D (observed)	D	Observer	Instrument
1903 July 9	2 53 p.m. 3 7 p.m.	21° 16′·5 W. 21 16 ·9	21° 16′·7 W.	Beattie Löwinger	73

Dip. θ.

Date	G.M.T.	Needle	θ (observed)	θ	Observer	Instrument
1903 July 9	11 31 a.m. 11 31 a.m.	1 4	57° 16′·6 S. 57 18 ·2	57° 17′·2 S.	Beattie	142

Horizontal Intensity. H.

Date	G.M.T.	H (observed)	H	Observer	Instrument
1903 July 9	1 35 p.m. V. 2 10 p.m. D. 2 42 p.m. V.	H_{30} ·19218 H_{40} ·19228	·19223	Beattie	73

365. WARRENTON. Lat. 28° 6′·9 S.; Long. 24° 52′·0 E. Right-hand side of railway, Kimberley to Vryburg. 120 yards from railway, starting from a point 218 yards from Vryburg end of platform, and going towards Vryburg.

Declination. D.

Date	G.M.T.	D (observed)	D	Observer	Instrument
1903 March 17	8 6 a.m.	24° 59′·8 W.	24° 51′·6 W.	Beattie	73
	8 22 a.m.	25 0·6			
	3 47 p.m.	24 51·0			
	4 2 p.m.	24 50·8			

Dip. θ.

Date	G.M.T.	Needle	θ (observed)	θ	Observer	Instrument
1903 March 17	10 22 a.m.	3	58° 47′·0 S.	58° 48′·9 S.	Beattie	142
	10 24 a.m.	4	58 45·8			
	10 24 a.m.	1	58 47·0			

Horizontal Intensity. H.

Date	G.M.T.	H (observed)	H	Observer	Instrument
1903 March 17	1 52 p.m. V.	H_{30} ·18820	·18798	Beattie	73
	2 35 p.m. D.	H_{40} ·18818			
	4 20 p.m. V.				

366. WASCHBANK. Lat. 28° 18′·8 S.; Long. 30° 8′·0 E. Right-hand side of railway, Newcastle to Ladysmith from a point at middle of platform, then 250 paces at right angles to the railway.

Declination. D.

Date	G.M.T.	D (observed)	D	Observer	Instrument
1903 Oct. 5	4 57 a.m.	23° 11′·4 W.	23° 16′·0 W.	Beattie	73
	5 5 a.m.	23 11·6			
	8 47 a.m.	23 16·3			
	8 54 a.m.	23 16·9			

Dip. θ.

Date	G.M.T.	Needle	θ (observed)	θ	Observer	Instrument
1903 Oct. 5	10 37 a.m.	1	59° 59′·6 S.	59° 59′·0 S.	Beattie	142
	10 37 a.m.	4_9	60 2·6			

Horizontal Intensity. H.

Date	G.M.T.	H (observed)	H	Observer	Instrument
1903 Oct. 5	6 46 a.m. V.	H_{30} ·18588	·18608	Beattie	73
	8 13 a.m. D.	H_{40} ·18592			
	9 7 a.m. V.				

367. WATERWORKS. Lat. 29° 4′·5 S.; Long. 26° 28′·0 E. Left-hand side of railway, Bloemfontein to Thaba'Nchu. 250 paces from line, starting from Thaba'Nchu side of station signboard.

Declination. D.

Date	G.M.T.	D (observed)	D	Observer	Instrument
1903 June 4	6 48 a.m. 7 20 a.m.	24° 30′·1 W. 24 29 ·5	24° 29′·6 W.	Beattie	73

Dip. θ.

Date	G.M.T.	Needle	θ (observed)	θ	Observer	Instrument
1903 June 4	8 24 a.m. 8 24 a.m. 8 25 a.m.	3 4 1	59° 37′·4 S. 59 36 ·4 59 36 ·4	59° 37′·2 S.	Beattie	142

Horizontal Intensity. H.

Date	G.M.T.	H (observed)	H	Observer	Instrument
1903 June 4	9 28 a.m. V. 10 42 a.m. D. 11 40 a.m. V.	H_{30} ·18506 H_{40} ·18504 H_{25} ·18493 H_{35} ·18499	·18500	Beattie	73

368. WELVERDIEND. Lat. 26° 22′·7 S.; Long. 27° 17′·0 E. Right-hand side of railway, Klerksdorp to Randfontein. Point reached by starting from dead end and going along railway towards Randfontein for 287 paces, then going at right angles to railway for 255 paces.

Declination. D.

Date	G.M.T.	D (observed)	D	Observer	Instrument
1903 June 22	6 25 a.m. 6 40 a.m. 6 57 a.m. 12 49 p.m. 1 2 p.m. 1 16 p.m.	22° 59′·3 W. 22 59 ·6 23 0 ·0 22 59 ·0 22 59 ·2 22 58 ·5	22° 59′·5 W.	Beattie Löwinger	73

Dip. θ.

Date	G.M.T.	Needle	θ (observed)	θ	Observer	Instrument
1903 June 22	8 58 a.m. 8 59 a.m. 9 0 a.m. 9 0 a.m.	3 4 1 4_9	58° 42′·7 S. 58 42 ·5 58 43 ·7 58 45 ·6	58° 43′·7 S.	Beattie	142

Horizontal Intensity. H.

Date	G.M.T.	H (observed)	H	Observer	Instrument
1903 June 22	9 53 a.m. V. 10 25 a.m. D. 12 5 p.m. D. 12 35 p.m. V.	H_{30} ·18988 H_{40} ·18983	·18985	Beattie	73

369. WEPENER. Lat. 29° 43′·6 S.; Long. 27° 3′·7 E. On right-hand side of road, Dewetsdorp to Wepener, just over the spruit at the entrance to the dorp. Place of observation back from road alongside spruit.

Declination. D.

Date	G.M.T.	D (observed)	D	Observer	Instrument
1904 Jan. 15	7 51 a.m.	25° 19′·9 W.	25° 23′·1 W.	Beattie	73

Dip. θ.

Date	G.M.T.	Needle	θ (observed)	θ	Observer	Instrument
1904 Jan. 15	9 49 a.m.	6	60° 10′·9 S.	60° 9′·9 S.	Beattie	142
	9 49 a.m.	5	60 16 ·5			

Horizontal Intensity. H.

Date	G.M.T.	H (observed)	H	Observer	Instrument
1904 Jan. 15	8 44 a.m. V.	·18263	·18303	Beattie	73

370. WILLISTON. Lat. 31° 20′·4 S.; Long. 20° 55′·2 E.

Declination. D.

Date	G.M.T.	D (observed)	D	Observer	Instrument
1905 Jan. 21	5 48 a.m.	26° 31′·0 W.	26° 39′·6 W.	Beattie	73
	6 2 a.m.	26 31 ·6		Brown	

Dip. θ.

Date	G.M.T.	Needle	θ (observed)	θ	Observer	Instrument
1905 Jan. 21	9 44 a.m.	3_9	59° 15′·0 S.	59° 2′·9 S.	Beattie	142
	9 43 a.m.	4_9	59 15 ·5			

Horizontal Intensity. H.

Date	G.M.T.	H (observed)	H	Observer	Instrument
1905 Jan. 21	7 38 a.m. V.	H_{30} ·18327	·18451	Beattie	73
	8 38 a.m. D.	H_{40} ·18325			

371. WILLOWMORE. Lat. 33° 9′·4 S.; Long. 23° 30′·0 E. Over river on road to Prince Albert. Left-hand side of road leaving Willowmore in fork of roads.

Declination. D.

Date	G.M.T.	D (observed)	D	Observer	Instrument
1900 July 14	1 44 p.m.	28° 21′·1 W.			
	1 53 p.m.	28 20·3			
	2 2 p.m.	28 21·0			
	2 23 p.m.	28 19·6	27° 51′·2 W.	Beattie Morrison	31
	2 37 p.m.	28 20·6			
	2 53 p.m.	28 20·8			
	3 0 p.m.	28 21·5			
1903 Feb. 20	5 33 a.m.	27 53·9			
	5 56 a.m.	27 54·2			
	8 15 a.m.	27 54·6	27 50·8	Beattie	31
	8 30 a.m.	27 55·0			
	2 27 p.m.	27 56·1			
	2 44 p.m.	27 57·5			
			27 51·0 (mean adopted)		

Dip. θ.

Date	G.M.T.	Needle	θ (observed)	θ	Observer	Instrument
1900 July 14	10 0 a.m.	1	60° 25′·2 S.			
	10 0 a.m.	2	60 22·5	60° 48′·0 S.	Beattie	9
	11 50 a.m.	1	60 25·8			
	11 50 a.m.	2	60 24·0			
1903 Feb. 20	10 18 a.m.	3	60 44·6			
	10 19 a.m.	4	60 44·3	60 49·6	Beattie	142
	9 58 a.m.	4	60 49·6			
	9 59 a.m.	1	60 48·2			
				60 48·8 (mean adopted)		

Horizontal Intensity. H.

Date	G.M.T.	H (observed)	H	Observer	Instrument
1900 July 15	9 8 a.m. V.				
	9 55 a.m. D.	H_{30} ·18068			
	10 36 a.m. V.	H_{40} ·18063	·17796	Morrison	31
	8 38 a.m. V.				
	9 21 a.m. D.	H_{30} ·18059			
	9 58 a.m. V.	H_{40} ·18059			
1903 Feb. 19	1 22 p.m. V.	H_{30} ·17830			
	2 58 p.m. D.	H_{40} ·17829	·17792	Beattie	31
	3 41 p.m. V.				
			·17794 (mean adopted)		

372. WINBURG. Lat. 28° 31′·2 S.; Long. 27° 3′·0 E. Right-hand side of railway, from Smaldeel to Winburg. 286 paces at right angles to it, starting from point 143 paces from Smaldeel end of platform, and going towards Smaldeel.

Declination. D.

Date	G.M.T.	D (observed)	D	Observer	Instrument
1903 June 8	7 28 a.m. 7 42 a.m.	24° 12′·5 W. 24 14 ·1	24° 12′·7 W.	Beattie	73

Dip. θ.

Date	G.M.T.	Needle	θ (observed)	θ	Observer	Instrument
1903 June 8	8 44 a.m. 8 45 a.m. 8 44 a.m.	3 4 1	59° 17′·2 S. 59 16 ·4 59 17 ·2	59° 17′·4 S.	Beattie	142

Horizontal Intensity. H.

Date	G.M.T.	H (observed)	H	Observer	Instrument
1903 June 8	9 32 a.m. V. 10 31 a.m. D. 11 0 a.m. V.	H_{30} ·18648 H_{40} ·18644	·18646	Beattie	73

373. WINKELDRIFT. Lat. 27° 10′·6 S.; Long. 27° 7′·6 E. Just across Rhenoster River on left-hand side of road, Reitzburg to Bothaville, on Bothaville side of river.

Declination. D.

Date	G.M.T.	D (observed)	D	Observer	Instrument
1904 Feb. 11	7 31 a.m. 7 41 a.m. 12 29 p.m. 12 38 p.m.	24° 13′·2 W. 24 13 ·3 24 5 ·6 24 5 ·5	24° 15′·3 W	Beattie	73

Dip. θ.

Date	G.M.T.	Needle	θ (observed)	θ	Observer	Instrument
1904 Feb. 11	10 46 a.m. 10 46 a.m.	1 4_9	58° 46′·3 S. 58 49 ·3	58° 43′·5 S.	Beattie	142

Horizontal Intensity. H.

Date	G.M.T.	H (observed)	H	Observer	Instrument
1904 Feb. 11	8 25 a.m. V. 8 55 a.m. D. 9 40 a.m. V.	H_{30} ·18939 H_{40} ·18944	·18988	Beattie	73

374. Witklip. Lat. 23° 16′·5 S.; Long. 29° 17′·0 E. 180 paces east of big white rock starting from north end of it.

Declination. D.

Date	G.M.T.	D (observed)	D	Observer	Instrument
1903 July 2	3 18 p.m. 3 30 p.m.	21° 58′·2 W. 21 58·4	21° 58′·0 W.	Beattie Löwinger	73

Dip. θ.

Date	G.M.T.	Needle	θ (observed)	θ	Observer	Instrument
1903 July 2	5 35 a.m. 5 35 a.m.	1 4	56° 30′·8 S. 56 31·0	56° 30′·9 S.	Beattie	142

Horizontal Intensity. H.

Date	G.M.T.	H (observed)	H	Observer	Instrument
1903 July 2	6 55 a.m. V. 7 25 a.m. D.	H_{30} ·19959 H_{40} ·19956	·19958	Beattie	73

375. Witmoss. Lat. 32° 33′·0 S.; Long. 25° 45′·0 E. On flat ground, 400 yards south west of railway station.

Declination. D.

Date	G.M.T.	D (observed)	D	Observer	Instrument
1902 July 15	7 48 a.m. 7 59 a.m. 8 7 a.m. 1 22 p.m. 1 32 p.m. 2 31 p.m. 2 38 p.m.	26° 37′·4 W. 26 37·5 26 36·9 26 35·4 26 35·5 26 36·0 26 36·5	26° 26′·8 W.	Beattie Morrison	31

Dip. θ.

Date	G.M.T.	Needle	θ (observed)	θ	Observer	Instrument
1902 July 15	10 29 a.m. 10 29 a.m. 10 29 a.m. 10 30 a.m.	1 2 3 4	61° 8′·7 S. 61 10·5 61 4·9 61 3·6	61° 14′·6 S.	Beattie	142

Horizontal Intensity. H.

Date	G.M.T.	H (observed)	H	Observer	Instrument
1902 July 15	12 58 p.m. V. 2 4 p.m. D. 2 53 p.m. V.	H_{30} ·17940 H_{40} ·17943	·17852	Morrison	31

376. WOLVEFONTEIN. Lat. 23° 19′·0 S.; Long. 24° 55′·0 E.

Dip. θ.

Date	G.M.T.	Needle	θ (observed)	θ	Observer	Instrument
1900 July 19	9 20 a.m.	1	60° 56′·9 S.	61° 19′·2 S.	Beattie	9
	9 20 a.m.	2	60 54 ·4			

Horizontal Intensity. H.

Date	G.M.T.	H (observed)	H	Observer	Instrument
1900 July 20	11 3 a.m. V.	H_{30} ·17848	·17600	Morrison	31
	11 50 a.m. D.	H_{40} ·17849			
	12 27 p.m. V.				

377. WOLVEHOEK. Lat. 26° 54′·9 S.; Long. 27° 50′·0 E. Right-hand side of railway, Heilbron to Pretoria. 250 paces at right angles to railway reckoned from a point 115 paces from Heilbron end of platform and towards Meyerton.

Declination. D.

Date	G.M.T.	D (observed)	D	Observer	Instrument
1903 June 13	6 13 a.m.	23° 1′·8 W.	23° 1′·4 W.	Beattie Löwinger	73
	6 25 a.m.	23 1 ·6			
	6 41 a.m.	23 1 ·5			
	6 55 a.m.	23 1 ·5			

Dip. θ.

Date	G.M.T.	Needle	θ (observed)	θ	Observer	Instrument
1903 June 13	8 42 a.m.	3	58° 38′·1 S.	58° 36′·2 S.	Beattie	142
	8 41 a.m.	4	58 34 ·0			
	8 43 a.m.	1	58 35 ·6			

Horizontal Intensity. H.

Date	G.M.T.	H (observed)	H	Observer	Instrument
1903 June 13	9 37 a.m. V.		·19040	Beattie	73
	10 10 a.m. D.	H_{30} ·19040			
	10 35 a.m. D.	H_{40} ·19040			
	11 0 a.m. V.				

378. North of Limpopo. Lat. 22° 7'·2 S.; Long. 29° 10'·0 E. Nine miles over river from police camp, near a ruined house.

Declination. D.

Date	G.M.T.	D (observed)	D	Observer	Instrument
1903 July 8	6 43 a.m. 6 55 a.m. 7 11 a.m.	19° 57'·7 W. 19 59 ·5 20 0 ·6	19° 59'·5 W.	Beattie Löwinger	73

Dip. θ.

Date	G.M.T.	Needle	θ (observed)	θ	Observer	Instrument
1903 July 8	10 8 a.m. 10 8 a.m.	1 4	55° 51'·6 S. 55 52 ·3	55° 52'·0 S.	Beattie	142

Horizontal Intensity. H.

Date	G.M.T.	H (observed)	H	Observer	Instrument
1903 July 8	8 8 a.m. V. 8 50 a.m. D.	H_{30} ·20292 H_{40} ·20289	·20291	Beattie	73

379. Wonderfontein. Lat. 25° 48'·3 S.; Long. 29° 53'·0 E. Right-hand side of railway, Belfast to Middelburg. 92 paces from Belfast end of platform towards Belfast, then 231 paces from the railway.

Declination. D.

Date	G.M.T.	D (observed)	D	Observer	Instrument
1903 Sept. 19	6 24 a.m. 6 54 a.m. 7 5 a.m.	23° 34'·6 W. 23 33 ·2 23 34 ·1	23° 35'·4 W.	Beattie	73

Dip. θ.

Date	G.M.T.	Needle	θ (observed)	θ	Observer	Instrument
1903 Sept. 19	9 57 a.m. 9 57 a.m.	1 4_9	58° 2'·1 S. 58 0 ·6	58° 0'·0 S.	Beattie	142

Horizontal Intensity. H.

Date	G.M.T.	H (observed)	H	Observer	Instrument
1903 Sept. 19	7 23 a.m. V. 8 17 a.m. D. 8 50 a.m. D.	H_{30} ·19833 H_{40} ·19837	·19847	Beattie	73

380. Woodville. Lat. 33° 56′·3 S.; Long. 22° 41′·0 E. In orchard alongside orange trees.

Declination. D.

Date	G.M.T.	D (observed)	D	Observer	Instrument
1903 Jan. 26	6 10 a.m.	28° 3′·3 W.			
	6 18 a.m.	28 3 ·6	27° 56′·8 W.	Beattie	31
	6 28 a.m.	28 3 ·4			

Dip. θ.

Date	G.M.T.	Needle	θ (observed)	θ	Observer	Instrument
1903 Jan. 26	9 10 a.m.	3	60° 33′·5 S.	60° 37′·3 S.	Beattie	142
	9 10 a.m.	1	60 34 ·6			

Horizontal Intensity. H.

Date	G.M.T.	H (observed)	H	Observer	Instrument
1903 Jan. 26	1 55 p.m. V.	H_{30} ·17927	·17881	Beattie	31
	2 47 p.m. D.	H_{40} ·17924			
	3 42 p.m. V.				

381. Worcester. Lat. 33° 39′·0 S.; Long. 19° 26′·0 E. On town commonage.

Declination. D.

Date	G.M.T.	D (observed)	D	Observer	Instrument
1902 April 15	3 12 p.m.	28° 38′·2 W.			
	3 22 p.m.	28 37 ·3	28° 34′·0 W.	Morrison	31
1902 April 16	5 30 a.m.	28 39 ·2			
	5 57 a.m.	28 39 ·3			
	6 7 a.m.	28 39 ·6			

Dip. θ.

Date	G.M.T.	Needle	θ (observed)	θ	Observer	Instrument
1899 Oct. 11	9 10 a.m.	1	58° 49′·9 S.	59° 22′·5 S.	Beattie	9
	9 10 a.m.	2	58 47 ·1			
1902 April 15	10 42 a.m.	1	59 10 ·6			
	10 42 a.m.	2	59 15 ·4	59 24 ·5	Morrison	9
1902 April 16	2 48 p.m.	1	59 11 ·1			
	2 52 p.m.	2	59 16 ·8			
				59 23 ·5 (mean adopted)		

Horizontal Intensity. H.

Date	G.M.T.	H (observed)	H	Observer	Instrument
1899 Oct. 9	1 3 p.m. V.	H_{30} ·18495	·18180	Morrison	31
	2 7 p.m. D.				
1902 April 16	9 4 a.m. V.	H_{30} ·18313	·18217	Morrison	31
	10 1 a.m. D.	H_{40} ·18308			
	10 49 a.m. V.				
			·18199 (mean adopted)		

382. Zak Rivier. Lat. 30° 30′·9 S.; Long. 20° 31′·0 E. On Williston side of river just at the drift.

Declination. D.

Date	G.M.T.	D (observed)	D	Observer	Instrument
1905 Jan. 24	6 3 a.m. 6 18 a.m.	26° 52′·4 W. 26 52·1	27° 0′·9 W.	Beattie Brown	73

Dip. θ.

Date	G.M.T.	Needle	θ (observed)	θ	Observer	Instrument
1905 Jan. 24	9 53 a.m. 9 53 a.m.	3_9 4_9	58° 31′·4 S. 58 ·27·9	58° 17′·2 S.	Beattie	142

Horizontal Intensity. H.

Date	G.M.T.	H (observed)	H	Observer	Instrument
1905 Jan. 24	7 38 a.m. V. 9 1 a.m. D.	H_{30} ·18581 H_{40} ·18581	·18707	Beattie	73

383. Zand River. Lat. 23° 3′·8 S.; Long. 29° 34′·0 E. Right-hand side of road on rise just before drift, on road Mara to Spelonken.

Declination. D.

Date	G.M.T.	D (observed)	D	Observer	Instrument
1903 July 13	7 36 a.m. 7 51 a.m.	21° 12′·9 W. 21 12·6	21° 13′·0 W.	Beattie Löwinger	73

Dip. θ.

Date	G.M.T.	Needle	θ (observed)	θ	Observer	Instrument
1903 July 13	5 33 a.m. 5 35 a.m.	1 4	56° 21′·3 S. 56 20·6	56° 20′·7 S.	Beattie	142

Horizontal Intensity. H.

Date	G.M.T.	H (observed)	H	Observer	Instrument
1903 July 13	6 20 a.m. V. 6 58 a.m. D.	H_{30} ·20234 H_{40} ·20234	·20234	Beattie	73

384. ZEEKOEGAT. Lat. 33° 3′·0 S.; Long. 22° 31′·0 E. On left-hand side of Beaufort West Road, 300 yards west of the hotel.

Declination. D.

Date	G.M.T.	D (observed)	D	Observer	Instrument
1903 Jan. 4	4 41 a.m.	28° 1′·3 W.			
	4 55 a.m.	28 1·4	27° 57′·2 W.	Beattie	31
	4 39 p.m.	28 0·3			

Dip. θ.

Date	G.M.T.	Needle	θ (observed)	θ	Observer	Instrument
1903 Jan. 4	8 45 a.m.	1	60° 36′·7 S.			
	8 44 a.m.	3	60 33·4	60° 38′·4 S.	Beattie	142
	8 44 a.m.	4	60 33·2			

Horizontal Intensity. H.

Date	G.M.T.	H (observed)	H	Observer	Instrument
1903 Jan. 4	12 40 p.m. V.	H_{30} ·17680	·17639	Beattie	31
	3 10 p.m. D.	H_{40} ·17687			

385. ZUURBRAAK. Lat. 34° 0′·3 S.; Long. 20° 39′·0 E. In field behind garden of boarding house.

Declination. D.

Date	G.M.T.	D (observed)	D	Observer	Instrument
1901 Jan. 31	5 48 a.m.	28° 37′·8 W.			
	6 7 a.m.	28 36·9			
	6 19 a.m.	28 37·0	28° 20′·2 W.	Beattie Morrison	31
	6 30 a.m.	28 37·0			
	6 42 a.m.	28 36·6			

Dip. θ.

Date	G.M.T.	Needle	θ (observed)	θ	Observer	Instrument
1901 Feb. 1	10 12 a.m.	1	59° 36′·7 S.	59° 58′·9 S.	Beattie	142
	10 12 a.m.	2	59 37·5			

Horizontal Intensity. H.

Date	G.M.T.	H (observed)	H	Observer	Instrument
1901 Feb. 1	7 51 a.m. V.	H_{30} ·18299			
	8 32 a.m. D.		·18092	Morrison	31
	9 6 a.m. V.	H_{40} ·18304			

386. Zuurfontein. Lat. 32° 51′·0 S.; Long. 18° 35′·0 E. In field in front of and N.N.E. of Zaack's Store.

Declination. D.

Date	G.M.T.	D (observed)	D	Observer	Instrument
1901 July 17	9 17 a.m.	28° 25′·2 W.	28° 16′·6 W.	Beattie Morrison	31
	9 25 a.m.	28 25 ·1			
	9 40 a.m.	28 26 ·7			

Dip. θ.

Date	G.M.T.	Needle	θ (observed)	θ	Observer	Instrument
1901 July 17	11 0 a.m.	1	58° 25′·4 S.	58° 44′·4 S.	Beattie	142
	11 0 a.m.	2	58 28 ·2			

Horizontal Intensity. H.

Date	G.M.T.	H (observed)	H	Observer	Instrument
1901 July 17	7 26 a.m. V.	H_{30} ·18614	·18426	Morrison	31
	8 28 a.m. D.	H_{40} ·18613			
	9 53 a.m. V.				

387. Zuurpoort. Lat. 32° 2′·9 S.; Long. 24° 8′·0 E. In field in front of post office. Right-hand side of road coming from Graaff Reinet, and in front of dam.

Declination. D.

Date	G.M.T.	D (observed)	D	Observer	Instrument
1900 July 9	9 21 a.m.	27° 28′·1 W.	26° 53′·0 W.	Beattie Morrison	31
	9 30 a.m.	27 27 ·9			
	9 43 a.m.	27 27 ·2			
	1 58 p.m.	27 20 ·1			
	2 9 p.m.	27 18 ·1			
	2 25 p.m.	27 18 ·6			

Dip. θ.

Date	G.M.T.	Needle	θ (observed)	θ	Observer	Instrument
1900 July 9	10 20 a.m.	1	59° 55′·6 S.	60° 23′·4 S.	Beattie	9
	10 20 a.m.	2	59 55 ·2			

Horizontal Intensity. H.

Date	G.M.T.	H (observed)	H	Observer	Instrumen
1900 July 9	12 24 p.m. V.	H_{30} ·18319	·18072	Morrison	31
	1 16 p.m. D.	H_{40} ·18317			
	1 56 p.m. V.				
	6 19 a.m. V.	H_{30} ·18333			
	7 17 a.m. D.	H_{40} ·18328			
	8 46 a.m. V.				

388. Lat. 23° 42′·7 S.; Long. 29° 44′·0 E. About 20 miles from Pietersburg on Birthday road.

Declination. D.

Date	G.M.T.	D (observed)	D	Observer	Instrument
1903 July 21	1 38 p.m. 1 49 p.m.	20° 47′·2 W. 20 47 ·3	20° 47′·7 W.	Beattie Löwinger	73

Dip. θ.

Date	G.M.T.	Needle	θ (observed)	θ	Observer	Instrument
1903 July 21	12 50 p.m. 12 48 p.m.	1 4	56° 34′·2 S. 56 37 ·0	56° 35′·2 S.	Beattie	142

Horizontal Intensity. H.

Date	G.M.T.	H (observed)	H	Observer	Instrument
1903 July 21	1 26 p.m. V.	·20114	·20114	Beattie	73

389. Lat. 24° 8′·0 S.; Long. 29° 28′·0 E. On road from Pietersburg to Chunie's Poort. At second outspan from Pietersburg on right-hand side of road, and just over the drift coming from Pietersburg.

Declination. D.

Date	G.M.T.	D (observed)	D	Observer	Instrument
1903 July 24	1 28 p.m. 1 45 p.m.	20° 23′·0 W. 20 23 ·2	20° 23′·9 W.	Beattie Löwinger	73

Dip. θ.

Date	G.M.T.	Needle	θ (observed)	θ	Observer	Instrument
1903 July 24	11 4 a.m. 11 4 a.m.	1 4	56° 55′·8 S. 56 56 ·0	56° 55′·4 S.	Beattie	142

Horizontal Intensity. H.

Date	G.M.T.	H (observed)	H	Observer	Instrument
1903 July 24	11 49 a.m. V. 12 45 p.m. D. 1 15 p.m. V.	H_{30} ·19905 H_{40} ·19910	·19908	Beattie	73

390. Lat. 25° 9′·8 S.; Long. 29° 4′·8 E. On Pietersburg to Middelburg road, about 28 miles from Pokwani, and on Middelburg side of latter.

Declination. D.

Date	G.M.T.	D (observed)	D	Observer	Instrument
1903 July 29	1 13 p.m. 1 27 p.m. 1 43 p.m.	21° 34′·0 W. 21 32 ·4 21 32 ·0	21° 33′·7 W.	Beattie Löwinger	73

Dip. θ.

Date	G.M.T.	Needle	θ (observed)	θ	Observer	Instrument
1903 July 29	11 2 a.m. 11 3 a.m.	1 4	58° 14′·9 S. 58 14 ·2	58° 14′·0 S.	Beattie	142

Horizontal Intensity. H.

Date	G.M.T.	H (observed)	H	Observer	Instrument
1903 July 29	11 54 a.m. V. 12 27 p.m. D. 12 54 p.m. V.	H_{30} ·19552 H_{40} ·19549	·19557	Beattie	73

391. Lat. 25° 47′·5 S.; Long. 29° 36′·0 E. Left Middelburg by Machadodorp road. Took first turn to right. Observed at place about 10 miles from Middelburg.

Declination. D.

Date	G.M.T.	D (observed)	D	Observer	Instrument
1903 Aug. 2	11 56 a.m. 12 10 p.m. 12 24 p.m.	21° 8′·2 W. 21 9 ·2 21 9 ·2	21° 10′·0 W.	Beattie	73

Dip. θ.

Date	G.M.T.	Needle	θ (observed)	θ	Observer	Instrument
1903 Aug. 2	10 35 a.m. 10 35 a.m.	1 4	58° 30′·3 S. 58 29 ·8	58° 29′·5 S.	Beattie	142

Horizontal Intensity. H.

Date	G.M.T.	H (observed)	H	Observer	Instrument
1903 Aug. 2	11 45 a.m. V	·19222	·19228	Beattie	73

392. Lat. 24° 47′·0 S.; Long. 30° 40′·0 E. On right-hand side of road, Kasper's Nek to Pilgrim's Rest, just before the second drift over the Blyde River coming from Kasper's Nek.

Dip. θ.

Date	G.M.T.	Needle	θ (observed)	θ	Observer	Instrument
1903 Sept. 5	12 49 p.m.	1	58° 10′·3 S.	58° 9′·0 S.	Beattie	142

Horizontal Intensity. H.

Date	G.M.T.	H (observed)	H	Observer	Instrument
1903 Sept. 5	12 34 p.m. V.	·19535	·19547	Beattie	73

393. Lat. 28° 54′·6 S.; Long. 27° 44′·1 E. Third farm after passing Stephen's Store, on Ladybrand road to Ficksburg. Just through gate on hill side. Right-hand side of road.

Declination. D.

Date	G.M.T.	D (observed)	D	Observer	Instrument
1904 Jan. 22	1 21 p.m.	23° 58′·3 W.	24° 5′·5 W.	Beattie	73

Dip. θ.

Date	G.M.T.	Needle	θ (observed)	θ	Observer	Instrument
1904 Jan. 22	10 6 a.m.	1	59° 58′·0 S.	59° 54′·2 S.	Beattie	142

Horizontal Intensity. H.

Date	G.M.T.	H (observed)	H	Observer	Instrument
1904 Jan. 22	12 39 p.m. V.	·18433	·18476	Beattie	73

394. Lat. 28° 31′·6 S.; Long. 27° 42′·3 E. Left-hand side of road, Ficksburg to Senekal. A cross road goes at right angles to the first. Observations made in hollow on right-hand side of second road going to Winburg, about 100 paces from road on Winburg side of sluit.

Declination. D.

Date	G.M.T.	D (observed)	D	Observer	Instrument
1904 Jan. 24	7 9 a.m.	23° 55′·1 W.			
	7 54 a.m.	23 53 ·1	23° 58′·1 W.	Beattie	73
	8 22 a.m.	23 55 ·3			

Dip. θ.

Date	G.M.T.	Needle	θ (observed)	θ	Observer	Instrument
1904 Jan. 24	9 30 a.m.	1	59° 34′·4 S.			
	9 31 a.m.	4_9	59 36 ·9	59° 31′·9 S.	Beattie	142
	9 31 a.m.	6	59 36 ·3			

Horizontal Intensity. H.

Date	G.M.T.	H (observed)	H	Observer	Instrument
1904 Jan. 24	8 37 a.m. V.	·18654	·18697	Beattie	73

395. Lat. 28° 6′·7 S.; Long. 29° 3′·1 E. Left-hand side of main road from Harrismith to Vrede, about 10 miles from Harrismith.

Declination. D.

Date	G.M.T.	D (observed)	D	Observer	Instrument
1904 Feb. 1	8 50 a.m.	22° 51′·9 W.			
	9 2 a.m.	22 51 ·8	22° 57′·9 W.	Beattie	73
	2 0 p.m.	22 51 ·7			
	2 8 p.m.	22 52 ·4			

Dip. θ.

Date	G.M.T.	Needle	θ (observed)	θ	Observer	Instrument
1904 Feb. 1	11 26 a.m.	1	59° 43′·6 S.	59° 39′·3 S.	Beattie	142
	11 26 a.m.	4_9	59 43 ·2			

Horizontal Intensity. H.

Date	G.M.T.	H (observed)	H	Observer	Instrument
1904 Feb. 1	9 19 a.m. V.				
	9 34 a.m. D.	H_{30} ·18587 H_{40} ·18589	·18634	Beattie	73
	10 43 a.m. V.				

396. Lat. 27° 22′·7 S.; Long. 29° 0′·0 E.

Declination. D.

Date	G.M.T.	D (observed)	D	Observer	Instrument
1904 Feb. 5	12 40 p.m. 12 48 p.m.	22° 24′·4 W. 22 25 ·0	22° 32′·3 W.	Beattie	73

Dip. θ.

Date	G.M.T.	Needle	θ (observed)	θ	Observer	Instrument
1904 Feb. 5	11 39 a.m. 11 39 a.m.	1 4_9	59° 11′·7 S. 59 13 ·1	59° 8′·2 S.	Beattie	142

Horizontal Intensity. H.

Date	G.M.T.	H (observed)	H	Observer	Instrument
1904 Feb. 5	12 22 p.m. V.	·18846	·18892	Beattie	73

397. Lat. 27° 30′·0 S.; Long. 26° 38′·3 E.

Dip. θ.

Date	G.M.T.	Needle	θ (observed)	θ	Observer	Instrument
1904 Feb. 13	11 25 a.m. 11 26 a.m.	1 4_9	59° 1′·1 S. 59 2 ·7	58° 57′·5 S.	Beattie	142

Horizontal Intensity. H.

Date	G.M.T.	H (observed)	H	Observer	Instrument
1904 Feb. 13	12 26 p.m. V.	·18727	·18777	Beattie	73

398. East of Komgha, 6 miles. Lat. 32° 32′·6 S.; Long. 27° 58′·3 E.

Declination. D.

Date	G.M.T.	D (observed)	D	Observer	Instrument
1906 Jan. 14	5 35 a.m.	25° 44′·3 W.	26° 7′·2 W.	Brown Morrison	31
	5 44 a.m.	25 44 ·4			
	5 52 a.m.	25 44 ·0			

Dip. θ.

Date	G.M.T.	Needle	θ (observed)	θ	Observer	Instrument
1906 Jan. 14	9 48 a.m.	1	61° 52′·6 S.	61° 32′·6 S.	Morrison	9
	9 44 a.m.	2	61 53 ·2			

Horizontal Intensity. H.

Date	G.M.T.	H (observed)	H	Observer	Instrument
1906 Jan. 14	7 34 a.m. V.	H_{30} ·17488	·17730	Morrison	31
	8 10 a.m. D.	H_{40} ·17486			
	8 56 a.m. V.				

399. Outspan. Lat. 32° 13′·4 S.; Long. 28° 10′·4 E. On public outspan half-way between Butterworth and Idutywa.

Declination. D.

Date	G.M.T.	D (observed)	D	Observer	Instrument
1906 Jan. 16	5 58 a.m.	25° 33′·1 W.	25° 56′·9 W.	Brown Morrison	31
	6 13 a.m.	25 34 ·0			
	6 24 a.m.	25 34 ·2			

Dip. θ.

Date	G.M.T.	Needle	θ (observed)	θ	Observer	Instrument
1906 Jan. 16	10 12 a.m.	1	61° 37′·5 S.	61° 18′·7 S.	Morrison	9
	10 11 a.m.	2	61 40 ·6			

Horizontal Intensity. H.

Date	G.M.T.	H (observed)	H	Observer	Instrument
1906 Jan. 16	7 50 a.m. V.	H_{30} ·17559	·17803	Morrison	31
	8 26 a.m. D.	H_{40} ·17558			
	9 9 a.m. V.				

400. BASHEE. Lat. 31° 42′·0 S.; Long. 28° 30′·0 E. About 26 miles east of Idutywa.

Declination. D.

Date	G.M.T.	D (observed)	D	Observer	Instrument
1906 Jan. 18	6 50 a.m.	25° 37′·0 W.			
	6 58 a.m.	25 34 ·5	25° 58′·4 W.	Brown Morrison	31
	7 8 a.m.	25 34 ·4			

Dip. θ.

Date	G.M.T.	Needle	θ (observed)	θ	Observer	Instrument
1906 Jan. 18	11 44 a.m.	1	61° 54′·2 S.	61° 35′·1 S.	Morrison	9
	11 46 a.m.	2	61 56 ·7			

Horizontal Intensity. H.

Date	G.M.T.	H (observed)	H	Observer	Instrument
1906 Jan. 18	9 8 a.m. V.				
	9 47 a.m. D.	H_{30} ·17543 H_{40} ·17540	·17786	Morrison	31
	10 4 a.m. V.				

401. Lat. 31° 26′·0 S.; Long. 29° 31′·5 E. On road from Port St Johns to Mount Ayliff. 30 miles from Port St Johns.

Dip. θ

Date	G.M.T.	Needle	θ (observed)	θ	Observer	Instrument
1906 Jan. 25	2 59 a.m.	1	61° 21′·6 S.	61° 2′·1 S.	Morrison	9
	2 59 a.m.	2	61° 23 ·7			

Horizontal Intensity. H.

Date	G.M.T.	H (observed)	H	Observer	Instrument
1906 Jan. 25	8 18 a.m. V.				
	8 57 a.m. D.	H_{30} ·18205 H_{40} ·18201	·18449	Morrison	31
	9 34 a.m. V.				

402. Lat. 31° 0′·6 S.; Long. 29° 30′·5 E. About 20 miles from Mount Ayliff.

Declination. D.

Date	G.M.T.	D (observed)	D	Observer	Instrument
1906 Jan. 27	10 18 a.m. 10 24 a.m. 10 33 a.m.	23° 51′·3 W. 23 52·5 23 52·7	24° 17′·9 W.	Brown Morrison	31

Dip. θ.

Date	G.M.T.	Needle	θ (observed)	θ	Observer	Instrument
1906 Jan. 27	9 30 a.m. 9 26 a.m.	1 2	61° 26′·9 S. 61 28·0	61° 6′·8 S.	Morrison	9

Horizontal Intensity. H.

Date	G.M.T.	H (observed)	H	Observer	Instrument
1906 Jan. 27	11 41 a.m. V. 12 12 p.m. D. 12 49 p.m. V.	H_{30} ·17726 H_{40} ·17724	·17973	Morrison	31

403. Lat. 30° 49′·7 S.; Long. 29° 15′·5 E. About nine miles along the road from Mount Ayliff to Mount Frere. On top of rise.

Declination. D.

Date	G.M.T.	D (observed)	D	Observer	Instrument
1906 Jan. 29	1 48 p.m. 1 56 p.m. 2 4 p.m.	23° 46′·4 W. 23 47·5 23 47·4	24° 14′·4 W.	Brown Morrison	31

Dip. θ.

Date	G.M.T.	Needle	θ (observed)	θ	Observer	Instrument
1906 Jan. 29	11 59 a.m. 11 56 a.m.	1 2	61° 37′·6 S. 61 39·4	61° 17′·8 S.	Morrison	9

Horizontal Intensity. H.

Date	G.M.T.	H (observed)	H	Observer	Instrument
1906 Jan. 29	9 32 a.m. V. 10 14 a.m. D. 10 54 a.m. V.	H_{30} ·17706 H_{40} ·17704	·17953	Morrison	31

404. Lat. 31° 6′·0 S.; Long. 28° 52′·0 E. 3 hours by ox wagon from Mount Frere, on road to Qumbu.

Dip. θ.

Date	G.M.T.	Needle	θ (observed)	θ	Observer	Instrument
1906 Jan. 31	9 38 a.m.	1	61° 40′·7 S.	61° 19′·4 S.	Morrison	9
	9 38 a.m.	2	61 39 ·4			

Horizontal Intensity. H.

Date	G.M.T.	H (observed)	H	Observer	Instrument
1906 Jan. 31	11 6 a.m. V.	H ·17548	·17793	Morrison	31

405. Ugie, on road to. Lat. 31° 8′·3 S.; Long. 28° 26′·2 E.

Declination. D.

Date	G.M.T.	D (observed)	D	Observer	Instrument
1906 Feb. 5	6 36 a.m.	24° 43′·5 W.	25° 7′·3 W.	Brown Morrison	31
	6 46 a.m.	24 43 ·8			
	6 54 a.m.	24 44 ·4			

Dip. θ.

Date	G.M.T.	Needle	θ (observed)	θ	Observer	Instrument
1906 Feb. 5	10 26 a.m.	1	61° 32′·6 S.	61° 12′·0 S.	Morrison	9
	10 28 a.m.	2	61 32 ·7			

Horizontal Intensity. H.

Date	G.M.T.	H (observed)	H	Observer	Instrument
1906 Feb. 5	8 13 a.m. V.	H_{30} ·17722	·17971	Morrison	31
	8 54 a.m. D.	H_{40} ·17720			
	9 21 a.m. V.				

APPENDIX F.

MAGNETIC DECLINATION (W.) OBSERVED AT NATAL OBSERVATORY, DURBAN.

SUPPLIED BY MR E. N. NEVILL.

(*Monthly Means.*)

1893. 25° +

Month	8 a.m.	Noon	3 p.m.	9 p.m.	Mean
July	28′·97	33′·46	28′·74	28′·82	30′·00
Aug.	26·82	33·67	27·67	28·84	29·25
Sept.	28·06	31·47	25·55	27·22	28·07
Oct.	30·78	27·33	22·97	26·00	26·77
Nov.	28·98	19·73	18·93	22·23	22·47
Dec.	28·47	19·90	18·62	21·10	22·02
July—Dec.	28·68	27·59	23·75	25·70	26·43

1894. 25° +

Month	8 a.m.	Noon	3 p.m.	9 p.m.	Mean
Jan.	29′·47	20′·84	18′·19	23′·50	23′·00
Feb.	30·91	21·03	18·49	[23·50]	23·48
March	27·33	22·22	17·96		22·50
April	22·28	20·12	17·73		20·04
May	18·52	21·10	16·78		18·80
June	18·54	20·50	17·15		18·73
July	19·14	24·80	17·87	21·60	20·85
Aug.	18·41	24·13	17·85	18·90	19·82
Sept.	17·98	21·15	16·54	19·10	18·69
Oct.	22·03	18·76	13·14	16·48	17·60
Nov.	23·29	15·64	12·72	15·20	16·71
Dec.	19·02	11·61	10·43	14·23	13·82
Jan.—June	24·51	20·97	17·72		21·07
July—Dec.	19·98	19·35	14·76	17·59	17·92
Year					19·50

1895. 25° +

Month	9 a.m.	Noon	3 p.m.	9 p.m.	Mean
Jan.	16′·60	10′·52	11′·30	11′·30	12′·43
Feb.	18·21	9·24	6·30	15·70	12·36
March	18·02	11·18			14·60
April	17·87	13·99			15·93
May	21·86	21·35			21·61
June	16·01	19·36	15·07		16·81
July	17·68	21·02			19·35
Aug.	17·69	20·73			19·21
Sept.	15·01	15·20			15·10
Oct.	14·81	10·90			12·85
Nov.	14·19	7·15			10·67
Dec.	10·94	3·60			7·27
Jan.—June	18·10	14·27			16·18
July—Dec.	15·05	13·10			14·08
Year					15·13

1896. 25° +

Month	9 a.m.	Noon	Mean
Jan.	10′·81	4′·76	7′·78
Feb.	8·45	4·03	6·24
April	8·26	5·15	6·70
May	4·99	4·05	4·52
Mean =	8·13	4·50	6·31

1897. 24° +

Month	8 a.m.
July	46′·59
Aug.	46·68
Sept.	46·33
Oct.	48·66
Nov.	47·74
Dec.	45·47
Mean =	46·91

1898. 24° +

Month	8 a.m.
Jan.	43′·81
Feb.	42·07
March	41·03
April	37·03
May	35·60
Mean =	39·91

1902. 24° +

Month	8 a.m.
Jan.	26′·92
Feb.	25·52
March	22·38
April	17·68
May	12·80
June	8·87
Mean =	19·03

1904. 23° +

Month	9 a.m.	Noon	3 p.m.	9 p.m.
April	28′·28	28′·68	29′·02	25′·77
May	25·82	25·36	25·64	22·18
June	24·11	26·78	25·48	22·71
Mean =	26·07	26·94	26·71	23·55
July 12—31	22·79	24·86		
Aug. 1—5	23·04	25·94		

INDEX

All figures refer to pages. The figures in italics refer to the Appendices

Abelsdam 25, 44, 63, 72, 119, *13*
Aberdeen (Cape Colony) 25, 44, 63, 72, 94, 95, *13*
Aberdeen (Transvaal) 44, 63, 72, 109, 112, *14*
Aberdeen Road 28, 30, 32, 44, 63, 72, 93, 94, *14*
Aberfeldy 26, 44, 63, 72, 84, 106, 109, *15*
Adelaide 44, 63, 72, 93, 94, 95, *15*
Albert Falls 26, 45, 63, 72, 83, 85, 88, *16*
Algoa Bay 5
Alicedale 45, 63, 72, 84, 92, *16*
Aliwal North 25, 45, 63, 72, *17*
Alma 45, 63, 72, 112, 114, 120, 121, *17*
Amabele Junction 45, 63, 72, 92, *18*
Amaranja 45, 72, *18*
Amatongas 45, 63, 72, *18*
Annual Variation 3
Anomaly (magnetic) 62, Tables of 63, 72
Aries 7
Ashton 45, 72, 97, 98, 100, *19*
Assegai Bosch 27, 45, 63, 72, 100, *19*
Avontuur 25, 45, 63, 72, 97, *19*
Ayrshire Mine 25, 45, 63, 72, 125, *20*
Azimuth, Determination of 14
  Specimen observation *2*

Balmoral 26, 45, 63, 72, 84, 106, 118, *20*
Bamboo Creek 45, 63, 72, *21*
Bankpan 45, 72, 112, *21*
Barberton 26, 45, 63, 72, 112, 114, *22*
Barrington 27, 45, 63, 72, 97, 100, 103, *22*
Bashee 51, 71, 81, 91, 92, *224*
Battlefields 25, 45, 63, 72, 125, *23*
Bavaria 45, 63, 72, 112, 114, 121, 122, *23*
Baviaanskrantz 45, 63, 72, *24*
Beaconsfield 45, *24*
Beaufort West 27, 28, 29, 30, 32, 45, 63, 72, 94, 95, *24*
Beaufort West—Cradock Valley 94, 95, 103
Bechuanaland 11, 122
Beira 7, 10, 25, 28, 30, 32, 45, 63, 72, *25*
Belleville 45, 72, *25*

Berg River Mouth 25, 45, 63, 72, *26*
Bethal 26, 45, 63, 72, *26*
Bethany 25, 45, 63, 72, *27*
Bethesda Road 25, 45, 63, 72, 83, 94, *27*
Bethlehem 45, 63, 72, 116, 121, *28*
Bethulie 25, 45, 63, 72, *28*
Biesjesdal 7
Biesjespoort 45, 63, 72, 105, *29*
Birthday 45, 63, 72, *29*
Blaauwbosch 45, 63, 72, *30*
Blaauwkrantz 27, 45, 63, 73, 97, 100, *30*
Bluff, The 45, 64, 73, 88, 90, *31*
Boane 10
Boschkopjes 45, 64, 73, *31*
Boschrand 25, 45, 64, 73, 114, *32*
Bosman, J. J. 4
Boston 45, 64, 73, 83, 85, 88, 89, 90, *32*
Botha's Berg 45, 64, 73, 106, 109, 118, *33*
Brak River 45, 64, 73, 111, *33*
Brandboontjes 45, 64, 73, 109, *34*
Bredasdorp 25, 45, 64, 73, 100, *34*
Breekkerrie 45, 64, 73, 106, *35*
Britstown 45, 64, 73, 105, 106, *35*
Buffelsberg 45, 64, 73, 109, *36*
Buffelshoek 45, 64, 73, 112, *36*
Buffelsklip 25, 27, 45, 64, 73, 103, *37*
Bulawayo 11, 30, 32, 45, 64, 73, 122, 124, 125, *37*
Bult and Baatjes 45, 73, *38*
Bulwer 45, 64, 73, 85, 88, *38*
Burghersdorp 29, 45, 64, 73, *38*
Bushmanskop 45, 64, 73, 95, *39*
Butterworth 45, 64, 73, 92, *39*

Caledon River 45, 64, 73, *40*
Calitzdorp 45, 64, 73, 97, 98, *40*
Camperdown 45, 73, 85, 88, 90, *41*
Cango 25, 45, 64, 73, 97, 101, 103, *41*
Cape Colony 4, 8, 11, 85
Cape of Good Hope 4, 5

Cape Town 3, 4, 5, 6, 7, 8, 10, 20, 25, 27, 28, 29, 30, 31, 32, 45, 64, 73, 98, 99, *42*
Cathcart 45, 73, 92, *45*
Centre of Attraction, Definition of 83
Centre of Repulsion, Definition of 83
Ceres Road 30, 32, 45, 64, 73, 98, 99, 100, *45*
Charlestown 26, 46, 64, 73, 112, 114, 119, 121, 122, *46*
Charts 11, 82, at end
Chaves, Major F. A. 34, 40, 41, 42
Clarkson 46, 73, *46*
Claxton, C. T. F. 20
Coerney 25, 46, 64, 73, 84, 92, *47*
Col, Definition of 83
Cold Bokkeveld 6
Colenso 26, 46, 64, 73, 83, 85, 88, 89, 116, *47*
Colesberg 25, 28, 30, 32, 46, 64, 73, 116, 118, *48*
Colesberg—Norval's Pont Valley 118
Connan's Farm 46, 64, 73, *49*
Cooke & Son 11
Cookhouse 28, 30, 32, 46, 64, 73, 91, 93, 94, *49*
Cotswold Hotel 46, 64, 73, 90, 91, *50*
Cradock 32, 46, 64, 73, 83, 94, 95, *50*
Cream of Tartarfontein 46, 64, 73, 111, *51*
Crocodile Pools 46, 64, 73, 123, *51*

Daily Variation 3, 20
Dalton 46, 73, 85, *52*
Dambiesfontein 46, 64, 73, 103, *52*
Dannhauser 26, 46, 64, 73, 85, 112, 114, 121, *53*
Dargle Road 46, 64, 73, 83, 85, 87, 88, *53*
Darling 25, 46, 64, 73, 103, *54*
De Aar 11, 25, 46, 64, 73, 105, 106, *54*
Declination
  calculated and observed diurnal differences 25
  changes in per degree of latitude and of longitude 53, 122, 123, 124
  daily or diurnal variation 20
  defined 1
  determination of 15
  diurnal variation at the Cape, St Helena and Mauritius 21
  diurnal variation at Matjesfontein 23
  early and miscellaneous observed values 5, 10
  errors of observation 15
  formula for calculation of diurnal variation in 22
  observed values 44
  observed and calculated values and anomalies 63
  properties of near ridges and valleys 83
  secular variation of at various stations 28
  specimen observation of *6*
  values of at intersections of degrees of latitude and of longitude 55
  values of at mean stations 52
De Doorns 46, 74, 100, *55*
Deelfontein 46, 74, 114, 116, *55*
Deelfontein Farm 46, 64, 74, 118, *55*
De Jager's Farm 25, 46, 65, 74, *56*
Dewetsdorp 46, 65, 74, *56*
"Discovery" comparison with instruments of 39
Districts 52, 124
Disturbances (magnetic) 82
Disturbing Horizontal Force 82
Draghoender 46, 65, 74, 106, *57*
Drew 46, 65, 74, 98, 100, *57*
Driefontein 46, 65, 74, 114, *58*
Driehoek 46, 65, 74, 112, 120, 121, *58*
Durban 28, 88, 90, 96
  Diurnal variation of Declination at 23, *227*
Dip
  changes in per degree of latitude and of longitude 53, 122, 123, 124
  defined 2
  determination of 16
  errors of observation 16
  observed values 44
  observed and calculated values and anomalies 72
  properties of near ridges and valleys 84
  secular variation of at various stations 30
  specimen observation of *7*
  values of at intersections of degrees of latitude and of longitude 56
  values of at mean stations 52, 124

Eastern Province Valley 92, 96
East London 27, 46, 65, 74, 92, 93, *59*
Elandshoek 46, 74, *59*
Elandskloof Farm 46, 74, *60*
Elim 25, 46, 65, 74, *60*
Ellerton 46, 74, *61*
Elliot 46, 65, 74, 91, 93, 97, *61*
Elliot Brothers 11
Elsburg 26, 46, 65, 74, 84, 106, 109, *62*
Emmasheim 46, 65, 74, 116, *62*
Epoch of Survey 44
Erman 10
Errors accidental and effective of Declination 15, of Dip 16, of Horizontal Intensity 19
Estcourt 46, 65, 74, 83, 85, 88, 90, *63*

Ferreira 25, 46, 65, 74, *63*
Fish River 25, 46, 65, 74, 83, 94, 95, 116, *64*
Forty-one mile siding 25, 46, 65, 74, 125, *64*
Fountain Hall 46, 65, 74, 83, 85, 88, *65*
Fourcade, H. G. 4
Francistown 46, 74, *65*
Fraserburg 46, 65, 74, 94, 95, 103, 104, *66*
Fraserburg Road 46, 74, 94, 97, *66*

Gahoe 7

Gamtoos River Bridge 11, 25, 27, 46, 65, 74, 84, *67*
Gemsbokfontein 46, 65, 74, 105, *68*
Geological map, Note on 126
George Town 46, 74, 101, *67*
Gilbert 7
Ginginhlovu 26, 46, 65, 74, 85, 88, 90, *1*, *68*
Glenallen 46, 65, 74, *69*
Glenconnor 46, 65, 74, 84, 92, *69*
Globe and Phœnix 46, 65, 74, 125, *70*
Goedgedacht 46, 65, 74, *70*
Gordon's Bay 46, 74, *70*
Graaf Reinet 46, 65, 74, 83, 94, 95, *71*
Graaf Reinet-Zuurpoort Ridge 94
Grahamstown 25, 27, 46, 65, 74, 84, 92, 93, 97, *71*
Grange 46, 74, 85, 88, 90, *72*
Graskop 46, 65, 74, 112, 114, 119, 121, 122, *72*
Great Karroo Ridge 93
Greylingstad 26, 46, 65, 74, *73*
Greytown 46, 65, 75, 83, 85, 87, 88, *73*
Griqualand East valley 91, 96
Griqualand East and South Natal valley 85, 88, 96
Griqualand East and South Natal Ridge 90, 96
Grobler's Bridge 46, 65, 75, 106, 109, 118, *74*
Groenkloof 47, 65, 75, 116, 118, *74*
Groenplaats 25, 47, 65, 75, 112, 114, 119, *75*
Groote Laagte 10
Grootfontein 47, 65, 75, 93, 94, *75*
Gwaai 25, 47, 65, 75, *76*
Gwelo 47, 65, 75, 125, *76*

Hamaan's Kraal 26, 47, 65, 75, *77*
Hankey 47, 65, 75, 84, 92, *77*
Harreeboom 7
Hartley 25, 47, 65, 75, 125, *78*
Hector Spruit 47, 65, 75, 112, 114, 118, *78*
Heidelberg 47, 65, 75, 98, 100, *79*
Heilbron 25, 47, 65, 75, 112, 114, 120, *79*
Helvetia 47, 65, 75, *80*
Hermanus 25, 47, 65, 75, 98, 100, *80*
Hermon 47, 65, 75, 98, 100, 101, *81*
Hex River 7, 98
Highlands 47, 65, 75, 112, 114, 121, *81*
Hlabisa 26, 47, 66, 75, *82*
Hluti 47, 66, 75, 112, 120, *82*
Hoetjes Bay 25, 47, 66, 75, *83*
Holfontein 47, 66, 75, 116, 118, *83*
Honey Nest Kloof 25, 47, 66, 75, 106, *84*
Honing Spruit 25, 47, 66, 75, 114, 120, *84*
Hopefield 47, 66, 75, *85*
Horizontal Intensity (or force)
  changes in per degree of latitude and of longitude 53, 122, 123, 124
  daily, or diurnal, variation 26
  determination of 17
  early and miscellaneous values 6, 10
  error of observation 19
  observed values 44
  observed and calculated values and anomalies 72
  secular variation of at various stations 32
  specimen observation of *9*
  values of at intersection of degrees of latitude and of longitude 57
  values of at mean stations 52, 124
Hout Bay 5
Houtman, C. 4
Howhoek 25, 47, 66, 75, 98, 100, *85*
Huguenot 47, 66, 75, 98, 99, 100, 103, *86*
Humansdorp 47, 66, 75, 84, 92, 100, 103, *86*
Hutchinson 25, 28, 30, 32, 47, 66, 75, 105, *87*

Ibisi Bridge 26, 47, 66, 75, 88, 90, 91, *88*
Idutywa 47, 66, 75, 92, *88*
Igusi 47, 66, 75, *89*
Illovo River 47, 66, 75, 88, 90, *89*
Imvani 47, 66, 75, 93, 94, *90*
Inclination, see Dip
Indowane 26, 47, 66, 75, 88, *90*
Indwe 25, 47, 66, 75, 91, 93, 94, *91*
Inhambane 10
Inoculation 47, 66, 75, *91*
Instruments, comparison of 34
  description of 11
Inyantué 47, 66, 75, *92*

Kaalfontein 47, 75, 84, 106, 109, *92*
Kaalfontein-Wonderfontein Ridge 106
Kaalkop Farm 47, 66, 75, 112, 114, *93*
Kaapmuiden 47, 66, 75, 112, *93*
Kalkbank 26, 47, 66, 75, 111, 112, *94*
Kaloombies 47, 66, 75, *94*
Karree 25, 47, 66, 75, 116, 118, *95*
Kathoek 47, 66, 75, 98, 100, *95*
Kenhardt 7, 8, 47, 66, 76, *96*
Kenilworth 7, 8, 47, *96*
Kimberley 27, 47, 76, *96*
King William's Town 47, 66, 76, 92, 93, *97*
Klaarstrom 47, 66, 76, 97, *97*
Klerksdorp 47, 66, 76, 114, *98*
Klerksdorp-Virginia-Saxony Ridge 114
Klipfontein (Cape Colony) 6, 47, 66, 76, *98*
Klipfontein (Spelonken) 26, 47, 66, 76, 109, *99*
Klippan 7
Klipplaat 25, 47, 66, 76, 93, 94, *99*
Knysna 27, 29, 47, 66, 76, 100, *100*
Kokstad 47, 66, 76, 88, 90, *100*
Komati Poort 47, 76, 112, 114, *101*
Komgha 47, 66, 76, 92, *101*
Komgha, East of 51, 71, 81, *223*
Kosi River 7
Kraal 26, 47, 66, 76, 112, *102*

Krantz Kloof 26, 47, 66, 76, 88, 90, *102*
Krantz Kop 47, 66, 76, *103*
Kromm River 47, 66, 76, 112, 114, 119, 120, *103*
Krugers 25, 48, 67, 76, 116, 118, *104*
Kruispad 48, 67, 76, 93, 94, *104*
Kwambonambi 26, 48, 67, 76, 88, *105*

Laat Rivier 48, 67, 76, 106, *105*
La Caille 4
Ladismith 11, 25, 48, 67, 76, 97, 98, *106*
L'Agulhas 5, 25, 48, 67, 76, *106*
Laingsburg 48, 67, 76, 93, 94, *107*
Lake Banagher 48, 67, 76, 112, 114, *107*
Langlaagte 48, 67, 76, 84, 106, 109, *108*
Lapanie 7
Latitude, Determination of 12
  of stations 44
  specimen observation *1*
Leeufontein 6
Letjesbosch 48, 76, *108*
Libode 48, 67, 76, *109*
Limpopo, North of 51, 71, 81, *213*
Little Karroo Valley 97, 98
Lobatsi 48, 67, 76, 123, 124, *109*
Lochard 25, 48, 67, 76, 125, *110*
Longitude, Determination of 12
  of stations 44
  specimen observation *1*
Lourenço Marques 7
Lydenburg 26, 48, 67, 76, 109, *110*

Machadodorp 48, 67, 76, 106, 109, 118, *111*
Macheke 48, 67, 76, *111*
Maclear, Sir T. 4
Mafeking 11, 25, 28, 29, 30, 32, 48, 67, 76, 123, *112*
Magalapye 25, 48, 67, 76, 123, *113*
Magnet Heights 48, 67, 76, 109, *113*
Magnetometer Constants 12
Makwiro 25, 48, 67, 76, 125, *114*
Malagas 48, 67, 76, 98, 100, *114*
Malenje Siding 25, 48, 67, 76, *115*
Malinde 25, 48, 67, 76, *115*
Malmesbury 25, 27, 30, 32, 48, 67, 77, 98, 100, *116*
Mandegos 48, 67, 77, *116*
Mapani Loep 48, 67, 77, 111, *117*
Mara 7, 48, 77, 112, *117*
Marandellas 25, 48, 67, 77, 125, *118*
Maribogo 48, 67, 77, 123, *118*
Mashonaland 11, 124
Matabeleland 11, 124
Matetsi 48, 67, 77, *119*
Matjesfontein 25, 28, 30, 32, 35, 38, 39, 40, 48, 67, 77, *119*
  Diurnal variation of Declination at 23
Mauritius, Diurnal variation of Declination at 21

Mean stations, values of elements at 52, 124
Meyerton 48, 67, 77, 112, 114, *126*
Meyerton-Pivaan's Poort Ridge 112
Middelberg 27, 48, 67, 77, 106, *126*
Middlepost 48, 67, 77, 103, 104, *127*
Middleton 25, 48, 67, 77, 93, 94, *127*
Mill River 48, 77, *128*
Miller Siding 28, 30, 32, 48, 67, 77, 93, 94, *128*
Miller's Point 48, 67, 77, 98, 99, *129*
Misgund 48, 67, 77, 97, 100, *129*
Mission Station 48, 67, 77, *130*
Modder Spruit 26, 48, 67, 77, 83, 85, 86, 88, 89, 90, 116, 121, 122, *130*
Mogweding 7, 8
Molteno 48, 67, 77, 116, *131*
Mooifontein 7
Moore, Rear-Admiral 39
Moorrees, A. 6, 7, 8
Mossel Bay 4, 25, 27, 48, 67, 77, *131*
Mount Ayliff 48, 67, 77, 90, 91, *132*
Mount Frere 48, 67, 77, 90, *132*
Mount Moreland 48, 67, 77, 88, *133*
Movene 48, 67, 77, 112, 114, *133*
M'Phatele's Location 48, 67, 77, 109, *134*

Naauwpoort 25, 30, 32, 48, 68, 77, 95, 116, 118, *134*
Naboomspruit 26, 48, 68, 77, 109, *135*
Nada 7
Natal 11, 85
Natal-Zululand Ridge 85, 88, 90, 116
Nelspoort 25, 48, 68, 77, 94, 95, *135*
Nelspruit 26, 48, 68, 77, 112, 114, 118, *136*
Neumayer 30
Nevill, E. N. 24
Newcastle 48, 68, 77, 112, 114, *136*
Newcastle (Transvaal) 26, 48, 68, 77, *137*
Nooitgedacht 48, 68, 77, 109, 118, *137*
Northerly Component (or Intensity)
  Changes in per degree of latitude and of longitude 54, 124
  defined 2
  observed and calculated values and anomalies 63
  Values of at intersections of degrees of latitude and of longitude 60
  Values of at mean stations 52, 124
Norval's Pont 48, 77, 116, 118, *137*
Novara 6
'Nqutu Road 48, 77, *138*
Nylstroom 26, 48, 68, 77, 109, *138*

Olvrida 10
Oo'kiep 6
Orange River 6, 30, 48, 68, 77, 105, 106, *139*
Orange River Colony 11, 106
Orjida 49, 68, 77, 97, *139*

Oro Point 7
Oudemuur 49, 68, 77, *140*
Oudtshoorn 25, 49, 68, 77, 97, 98, 103, *140*
Outspan 51, 71, 81, 92, *223*

"P" constant, values of 19
Paardevlei 49, 68, 77, 106, *141*
Paarl 49, 77, *141*
Palapye 25, 30, 32, 49, 68, 77, 123, *142*
Pampoenpoort 49, 68, 77, 105, *142*
Payne's Farm 49, 68, 77, 91, *143*
Peak (magnetic), Definition of 83
Picene 28, 30, 32, 49, 68, 78, 112, 114, 118, *143*
Pienaar's River 49, 68, 78, 109, *144*
Pietersburg 26, 49, 68, 78, 111, *144*
Piet Potgietersrust 26, 49, 68, 78, 109, *145*
Piet Retief 49, 68, 78, 112, 114, *145*
Pilgrim's Rest 49, 68, 78, 109, *146*
Piquetberg 49, 68, 78, *146*
Pivaan's Poort 49, 68, 78, 112, 120, 121, 122, *147*
Platrand 49, 68, 78, 119, 121, *147*
Plettenberg Bay 25, 49, 68, 78, 97, 100, *148*
Plumtree 49, 78, *148*
Pokwani 49, 68, 78, 109, *149*
Port Alfred 25, 49, 68, 78, 84, 92, 93, *149*
Port Beaufort 49, 68, 78, 98, 100, *150*
Port Elizabeth 49, 68, 78, *150*
Port Nolloth 6
Port Shepstone 49, 68, 78, 90, 91, *151*
Port St Johns 49, 68, 78, 92, 97, *151*
Port St Johns—Transkei Ridge 92, 93, 97
Portuguese East Africa 11
Potchefstroom 49, 68, 78, 114, *152*
Potfontein 25, 49, 68, 78, 105, 106, *152*
Pretoria 49, 68, 78, 84, 106, 109, 118, *153*
Prince Albert 25, 27, 49, 68, 78, 97, *153*
Prince Albert Road 25, 28, 30, 49, 68, 78, 93, 94, *154*
Probable error of Azimuth 14

Queenstown 25, 49, 68, 78, 93, 94, 95, *154*
Quelimane 10

Randfontein 26, 49, 68, 78, 84, 106, 109, *155*
Rateldraai 49, 69, 78, 106, *155*
Rateldrift 49, 69, 78, 103, 104, *156*
Reid and Son 11
Repeat Stations 28, 30, 32
Reuben point 7, 28
Revué 49, 69, 79, *163*
Rhodesia 122
Richmond 49, 78, 85, 88, 89, 90, *156*
Richmond Road 49, 69, 78, 105, *157*
Ridge, definition of 83
  Graaf Reinet-Zuurpoort 94
  Great Karroo 93
  Griqualand East and South Natal 90, 96
  Kaalfontein-Wonderfontein 106
  Klerksdorp-Virginia-Saxony 114
  Meyerton-Pivaan's Poort 112
  Natal-Zululand 85, 88, 90, 116
  Port St John's Transkei 92, 93, 97
  Stormberg-Molteno 118
  Underberg-Boston 89
Rietkuil Farm 49, 69, 78, 88, *157*
Rietpoort 49, 69, 78, 103, 104, *158*
Rietvlei 49, 69, 78, 109, 111, *158*
Rietvlei (Cape Colony) 49, 69, 78, 97, *159*
Riversdale 25, 49, 69, 78, 98, 100, *159*
Rivierplaats 49, 69, 78, 94, 95, 103, 104, *160*
Roadside 49, 69, 78, *160*
Robben Island 5
Robertson 25, 49, 69, 78, 97, 98, *161*
Rodekrantz 49, 69, 78, 109, *161*
Rhodesia 122
Roodepoort 49, 69, 78, 105, 106, *162*
Rogers, A. W. 126
Rooidam 49, 69, 78, 106, *162*
Rooipüts 49, 69, 79, 106, *163*
Rooival 49, 69, 79, 95, *164*
Rosmead Junction 25, 28, 30, 32, 49, 69, 79, 94, 95, 116, 118, *164*
Rosmead-Naauwpoort Valley 118
Rosmead-Steynsburg Valley 118
Rouxville 49, 69, 79, 116, 118, *165*
Royal Observatory, see Cape Town
Rücker, Sir A. 15, 16, 82
Rusapi 49, 69, 79, *165*
Rustplaats 49, 69, 79, 109, *166*
Ruyterbosch 49, 69, 79, 100, 103, *166*

Sabie River 49, 69, 79, 106, 109, *167*
Sabine, Sir E. 4
Saldanha 5
Salisbury 25, 49, 69, 79, 125, *167*
Salt River 7
Saxony 25, 49, 69, 79, 114, 116, *168*
Schietfontein 49, 69, 79, 93, 94, *168*
Schikhoek 50, 69, 79, 112, 120, 121, *169*
Schoemanshoek 50, 69, 79, 106, 109, *169*
Schuilplaats 50, 69, 79, *170*
Secocoeni's Stad 50, 69, 79, 109, *170*
Secular variation 3, 10, 27
Seruli 50, 69, 79, 122, 123, 124, *171*
Shangani 25, 50, 69, 79, 125, *171*
Shashi 50, 69, 79, *172*
Shela River 50, 69, 79, 112, *172*
Shoshong Road 25, 50, 69, 79, 123, *173*
Signal Hill 25, 27, 50, 69, 79, 98, 99, *173*
Simon's Bay 5
Simonstown 6, 25, 39, 50, 69, 79, 98, 99, *174*, *175*

Sir Lowry's Pass 25, 30, 50, 69, 79, 98, 100, 101, 103, *176*
Smaldeel 25, 50, 69, 79, 114, 116, *176*
Spitzkopje 50, 69, 79, 106, 109, *177*
Springfontein 25, 50, 69, 79, 118, *177*
Springs 50, 69, 79, 84, 106, 109, *178*
Stanford 25, 50, 69, 79, 98, 100, *178*
Stanger 26, 50, 69, 79, 85, 88, *179*
Stations, Distribution of 11; List of 44
Steekdoorns 7
Steenkampspoort 50, 69, 79, 94, 95, 103, 104, *179*
Stellenbosch 25, 27, 32, 50, 70, 79, 98, 100, *180*
Sterkstroom 50, 70, 79, 95, *181*
Steynsburg 50, 70, 79, 95, 116, *181*
St Helena, Diurnal variation of Declination at 21
Still Bay 25, 50, 70, 79, 98, 103, *182*
Stormberg Junction 50, 70, 79, 95, 116, *182*
Stormberg-Molteno Ridge 118
Storms River 27, 50, 70, 79, *183*
Strandfontein 25, 35, 50, 70, 79, 99, *183*
Sutherland 50, 70, 79, 94, 95, 103, *184*
Swellendam 25, 50, 70, 79, 98, 100, *184*

Table Bay 5
Tafelberg 6
Taungs 50, 70, 79, 123, 124, *185*
Thaba 'Nchu 25, 50, 70, 79, 116, *185*
Thirtyfirst 50, 70, 80, 118, *191*
Thorpe, T. E. 15, 16, 82
Tinfontein 50, 70, 79, 116, *186*
Tnooi 7
Toeslaan 7
Toise River 50, 70, 80, 91, 92, *186*
Total Intensity (or Force)
  changes in per degree of latitude and of longitude 54, 124
  early and miscellaneous values 5
  observed and calculated values and anomalies 72
  values of at intersections of degrees of latitude and of longitude 58
  values of at mean stations 52, 124
Touws River 50, 70, 80, 98, 100, *187*
Transkei 11, 86
Transvaal 106
Tsolo 50, 70, 80, 91, 94, *187*
Tugela, 50, 80, 85, 88, *188*
Tulbagh 6, 8, 28,
Tulbagh Road 25, 50, 70, 80, 98, 99, *188*
Tweepoort 50, 70, 80, 112, *189*
Twee Rivieren 50, 70, 80, 100, *190*
Twelfelhoek 50, 70, 80, 112, 114, 119, 121, *189*
Tygerfontein 25, 50, 70, 80, 98, 100, 103, *190*
Tygerkloof Drift 50, 70, 80, 116, 121, *191*

Ugie 51, 71, 81, 91, 94, 97, *226*
Uitenhage 30, 32, 50, 70, 80, 84, *192*
Uitkyk 26, 50, 70, 80, 84, 106, 109, *192*
Uitspan Farm 50, 70, 80, 103, 104, *193*
Umhlatuzi 50, 70, 80, 85, 88, 90, *193*
Umhlengana Pass 50, 70, 80, 92, *194*
Umtali 25, 50, 70, 80, 124, *194*
Umtata 50, 70, 80, 91, 92, *195*
Umtwalumi 50, 80, 90, 91, *195*
Umzinto 50, 70, 80, 88, 90, *196*
Underberg Hotel 26, 50, 70, 80, 85, 88, *196*
Underberg-Boston Ridge 89
Upington 50, 70, 80, *197*
Utrecht 50, 70, 80, 112, *197*

Val Joyeux 40
Valley, definition of 83
  Beaufort West-Cradock 94, 95, 103
  Colesberg-Norval's Pont 118
  Eastern Province 92, 96
  Griqualand East 91, 96
  Griqualand East and South Natal 85, 88, 96
  Little Karroo 97, 98
  Rosmead-Naauwpoort 118
  Rosmead-Steynsburg 118
Van Reenen 50, 70, 80, 88, 116, *198*
Van Roois Vlei 7
Van Wyk's Farm 25, 50, 70, 80, 97, 98, *198*
Van Wyk's Vlei 50, 70, 80, 106, *199*
Vertical Intensity (or Force)
  changes in per degree of latitude and of longitude 54, 124
  observed and calculated values and anomalies 72
  values of at intersection of degrees of latitude and of longitude 59
  values of at mean stations 52, 124
Victoria Falls 11, 50, 70, 80, 124, *199*
Villiersdorp 25, 51, 70, 80, 99, *200*
Virginia 51, 70, 80, 114, *200*
Vlaklaagte 26, 51, 70, 80, *201*
Vogelvlei 51, 71, 80, 114, 116, *201*
Vondeling 25, 51, 71, 80, 97, *202*
Vredefort 25, 51, 71, 80, 114, 119, 120, *202*
Vredefort Road 51, 80, *203*
Vryburg 51, 71, 80, 122, *203*

Wakkerstroom 51, 71, 80, 112, 120, *204*
Walfisch Bay 7, 28, 29
Wankie 51, 71, 80, *204*
Warmbad (Waterberg) 51, 71, 80, *205*
Warmbad (Zontpansberg) 51, 71, 80, 111, *205*
Warrenton 25, 51, 71, 81, 122, *206*
Waschbank 26, 51, 71, 81, 85, 86, 88, 121, *206*
Waterworks 51, 71, 81, 116, 118, *207*
Welverdiend 26, 51, 71, 81, *207*

Wepener 7, 8, 28, 51, 71, 81, 116, *208*
Westerly component (or Intensity)
  changes in per degree of latitude and of longitude 54
  defined 2
  observed and calculated values and anomalies 63
  values of at intersections of degrees of latitude and of longitude 61
  values of at mean stations 52, 124
Williston 51, 71, 81, 103, 104, *208*
Willowmore 27, 28, 30, 32, 51, 71, 81, 93, 94, *209*
Winburg 51, 71, 81, 114, 116, *210*
Winkeldrift 25, 51, 71, 81, 114, 119, 120, *210*
Witklip 51, 71, 81, 111, 112, *211*
Witmoss 25, 51, 71, 81, 83, 91, 93, 94, *211*
Wolvefontein 51, 81, 84, 92, *212*
Wolvehoek 51, 71, 81, 112, 114, *212*
Wonderfontein 51 71, 81, 106, 109, 118, *213*
Woodville 51, 71, 81, 97, 100, *214*
Worcester 7, 25, 28, 30, 51, 71, 81, 98, 99, *214*

Zak Rivier 51, 71, 81, *215*
Zand River 51, 71, 81, 111, *215*
Zeekoegat 25, 51, 71, 81, 93, 94, 97, *216*
Zuurbraak 51, 71, 81, 100, *216*
Zuurfontein 51, 71, 81, *217*
Zuurpoort 25, 27, 51, 71, 81, 83, 94, *217*

Printed in the United States